US Military Recipes
Volume 2 Armed Forces Recipe Service
Great for Cooking for Large Groups

Army TM 10-412
Navy NAVSUP Pub 7
USAF AFJMAN 34-606
USMC MCO P10110 42B

Contents
Basic Information
Conversion Charts
Equipment Guidelines
Ingredients
Food Safety

Sections
M - Salads
N - Sandwiches
O - Dressings, Gravies, and Sauces
P - Soups
Q - Vegetables

edited by
Brian Greul

All branches of the US Military use this standardized set of recipes. This is the 2003 edition. The full collection is 1691 pages. This book is the second 684 pages of the full 1600 page collection. For reasons related to the maximum size of a book, the collection has been split into two books. The front 77 pages are repeated in each volume because the contain the instructions that are common to all recipes. This allows the books to be used independently. The editor recommends that you use flags to mark your favorite recipes. The recipes are fully scalable up or down and the instructions make the calculations as easy as one of the many pie recipes.

An 8.5x11 3 hole punched loose leaf copy may be purchased for your 3 ring binder. Email books@ocotillopress.com for current information.

Edited 2021 Ocotillo Press
ISBN 978-1-954285-29-3

No rights reserved. This content of this book is in the public domain as it is a work of the US Government. It is reproduced by the publisher as a convenience to enthusiasts and others who may wish to own a quality copy of it. It has been adjusted to accomodate the printing and binding process.

Printed in the United States of America for domestic orders and overseas for foreign orders.

Ocotillo Press
Houston, TX 77017
Books@OcotilloPress.com

Disclaimer: The user of this book is responsible for following safe and lawful practices at all times. The publisher assumes no responsibility for the use of the content of this book. The publisher has made an effort to ensure that the text is complete and properly typeset, however omissions, errors, and other issues may exist that the publisher is unaware of.

Armed Forces Recipe Service

UNITED STATES ARMY
TM 10-412

UNITED STATES NAVY
NAVSUP Publication 7

UNITED STATES AIR FORCE
AFJMAN 34-606 Volume I and Volume II

UNITED STATES MARINE CORPS
MCO P10110.42B

Stock No. 0530-LP-188-7302

DEPARTMENTS OF THE ARMY, THE NAVY AND THE AIR FORCE
Washington, DC

ARMED FORCES RECIPE SERVICE

The Armed Forces Recipe Service has been revised and updated and is issued for the purpose of standardizing and improving food prepared and served in military food service operations.

This recipe service is available for use by all Military Services and was coordinated and developed by technical representatives from each of the following:

- Army Center of Excellence, Subsistence, Fort Lee, VA   23801-6020
- Navy Supply Systems Command, Mechanicsburg, PA   17055-0791
- Headquarters, Air Force Services Agency, San Antonio, TX   78216-4138
- Headquarters, U. S. Marine Corps, Washington, DC   20380-1775
- U. S. Army Soldier and Biological Chemical Command, Natick Soldier Center, Natick, MA   01760-5018

DEPARTMENTS OF THE ARMY, THE NAVY, AND THE AIR FORCE Washington, DC June 2003	USA, TM-10-412 NAVSUP Publication 7 AFJMAN 34-606 MCO 10110.42B

ARMED FORCES RECIPE SERVICE

This 2003 revision replaces the 1993 (original), 1997 (Change 1) and 1999 (Change 2) Armed Forces Recipe Service cards. This printing includes all new recipe development and revisions to the 1993, 1997 and 1999 sets of Armed Forces Recipe Service cards. This update contains a new recipe format which reflects the nutritional analysis per serving, located at the top of each recipe card.

New recipes have been incorporated. Some of the recipes are designated with a number 800 and above which represent recipes that include speed scratch items, convenience prepared foods and additional new recipes. Some recipes have been deleted while other recipes that were printed on cards as variations are now individual recipes. Sources of recipes include the U.S. Army Solider and Biological Command, commercial quantity food cookbooks and food product manufacturers.

Replace current recipe cards sets with this 2003 Update. Replace the Index of Recipes, NAVSUP Publication P-7, dated July 1999 with Index of Recipes, dated 2003. An index of recipes is issued to assist food service personnel to easily locate recipes by category to ensure a varied menu. This card should be retained and inserted in the front of the publication.

BY ORDER OF THE SECRETARIES OF THE ARMY, THE NAVY, AND THE AIR FORCE:

JOHN P. JUMPER
General, USAF
Chief of Staff

J. D. MCCARTHY
Rear Admiral, SC, United States Navy
Commander, Naval Supply Systems Command

ERIC K. SHINSEKI
General, United States Army
Chief of Staff

R. L. KELLY
Lieutenant General, U. S. Marine Corps
Deputy Chief of Staff for Installations & Logistics

Distribution:
    Navy: NAVSUP 51 Distribution

A. GENERAL INFORMATION No. 0

INDEX

Card No.
Basic Information
A-19 Handling Frozen Foods, Guidelines for
A-4   Measuring Equivalents, Table of
A-2   Terms Used in Food Preparation, Definitions of
A-35 Use of Convenience Prepared Foods, Guidelines for

Conversion Charts
A-5   Can Sizes, Table of Weights and Measures for
A-9   Container Yields, Canned Fruits, Guidelines for
A-13 Fruit Bars, Guidelines for
A-16 Measure Conversion, Guidelines for
A-27 Metric Conversion, Guidelines for
A-1   Information for Standardized Recipes
A-15 Weight Conversion

Equipment, Guidelines for
A-33 Combi-Ovens
A-23 Convection Ovens
A-34 Skittles
A-21 Steam Cookers
A-25 Steam Table, Baking and Roasting Pans, Capacities for
A-24 Tilting Fry Pans

Ingredients
A-20 Antibrowning Agent, Use of

Card No.
A-28 Dehydrated Cheese, Use of
A-11 Dehydrated Green Peppers, Onions, and Parsley Use of
A-8   Egg Equivalents, Table of
A-30 Herbs, Guide to Cooking with Popular
A-10 Milk, Nonfat, Dry, Reconstitution Chart

Safety
A-32 "HACCP" (Hazard Analysis Critical Control Point) Guidelines

GUIDELINE CARDS

SECTION C - BEVERAGES
C-G-1 Brewing Coffee, Guidelines for
C-G-4 Coffee Urn Capacities, Guidelines for

SECTION D - BREADS AND SWEET DOUGHS
D-G-1 Recipe Conversion
D-G-2 Preparation of Yeast Doughs, Guidelines for
D-G-3 Retarded Sweet Dough Methods
D-G-4 Good Quality Bread Products and Rolls, Characteristics of
D-G-5 Poor Quality Bread Products and Rolls, Characteristics of

A. GENERAL INFORMATION No. 0

INDEX

Card No.
GUIDELINE CARDS - CONTINUED

SECTION G - CAKES
G-G-2   Batter Cakes, Characteristics of Good Quality/Bad Quality
G-G-5   Cutting Cakes, Guidelines for
G-G-7   High Altitude Baking
G-G-4   Scaling Cake Batter, Guidelines for
G-G-1   Successful Cake Baking, Guidelines for

SECTION G - FROSTINGS
G-G-6   Prepared Frostings and Frosting Cakes, Guidelines for

SECTION H - COOKIES
H-G-1   General Information Regarding Cookies

SECTION I - PASTRY AND PIES
I-G-1   Making One Crust Pies
I-G-2   Making Two Crust Pies

SECTION L - POULTRY
L-G-5   Timetable for Roasting Turkeys

SECTION M - SALADS
M-G-1   Trays or Salad Bars, Guidelines for

SECTION O - DRESSINGS, GRAVIES AND SAUCES
O-G-1   Sauces and Gravies, Guidelines for Preparing

Card No.

SECTION Q - VEGETABLES
Q-G-4   Potato Bar, Guidelines for
Q-G-5   Dehydrated Vegetables, Guidelines for
Q-G-6   Steam Cooking Vegetables, Guidelines for

A. GENERAL INFORMATION No. 1 (1)

INFORMATION FOR STANDARDIZED RECIPES

Standardized recipes are a necessity for a well-run food service operation. All of the recipes have been developed, tested and standardized for product quality, consistency and yield. Recipes are the most effective management tool for guiding the requisitioning of supplies and controlling breakouts and inventory. The U. S. Dietary Guidelines were among the many considerations in both the selection and development of the recipes included in the file. Many of the recipes have been modified to reduce fat, salt and calories. For new and experienced cooks, consistent use of standardized recipes is essential for quality and economy. The **Armed Forces Recipe Service** contains over 1600 tested recipes yielding 100 portions printed on cards.

Yield - The quantity of cooked product a recipe produces. The yield for each recipe in the Armed Forces Recipe is generally given as 100 portions and in some recipes in count or volume, e.g., 2 pans, 8 loaves, 6-1/2 gallons. Portion size is key to determining the quantity of food to be prepared. Many recipes also specify the weight per portion. For example, 3/4 cup (6-1/2 ounces) Beef Stroganoff.

Ingredients Column – Ingredients are listed in the order used. The specific form or variety of each ingredient is indicated. For example:

Flour, wheat, general purpose	Eggs, whole	Sugar, granulated
Flour, wheat, bread	Egg whites	Sugar, brown

Measure, Weights, and Issue Columns – Measures and Weights indicate the Edible Portion (E.P.) quantity of the ingredient required to prepare the recipe for 100 portions. The issue column represents the As Purchased (A.P.) quantity required if this amount is different from the E.P. quantity.

Method Column - Describes how the ingredients are to be combined and cooked. For example, the method will describe the order in which to sift dry ingredients, to thicken a sauce, or to fold in beaten egg whites. The method contains directions for the most efficient order of work, eliminating unnecessary tools and equipment and unnecessary steps in preparation.

A. GENERAL INFORMATION No. 1(2)

INFORMATION FOR STANDARDIZED RECIPES
RECIPE CONVERSION

Since few dining facilities serve exactly 100 persons, and, in some instances, the acceptable size portion may be smaller or larger, it is often necessary to reduce or increase a recipe. You may adjust the recipe to yield the number of portions needed, or to use the amount of ingredients available, or to produce a specific number of smaller portions. When increasing or decreasing a recipe, the division or multiplication of pounds and ounces is simplified when decimals are used.

1. To convert the quantities to decimals, use this table:

Weight in Ounces	Decimal of Pound	Weight in Ounces	Decimal of Pound
1	.06	9	.56
2	.13	10	.63
3	.19	11	.69
4 (1/4 lb)	.25	12 (3/4 lb)	.75
5	.31	13	.81
6	.38	14	.88
7	.44	15	.94
8 (1/2 lb)	.50	16 (1 lb)	1.00

For example: 1 lb 4 oz is converted to 1.25 lb; 2 lb 10 oz is converted to 2.63 lb.

A. GENERAL INFORMATION No. 1(2)

2. To adjust the recipe to yield a specific number of portions:

> First -- Obtain a working factor by dividing the number of portions needed by 100. For example:
>
> 348 (portions needed) ÷ 100 = 3.48 (Working Factor)
>
> Then -- Multiply the quantity of each ingredient by the working factor. For example:
>
> 1.25 lb (recipe) X 3.48 (Working Factor) = 4.35 lb (quantity needed).
>
> The part of the pound is converted to ounces by multiplying the decimal by 16. For example:
>
> .35 lb X 16 ounces = 5.60 ounces
>
> After the part of the pound has been converted to ounces, use the following scale to "round off":
>
> | | | | | | |
> |---|---|---|---|---|---|
> | .00 to .12 | = | 0 | .63 to .87 | = | 3/4 ounce |
> | .13 to .37 | = | 1/4 ounce | .88 to .99 | = | 1 ounce |
> | .38 to .62 | = | 1/2 ounce | | | |
>
> Thus 5.60 ounces will be "rounded off" to 5 1/2 ounces, and 4 lb 5 1/2 ounces will be the quantity needed (equal to 4.35 lb).

A. GENERAL INFORMATION No. 1(3)

INFORMATION FOR STANDARDIZED RECIPES RECIPE CONVERSION

3. To adjust the recipe for volume:

   First -- Obtain a working factor by dividing the number of portions needed by 100 as shown in Step 2 of A.1, Recipe Conversion.

$$333/100 = 3.33$$

   Then -- Multiply the quantity of each ingredient by the working factor. You will round off to the nearest 1/4 teaspoon. For example, the recipe calls for 6 gallons of water per 100 portions. Portions to prepare are 333.

$$333/100 = 3.33 \text{ Working Factor (W/F)}$$

1. W/F x No. of gallons	= gallon	3.33 W/F x 6	= 19.98 GL
2. Decimal (of gal) x 4	= quart (QT)	.98 GL x 4	= 3.92 QT
3. Decimal (of quart) x 2	= pint (PT)	.92 QT x 2	= 1.84 PT
4. Decimal (of pint) x 2	= cup (C)	.84 PT x 2	= 1.68 C
5. Decimal (of tbsp) x 16	= tablespoon (TBSP)	.68 C x 16	= 10.88 TBSP
6. Decimal (of tbsp) x 3	= teaspoon (TSP)	.88 TBSP x 3	= 2.64 TSP
7. Round off decimal portion (see paragraph 2)		.64 TSP	= 3/4 TSP

A. GENERAL INFORMATION No. 1(3)

The amount of water needed for 333 portions is: 19 GL, 3 QT, 1 PT, 1 C, 10 TBSP and 2 3/4 TSP.

NOTE: 4 QT = 1 GL   2 C = 1PT   3 TSP = 1 TBSP
        2 PT = 1 QT   16 TBSP = 1C

4. To adjust the recipe on the basis of a quantity of an ingredient to be used:

    First -- Obtain a Working Factor by dividing the pounds you have to use by the pounds required to yield 100 portions.

    For example:

    102 lb ÷ 30 (lb per 100 servings) = 3.40 (Working Factor)

    Then -- Multiply the quantity of each ingredient in the recipe by the Working Factor.

5. To adjust the recipe to yield a specific number of portions of a specific size:

    First -- Divide the desired portion size by the standard portion of the recipe.
         3 oz (desired size) ÷ 4 oz (standard portion) = .75
         348 (servings needed) x .75 = 261
         261 ÷ 100 = 2.61 (Working Factor)

    Then -- Multiply the quantity of each ingredient in the recipe by the Working Factor.

DEFINITION OF TERMS USED IN FOOD PREPARATION

Bake	To cook by dry heat in an oven, either covered or uncovered.
Barbecue	To roast or cook slowly, basting with a highly seasoned sauce.
Baste	To moisten food with liquid or melted fat during cooking to prevent drying of the surface and to add flavor.
Batch Preparation	A predetermined quantity or number of servings of food that is to be prepared at selected time intervals in progressive cookery for a given meal period to ensure fresh, high quality cooked food to customers.
Beat	To make a mixture smooth by using a fast regular circular and lifting motion which incorporates air into a product.
Blanch	To partially cook in deep fat, boiling water or steam.
Blend	To mix two or more ingredients thoroughly.
Boil	To cook in liquid at boiling point (212° F.) in which bubbles rise and break at the surface.
Braise	To brown in small amount of fat, then to cook slowly in small amount of liquid below the boiling point in a covered utensil.
Bread	To cover with crumbs or other suitable dry coating ingredient; or to dredge in a mixture of flour seasonings, and/or condiments, dip in a mixture of milk and slightly beaten eggs and then dredge in crumbs.
Broil	To cook by direct exposure to heat.
Brown	To produce a brown color on the surface of food by subjecting it to heat.

A. GENERAL INFORMATION No. 2 (1)

Chop	To cut food into irregular small pieces.
Cream	To mix until smooth, so that the resulting mixture is softened and thoroughly blended.
Crimp	To pinch together in order to seal.
Cube	To cut any food into square-shaped pieces.
Dice	To cut into small cubes or pieces.
Dock	To punch a number of vertical impressions in a dough with a smooth round stick about the size of a pencil to allow for expansion and permit gases to escape during baking.
Dredge	To coat with crumbs, flour, sugar or corn meal.
Fermentation	The process by which yeast acts on the sugar and starches in the dough to produce carbon dioxide gas and alcohol, resulting in expansion of the dough. During this period, the dough doubles in bulk.
Flake	To break lightly into small pieces.
Fold	To blend two or more ingredients together with a cutting and folding motion.
Fry	To cook in hot fat.
Garnish	To decorate with small pieces of colorful food.

A. GENERAL INFORMATION No. 2 (2)

Glaze	A glossy coat given to foods, as by covering with a sauce or by adding a sugary syrup, icing, etc.
Gluten	A tough elastic protein that gives dough its strength and ability to retain gas.
Grate	To rub food on a grater and thus break it into tiny pieces.
Grill	To cook, uncovered, on a griddle, removing grease as it accumulates. No liquid is added.
Knead	To work dough by folding and pressing firmly with palms of hands, turning between foldings.
Marinade	A preparation containing spices, condiments, vegetables, and aromatic herbs, and a liquid (acid or oil or combination of these) in which a food is placed for a period of time to enhance its flavor or to increase its tenderness.
Marinate	To allow to stand in a marinade to add flavor or tenderness.
Mince	To cut or chop into very small pieces.
Panbroil	To cook uncovered in a hot frying pan, pouring off fat as it accumulates.
Pare	To cut away outer covering.
Peel	To remove the outer layer of skin of a vegetable or fruit, etc.

Progressive Cookery	The continuous preparation of food in successive steps during the entire serving period (i.e., continuous preparation of vegetables, cook-to-order hamburgers, steaks, fried eggs, pancakes). This procedure ensures fresh, high quality cooked food to customers on a continuous basis. See Batch Preparation.
Proof	To allow shaped and panned yeast products like bread and rolls to double in size under controlled atmospheric conditions.
Reconstitute	To restore to liquid state by adding water. Also to reheat frozen prepared foods.
Rehydrate	To soak, cook, or use other procedures with dehydrated foods to restore water lost during drying.
Roast	To cook by dry heat; usually uncovered, in an oven.
Roux	Roux is a French word for a mixture of flour and fat, cooked to eliminate the raw, uncooked taste of flour.
Sauté	To brown or cook in small amount of fat.
Scald	To heat a liquid over hot water or direct heat to a temperature just below the boiling point.
Scale	To measure a portion of food by weighing.
Scant	Not quite up to stated measure.
Score	To make shallow cuts across top of a food item.
Seasoned Flour or Crumbs	A mixture of flour or crumbs with seasonings.

Shred	To cut or tear into thin strips or pieces using a knife or a shredder attachment.
Sift	To put dry ingredients through a sieve.
Simmer	To cook gently in a liquid just below the boiling point (190° F. - 210° F.); bubbles will form slowly and break at the surface.
Slurry	A lump-free mixture made by whipping cornstarch or flour into cold water or other liquids
Steam	To cook over or surrounded by steam.
Stew	To simmer in enough liquid to cover solid foods.
Stir	To mix two or more ingredients with a circular motion.
Temper	To remove from freezer and place under refrigeration for a period of time sufficient to facilitate separation and handling of frozen product. Internal temperature of the food should be approximately 26° F. to 28° F.
Thaw	To remove from freezer and place under refrigeration approximately 18-48 hours. Internal temperature should be above 30° F.
Toss	To mix ingredients lightly.
Wash	The liquid brushed on the surface of unbaked pies or turnovers to give a golden brown color to the crust or on the surface of proofed breads and rolls before baking and on baked bread and rolls to give a shine to the crust.
Whip	To beat rapidly with wire whip to increase volume by incorporating air.

A. GENERAL INFORMATION No. 4

TABLE OF MEASURING EQUIVALENTS

TSP	TBSP	FLUID OUNCES	CUPS	SCOOPS	LADLES	FLUID MEASURE
3	1	1/2		1-No. 40		
	1-1/2	3/4		1-No. 30	Size 0	
	2	1		1-No. 24		
	2-2/3	1-1/3		1-No. 20		
	3	1-1/2		1-No. 16	Size 1	
	4	2	1/4	1-No. 12		
	5-1/3	2-2/3	1/3	1-No. 10		
	6	3	3/8	1-No. 8	Size 2	
	8	4	1/2	1-No. 6		
	10-2/3	5-1/3	2/3			
	12	6	3/4			
	14	7	7/8			
	16	8	1		Size 3	1/2 pt
	18	9	1-1/8			
		12	1-1/2		Size 4	3/4 pt
		16	2			1 pt
		24	3			1-1/2 pt
		32	4			1 qt
		64	8			2 qt
		128	16			1 gal

NOTE: 1. Use ladles to serve individual portions of liquid or semi-liquid foods.
2. Scoop number indicates the number of portions per quart.

A. GENERAL INFORMATION No. 5

TABLE OF WEIGHTS AND MEASURES FOR CAN SIZES

CAN SIZE	AVERAGE NET WEIGHT OR FLUID MEASURE PER CAN (SEE NOTE)	AVERAGE CUPS PER CAN	APPROX. CANS PER CASE	NO. CANS EQUIV. NO. 10 CN
No. 10	6 lb 8 oz	12-1/2	6	1
No. 3 cyl	3 lb 2 oz (46 fl oz)	5-3/4	12	2
No. 3 (vacuum)	1 lb 7 oz	2-3/4	24	4-1/2
No. 2-1/2	1 lb 12 oz	3-1/2	24	4
No. 2	1 lb 4 oz	2-1/3	24	5
No. 303	1 lb	2	24	7
No. 300	14 oz	1-3/4	24	7
No. 2 (vacuum)	12 oz	1-1/2	24	8
No. 1 picnic	11 oz	1-1/4	48	10

NOTE: The net weight on can or jar labels differs among foods due to different densities of foods. For example: A No. 10 cn contains 6 lb 3 oz sauerkraut or 7 lb 5 oz cranberry sauce.

A. GENERAL INFORMATION No. 8(1)

TABLE OF EGG EQUIVALENTS

FRESH WHOLE EGGS (SHELLED)			DEHYDRATED EGG MIX		
Medium Size	Weight	Volume	Weight	Volume (Approx.)	Water to be Added
1 egg	1.6 oz	3 tbsp	1/2 oz	2 tbsp	2-1/2 tbsp
2 eggs	3.2 oz	6 tbsp	1 oz	1/4 cup	5 tbsp
10 eggs*	1 lb	1-7/8 cups	5 oz	1-1/4 cups	1-1/2 cups
12 eggs	1 lb 3.2 oz	2-1/4 cups	6 oz	1-1/2 cups	scant - 2 cups
20 eggs	2 lb	3-3/4 cups	10 oz	2-1/2 cups	3 cups
40 eggs	4 lb	7-1/2 cups	20 oz	1-1/4 qt (1-No. 3 cyl can)	1-1/2 qt

* 10 large eggs = 1 lb 2 oz

NOTES:
1. Frozen Whole Eggs and Frozen Egg Whites may be used in equivalent weights to shelled fresh whole eggs.
2. Dehydrated Egg Mix may be used in most recipes requiring whole eggs as shown in the table above. DO NOT USE RECONSTITUTED EGGS IN UNCOOKED SALAD DRESSINGS OR OTHER RECIPES WHICH DO NOT REQUIRE COOKING. RECONSTITUTED DEHYDRATED EGG MIX SHOULD BE USED WITHIN ONE HOUR UNLESS REFRIGERATED. DO NOT HOLD OVERNIGHT. For greater accuracy, weigh dehydrated egg mix.

3. *Reconstitution Methods for Dehydrated Egg Mix*

   a. Method 1. Place dehydrated egg mix in bowl; stir with a wire whip; add 1/2 of the water; whip until a smooth paste is formed; add remaining water; whip until mixture is blended.

   b. Method 2. Add dehydrated egg mix to water; stir to moisten; let stand 5 minutes; whip until smooth.

For Baked Products

   a. Method 1. Reconstitute dehydrated egg mix; substitute for eggs in recipe.

   b. Method 2. Sift dehydrated egg mix with dry ingredients; add water in step in Method column where whole eggs are incorporated.

For Batter Dips. Dehydrated egg mix may be reconstituted and used.

A. GENERAL INFORMATION No. 9(1)

GUIDELINES FOR CONTAINER YIELDS FOR CANNED FRUITS

TYPE OF FRUIT	PORTION SIZE (Approximate)	CAN SIZE	NO. OF CANS FOR 100 PORTIONS
Applesauce	1/2 cup	No. 303 cn	25
		No. 10 cn	4
Applesauce, Instant	1/2 cup	No. 2-1/2 cn	4
Apricots, halved	3 to 5 halves	No. 2-1/2 cn	16
		No. 10 cn	4
Blueberries	1/2 cup	No. 10 cn	4
Cherries, sweet, dark or light, pitted or unpitted	1/2 cup	No. 303 cn	25
		No. 10 cn	4
Cranberry Sauce, strained	1/4 cup	No. 303 cn or 300 cn	13
Cranberry Sauce, whole	1/4 cup	No. 10 cn	2
Figs, Kadota	3 to 4 figs	No. 303 cn	25
Fruit Cocktail	1/2 cup	No. 2-1/2 cn	16
		No. 10 cn	4
Fruit Mix, chunks	1/2 cup	No. 303 cn	25
		No. 10 cn	4
Grapefruit	1/2 cup	No. 303 cn	25
		No. 3 cyl cn	8

A. GENERAL INFORMATION No. 9(1)

TYPE OF FRUIT	PORTION SIZE (Approximate)	CAN SIZE	NO. OF CANS FOR 100 PORTIONS
Peaches, halves	2 halves	No. 2-1/2 cn	16
		No. 10 cn	4
Peaches, quarters or slices	1/2 cup	No. 2-1/2 cn	16
		No. 10 cn	4
Pears, halves	2 halves	No. 2-1/2 cn	16
		No. 10 cn	4
Pears, quarters or slices	1/2 cup	No. 2-1/2 cn	16
		No. 10 cn	4
Pineapple, chunks or tidbits	1/2 cup	No. 2 cn	20
		No. 10 cn	4
Pineapple slices	1 large or 2 small slices	No. 2 cn	20
		No. 10 cn	4
Plums, whole	2 to 3 plums	No. 2-1/2 cn	16
		No. 10 cn	4
Prunes, whole, unpitted	3 prunes	No. 10 cn	1-1/2

A. GENERAL INFORMATION No. 10(1)

NONFAT DRY MILK
RECONSTITUTION CHART FOR COOKING

Nonfat Dry Milk (Conventional) +	Water =	Fluid Skim Milk
1-2/3 tbsp	1/2 cup	1/2 cup
3 tbsp	1 cup	1 cup
1-2/3 oz (6 tbsp)	1-7/8 cups	2 cups
3-1/4 oz (3/4 cup)	3-3/4 cups	1 qt
5 oz (1-1/8 cups)	5-3/4 cups	1-1/2 qt
6-1/2 oz (1-1/2 cups)	7-1/2 cups	2 qt
8 oz (1-7/8 cups)	9-1/2 cups	2-1/2 qt
10 oz (2-1/4 cups)	11-1/2 cups	3 qt
11-1/4 oz (2-2/3 cups)	3-1/3 qt	3-1/2 qt
13 oz (3 cups)	3-3/4 qt	1 gal
1 lb 10 oz (1-1/2 qt)	1-7/8 gal	2 gal
2 lb 7 oz (2-1/4 qt)	2-7/8 gal	3 gal
4 lb 2 oz (3-3/4 qt)	4-3/4 gal	5 gal
5 lb 2 oz (4-3/4 qt)	6 gal	6-1/4 gal

A. GENERAL INFORMATION No. 10(1)

NOTE:
1. Recipes in this file use conventional nonfat dry milk.
2. Instant nonfat dry milk may be substituted on a pound for pound basis for the nonfat dry milk specified in any recipe. It should be weighed because the measures for instant nonfat dry milk are different from measures for nonfat dry milk (conventional). Nonfat dry milk, instant settles. If instant milk must be measured, follow directions on the container.
3. For best results, nonfat dry milk should be weighed instead of measured. Measures vary from one manufacturer to another. However, as a general rule, 1 ounce of nonfat dry milk will measure 3-2/3 tablespoons, and 4-1/2 ounces of nonfat dry milk will measure 1 cup.
4. Dry milk must be reconstituted in clean containers using clean utensils and must be treated like fresh milk after it is reconstituted. It must be refrigerated and protected from contamination.
5. Dry milk reconstitutes more easily in warm water. It should be stirred into the water with a circular motion using a whip or slotted spoon. It may also be reconstituted in a mixer if a large quantity is being prepared. However, it should be mixed at low speed to prevent excessive foaming.
6. If nonfat dry milk is to be used for a beverage, it should be weighed using 1 lb dry milk and 3-3/4 qt water per gallon. Chill thoroughly before serving. For 100 portions (8 oz), use 6 lb 4 oz nonfat dry milk and 23-1/2 qt water.

A. GENERAL INFORMATION No. 11(1)

GUIDELINES FOR USE OF DEHYDRATED ONIONS, GREEN PEPPERS, AND PARSLEY

ONIONS

Dehydrated, chopped and dehydrated compressed, chopped onions may be used in any recipe which specifies "onions, fresh, chopped or sliced."

REHYDRATION GUIDE:	Dehydrated Onions +	Water = (70-90° F.)	Rehydrated Onions **OR**	Fresh Onion Equivalent*
Dehydrated chopped onions	2 oz (9-2/3 tbsp)	1-1/2 cups	8 oz (1-1/4 cups)	1 lb (3 cups) (1 lb 1-3/4 oz A.P.)
	3-1/3 oz (1 cup)	2-1/2 cups	13 oz (2 cups)	1 lb 10 oz (4-3/4 cups) (1 lb 13 oz A.P.)
	1 lb (4-7/8 cups)	3 qt	4 lb (2-1/2 qt)	8 lb (1-1/2 gal) (8 lb 14 oz A.P.)
	2 lb 8 oz (3 qt-1 No. 10 cn)	7-1/2 qt	10 lb (6-1/4 qt)	20 lb (3-3/4 gal) (22 lb 3 oz A.P.)
Dehydrated, compressed chopped onions	1-3/4 oz	1-1/2 cups	8 oz (1-1/8 cups)	1 lb (3 cups) (1 lb 1-3/4 oz A.P.)
	2-1/3 oz	2 cups	10-1/2 oz (1-3/8 cups)	1 lb 5 oz (1 qt) (1 lb 7 oz A.P.)
	1 lb	3 qt	4 lb 8 oz (2-1/2 qt)	9 lb (6-3/4 qt) (10 lb A.P.)
	1 lb 3 oz (1 No. 2-1/2 cn)	3-1/2 qt	5 lb 5-1/2 oz (3 qt)	10 lb 11 oz (2 gal) (11 lb 14 oz A.P.)

* Volume is for chopped onions.

FOR RECIPES WITH SMALL AMOUNTS OF LIQUID: Cover dehydrated onions with 70° F. to 90 F. water. Stir dehydrated compressed onions occasionally to break apart. Let dehydrated onions stand 30 minutes; compressed dehydrated onions 1 hour or more. Drain. Note: Weight of rehydrated onions will be less than weight of dry onions but appearance and flavor will be similar.

FOR SOUPS, STEWS, SAUCES OR RECIPES WITH A LOT OF LIQUID: Add dehydrated chopped or dehydrated compressed onions directly.

A. GENERAL INFORMATION No. 11(1)

GREEN PEPPERS

Dehydrated green peppers may be used in any recipe which specifies "peppers, sweet, diced or chopped."

REHYDRATION GUIDE:	Dehydrated Peppers +	Cold Water = (35-55° F.)	Rehydrated Peppers **OR**	Sweet Peppers Equivalent*
	1 oz (2/3 cup)	2 cups	6-1/2 oz (1-1/3 cups)	6-1/2 oz (1-1/4 cups) (8 oz A.P.)
	1 lb (2-1/2 qt)	2 gal	6 lb 8 oz (5-1/2 qt)	6 lb 8 oz (1-1/4 gal)(7 lb 15 oz A.P.)

* Volume is for chopped peppers

FOR SALADS OR UNCOOKED DISHES: Cover with cold water. Refrigerate 1 hour or overnight. Drain.
FOR RECIPES WITH SMALL AMOUNTS OF LIQUID: Cover with cold water. Let stand 30 minutes. Drain.
FOR SOUPS, STEWS, SAUCES OR RECIPES WITH A LOT OF LIQUID: Add dehydrated peppers directly.

PARSLEY

Dehydrated parsley may be used in any recipe which specifies "chopped, fresh parsley."

REHYDRATION GUIDE:	Dehydrated Parsley +	Cold Water = (30-35° F.)	Rehydrated Parsley **OR**	Fresh Parsley Equivalent*
	1 oz (1-2/3 cup)	3-1/3 cups	8 oz (1-3/4 cups)	9 oz (4-1/4 cups) (9-1/2 oz A.P.)

* Volume is for chopped parsley

FOR SALADS OR UNCOOKED DISHES: Cover with ice cold water. Let stand 3 to 5 minutes. Drain.
FOR SOUPS, STEWS, SAUCES OR RECIPES WITH A LOT OF LIQUID: Add dehydrated parsley directly.

A. GENERAL INFORMATION No. 13(1)

GUIDELINES FOR FRUIT BARS

Fruit bars provide important sources of nutrients such as Vitamins A and C, and fiber. All fruits are low in fat and calories and none contain cholesterol. They may be set up for service at breakfast, lunch, dinner and brunch meals. A variety of fresh, canned and frozen fruits may be used.

Preparation: Wash all fresh fruits except bananas. Drain well. Refrigerate until ready to serve. Keep bananas in a cool, dry place until ready to serve.

ITEM	PORTION SIZE	100 PORTIONS A.P. WEIGHT OR CONTAINER	E.P.
Apples, canned, drained	1/4 cup (1-1/2 oz)	13 lb 8 oz (2-No. 10 cn)	12 lb
Apples, fresh, eating	1 apple (6 oz)	37 lb 8 oz	
Applesauce, canned	1/4 cup (2 oz)	14 lb 10 oz (2-1/6-No. 10 cn)	
Apricots, canned, halves, drained	3 halves (1-1/2 oz)	20 lb 4 oz (3-No. 10 cn)	11 lb 10 oz
Apricots, fresh	2 apricots (2-1/2 oz)	16 lb 11 oz	
Bananas, fresh, peeled, sliced	1/2 cup (2-1/2 oz)	28 lb	18 lb 3 oz
Bananas, fresh	1 banana (6 oz)	40 lb	

A. GENERAL INFORMATION No. 13(1)

		100 PORTIONS	
ITEM	PORTION SIZE	A.P. WEIGHT OR CONTAINER	E.P.
Blueberries, canned drained	1/2 cup (4-1/2 oz)	52 lb 10 oz (8-1/4-No. 10 cn)	28 lb 6 oz
Cantaloupe, fresh, quartered, unpared	1/4 small cantaloupe (3 oz)	21 lb 14 oz	
Cantaloupe, fresh, pared, 1 inch pieces	1/2 cup (2-1/2 oz)	35 lb	17 lb 14 oz
Casaba melons, fresh, unpared, sliced	1/10 melon (4 oz)	31 lb 4 oz	
Casaba melons, fresh, pared 1 inch pieces	1/2 cup (2-1/2 oz)	29 lb 11 oz	17 lb 12 oz
Cherries, canned, sweet, drained	1/2 cup (3-1/2 oz)	38 lb 13 oz (5-3/4 No.-10 cn)	23 lb 14 oz
Cherries, fresh, sweet	1/2 cup (2-1/2 oz)	17 lb 10 oz	

GUIDELINES FOR FRUIT BARS - CONTINUED

ITEM	PORTION SIZE	100 PORTIONS	
		A.P. WEIGHT OR CONTAINER	E.P.
Coconut, prepared, sweetened, flakes	1 tbsp	1 lb 5 oz	
Fruit cocktail, canned, drained	1/2 cup (4 oz)	42 lb 3 oz (6-1/4-No. 10 cn)	27 lb 12 oz
Fruits, chunks, mixed, canned, drained	1/2 cup (3 oz)	39 lb 2 oz (5-3/4-No. 10 cn)	26 lb 3 oz
Grapefruit, canned, drained	1/2 cup (4 oz)	46 lb 14 oz (15-No. 3 cyl cn or 47-No. 303 cn)	25 lb 10 oz
Grapefruit, fresh, halved	1/2 grapefruit (8-3/4 oz)	54 lb 11 oz	
Grapefruit, fresh, segments	1/2 cup (4 oz)	48 lb	25 lb
Grapes, fresh	1/2 cup (2-1/2 oz)	16 lb 11 oz	
Honeyball melons, fresh, unpared, sliced	1/10 melon (3 oz)	40 lb 15 oz	
Honeyball melons, fresh, pared, 1 inch pieces	1/2 cup (2-1/2 oz)	37 lb 14 oz	17 lb 7 oz
Honeydew melons, fresh, unpared, sliced	1/10 melon (3 oz)	40 lb 15 oz	

A. GENERAL INFORMATION No. 13(2)

ITEM	PORTION SIZE	100 PORTIONS	
		A.P. WEIGHT OR CONTAINER	E.P.
Honeydew melons, fresh, pared, 1 inch pieces	1/2 cup (2-1/2 oz)	37 lb 14 oz	17 lb 7 oz
Kiwifruit, fresh, pared, sliced	2 slices (1/2 oz)	5 lb 14 oz	5 lb 1 oz
Mangoes, fresh, pared, diced	1/2 cup (3 oz)	27 lb 12 oz	19 lb 3 oz
Mangoes, fresh, pared, sliced	4 slices (2 oz)	18 lb 9 oz	12 lb 12 oz
Nectarines, fresh	1 nectarine (4-1/2 oz)	28 lb 2 oz	
Oranges, fresh, peeled, sliced	3 slices (2 oz)	20 lb 9 oz	14 lb 9 oz
Oranges, fresh	1 orange (6 oz)	37 lb 8 oz	
Oranges, Mandarin, canned, drained	1/4 cup (1-1/2 oz)	20 lb 4 oz (3 No. 10 cn)	10 lb 15 oz
Papaya, fresh, pared, seeded, cubed	1/2 cup (2-1/2 oz)	24 lb	15 lb 11 oz

GUIDELINES FOR FRUIT BARS - CONTINUED

ITEM	PORTION SIZE	100 PORTIONS A.P. WEIGHT OR CONTAINER	E.P.
Papaya, fresh, pared, sliced	3 slices (2 oz)	22 lb 8 oz	14 lb 11 oz
Peaches, canned, halves, drained	2 halves (4 oz)	45 lb 9 oz (6-3/4-No 10 cn)	27 lb 7 oz
Peaches, canned, quarters/slices, drained	1/2 cup (4 oz)	43 lb 14 oz (6-1/2-No. 10 cn)	27 lb
Peaches, fresh	1 peach (4 oz)	25 lb	
Peaches, frozen	1/2 cup (4 oz)	27 lb 13 oz (4-1/4-No. 10 cn)	
Pears, canned, halves, drained	2 halves (3-1/2 oz)	41 lb 7 oz (6-1/4-No. 10 cn)	25 lb
Pears, canned, quarters/slices, drained	1/2 cup (3-1/2 oz)	36 lb 7 oz (5-1/2-No. 10 cn)	22 lb 8 oz
Pears, fresh	1 pear (5-1/2 oz)	36 lb	
Persian melons, fresh, unpared, sliced	1/10 melon (3 oz)	45 lb 13 oz	
Persian melons, fresh, pared, diced	1/2 cup (2-1/2 oz)	41 lb 4 oz	17 lb 5 oz
Pineapple, canned, chunks/tidbits, drained	1/2 cup (3-1/2 oz)	37 lb 2 oz (5-1/2-No. 10 cn)	22 lb 10 oz

Pineapple, canned, slices, drained	2 slices (2 oz)	25 lb 5 oz (3-3/4 No. 10 cn)	14 lb 7 oz
Pineapple, fresh, pared, cored, 1 inch pieces	1/2 cup (2-1/2 oz)	33 lb 4 oz	17 lb 5 oz
Plums, canned, drained	3 plums (2-1/2 oz)	32 lb 1 oz (4-3/4-No. 10 cn)	17 lb 13 oz
Plums, fresh	1 plum (2-1/2 oz)	15 lb 10 oz	
Prunes, whole, canned, drained	3 prunes (1-1/2 oz)	10 lb 1 oz (1-2/5-No. 10 cn)	9 lb 10 oz
Raisins	1 tbsp	2 lb 4 oz (1/2-No. 10 cn)	
Raspberries, frozen	1/2 cup (4 oz)	27 lb 13 oz (4-1/4-No. 10 cn)	
Strawberries, fresh, sliced	1/2 cup (2-1/2 oz)	18 lb 4 oz	17 lb 3 oz
Strawberries, fresh, whole	1/2 cup (2-1/2 oz)	16 lb 9 oz	15 lb 10 oz
Strawberries, frozen, sliced	1/2 cup (4 oz)	27 lb 13 oz (4-1/4-No. 10 cn)	
Tangelos, fresh	1 tangelo (6 oz)	37 lb 8 oz	
Tangerines, fresh	1 tangerine (3-1/2 oz)	22 lb 15 oz	
Watermelons, fresh, unpared, wedge (1 inch by 4 inches)	1 wedge (4 oz)	51 lb	
Watermelons, fresh, pared, 1 inch pieces	1/2 cup (2-1/2 oz)	34 lb	17 lb 11 oz

A. GENERAL INFORMATION No. 15(1)

CONVERSION OF QUANTITIES IN RECIPES
Weight Conversion Chart

The following chart for weights permit easy adjustment of recipes to yield the number of portions actually needed. Since recipes are based on 100 portions, find the amount as specified in the recipe under the column headed 100 portions, and then use the amount shown in the column with the heading for the number of portions to be prepared, i.e., if a recipe for 100 uses 1 pound of flour, find 1 pound under the column headed 100 portions and then look in the column under 125 portion and you will see that your should use 1 pound 4 ounces to prepare 125 portions of the item.

Oz = ounce   Lb = pound

10 Portions	25 Portions	50 Portions	75 Portions	100 Portions	125 Portions	150 Portions	175 Portions	250 Portions	275 Portions	300 Portions
1/10 oz	1/4 oz	1/2 oz	3/4 oz	1 oz	1 1/4 oz	1 1/2 oz	1 3/4 oz	2 1/2 oz	2 3/4 oz	3 oz
1/5 oz	1/2 oz	1 oz	1 1/2 oz	2 oz	2 1/2 oz	3 oz	3 1/2 oz	5 oz	5 1/2 oz	6 oz
3/10 oz	3/4 oz	1 1/2 oz	2 1/4 oz	3 oz	3 3/4 oz	4 1/2 oz	5 1/4 oz	7 1/2 oz	8 1/4 oz	9 oz
2/5 oz	1 oz	2 oz	3 oz	4 oz	5 oz	6 oz	7 oz	10 oz	11 oz	12 oz
1/2 oz	1 1/4 oz	2 1/2 oz	3 3/4 oz	5 oz	6 1/4 oz	7 1/2 oz	8 3/4 oz	12 1/2 oz	13 3/4 oz	15 oz
3/5 oz	1 1/2 oz	3 oz	4 1/2 oz	6 oz	7-1/2 oz	9 oz	10 1/2 oz	15 oz	1 lb	1 lb 2 oz
7/10 oz	1 3/4 oz	3 1/2 oz	5 1/4 oz	7 oz	8 3/4 oz	10 1/2 oz	12 1/4 oz	1 lb 2 oz	1 lb 4 oz	1 lb 5 oz

A. GENERAL INFORMATION No. 15(1)

10 Portions	25 Portions	50 Portions	75 Portions	100 Portions	125 Portions	150 Portions	175 Portions	250 Portions	275 Portions	300 Portions
4/5 oz	2 oz	4 oz	6 oz	8 oz	10 oz	12 oz	14 oz	1 lb 4 oz	1 lb 6 oz	1 lb 8 oz
7/8 oz	2 1/4 oz	4 1/2 oz	6 3/4 oz	9 oz	11 1/4 oz	13 1/2 oz	15 3/4 oz	1 lb 6 oz	1 lb 8 oz	1 lb 11 oz
1 oz	2 1/2 oz	5 oz	7 1/2 oz	10 oz	12 1/2 oz	15 oz	1 lb 2 oz	1 lb 10 oz	1 lb 12 oz	1 lb 14 oz
1 1/8 oz	2 3/4 oz	5 1/2 oz	8 1/4 oz	11 oz	13 3/4 oz	1 lb	1 lb 4 oz	1 lb 12 oz	1 lb 14 oz	2 lb 2 oz
1 1/4 oz	3 oz	6 oz	9 oz	12 oz	15 oz	1 lb 2 oz	1 lb 5 oz	1 lb 14 oz	2 lb 2 oz	2 lb 4 oz
1 1/3 oz	3 1/4 oz	6 1/2 oz	9 3/4 oz	13 oz	1 lb	1 lb 4 oz	1 lb 6 oz	2 lb	2 lb 4 oz	2 lb 8 oz
1 3/8 oz	3 1/2 oz	7 oz	10 1/2 oz	14 oz	1 lb 2 oz	1 lb 5 oz	1 lb 8 oz	2 lb 4 oz	2 lb 6 oz	2 lb 10 oz
1 1/2 oz	3 3/4 oz	7 1/2 oz	11 oz	15 oz	1 lb 2 oz	1 lb 6 oz	1 lb 10 oz	2 lb 5 oz	2 lb 10 oz	2 lb 14 oz
1 5/8 oz	4 oz	8 oz	12 oz	1 lb	1 lb 4 oz	1 lb 8 oz	1 lb 12 oz	2 lb 8 oz	2 lb 12 oz	3 lb
2 oz	5 oz	10 oz	15 oz	1 lb 4 oz	1 lb 10 oz	1 lb 14 oz	2 lb 4 oz	3 lb 2 oz	3 lb 8 oz	3 lb 12 oz
2 2/5 oz	6 oz	12 oz	1 lb 2 oz	1 lb 8 oz	1 lb 14 oz	2 lb 4 oz	2 lb 10 oz	3 lb 12 oz	4 lb 2 oz	4 lb 8 oz
2 4/5 oz	7 oz	14 oz	1 lb 5 oz	1 lb 12 oz	2 lb 4 oz	2 lb 10 oz	3 lb 2 oz	4 lb 6 oz	4 lb 14 oz	5 lb 4 oz
3 1/5 oz	8 oz	1 lb	1 lb 8 oz	2 lb	2 lb 8 oz	3 lb	3 lb 8 oz	5 lb	5 lb 8 oz	6 lb
3 3/5 oz	9 oz	1 lb 2 oz	1 lb 11 oz	2 lb 4 oz	2 lb 14 oz	3 lb 6 oz	4 lb	5 lb 10 oz	6 lb 4 oz	6 lb 12 oz
4 oz	10 oz	1 lb 4 oz	1 lb 14 oz	2 lb 8 oz	3 lb 2 oz	3 lb 12 oz	4 lb 6 oz	6 lb 4 oz	6 lb 14 oz	7 lb 8 oz

A. GENERAL INFORMATION No. 15(2)

CONVERSION OF QUANTITIES IN RECIPES
Weight Conversion Chart

10 Portions	25 Portions	50 Portions	75 Portions	100 Portions	125 Portions	150 Portions	175 Portions	250 Portions	275 Portions	300 Portions
4 2/5 oz	11 oz	1 lb 6 oz	2 lb 2 oz	2 lb 12 oz	3 lb 8 oz	4 lb 2 oz	4 lb 14 oz	6 lb 14 oz	7 lb 10 oz	8 lb 4 oz
4 4/5 oz	12 oz	1 lb 8 oz	2 lb 4 oz	3 lb	3 lb 12 oz	4 lb 8 oz	5 lb 4 oz	7 lb 8 oz	8 lb 4 oz	9 lb
5 1/5 oz	13 oz	1 lb 10 oz	2 lb 8 oz	3 lb 4 oz	4 lb 2 oz	4 lb 14 oz	5 lb 11 oz	8 lb 2 oz	9 lb	9 lb 12 oz
5 3/5 oz	14 oz	1 lb 12 oz	2 lb 10 oz	3 lb 8 oz	4 lb 6 oz	5 lb 4 oz	6 lb 2 oz	8 lb 12 oz	9 lb 10 oz	10 lb 8 oz
6 oz	15 oz	1 lb 14 oz	2 lb 14 oz	3 lb 12 oz	4 lb 11 oz	5 lb 10 oz	6 lb 10 oz	9 lb 6 oz	10 lb 5 oz	11 lb 4 oz
6 2/5 oz	1 lb	2 lb	3 lb	4 lb	5 lb	6 lb	7 lb	10 lb	11 lb	12
8 oz	1 lb 4 oz	2 lb 8 oz	3 lb 12 oz	5 lb	6 lb 4 oz	7 lb 8 oz	8 lb 12 oz	12 lb 8 oz	13 lb 12 oz	15 lb
9 3/5 oz	1 lb 8 oz	3 lb	4 lb 8 oz	6 lb	7 lb 8 oz	9 lb	10 lb 8 oz	15 lb	16 lb 8 oz	18 lb
11 1/5 oz	1 lb 12 oz	3 lb 8 oz	5 lb 4 oz	7 lb	8 lb 12 oz	10 lb 8 oz	12 lb 4 oz	17 lb 8 oz	19 lb 4 oz	21 lb
12 4/5 oz	2 lb	4 lb	6 lb	8 lb	10 lb	12 lb	14 lb	20 lb	22 lb	24 lb
1 lb	2 lb 8 oz	5 lb	7 lb 8 oz	10 lb	12 lb 8 oz	15 lb	17 lb 8 oz	25 lb	27 lb 8 oz	30 lb
1 lb 4 oz	3 lb	6 lb	9 lb	12 lb	15 lb	18 lb	21 lb	30 lb	33 lb	36 lb
1 lb 8 oz	3 lb 12 oz	7 lb 8 oz	11 lb 4 oz	15 lb	18 lb 12 oz	22 lb 8 oz	26 lb 4 oz	37 lb 8 oz	41 lb 4 oz	45 lb
2 lb	5 lb	10 lb	15 lb	20 lb	25 lb	30 lb	35 lb	50 lb	55 lb	60 lb
3 lb	7 lb 8 oz	15 lb	22 lb 8 oz	30 lb	37 lb 8 oz	45 lb	52 lb 8 oz	75 lb	82 lb 8 oz	90 lb

A. GENERAL INFORMATION No. 16(1)

CONVERSION OF QUANTITIES IN RECIPES
Measure Conversion Chart

The following chart for measures permits easy adjustments of recipes to yield the number of portions actually needed. Since recipes are based on 100 portions, find the amount as specified in the recipe under column headed 100 portions and then use the amount shown in the column with the heading for the number of portions to be prepared, i.e., if a recipe for 100 uses 3 cups of flour, find 3 cups under the column headed 100 portions and then look in the column under 125 portions and you will see that you should use 3 ¾ cups to prepare 125 portions of the item.

tsp – teaspoon   tbsp – tablespoon   qt – quart   gal - gallon

10 Portions	25 Portions	50 Portions	75 Portions	100 Portions	125 Portions	150 Portions	175 Portions	250 Portions	275 Portions	300 Portions
……	¼ tsp	½ tsp	¾ tsp	1 tsp	1 ¼ tsp	1 ½ tsp	1 ¾ tsp	2 ½ tsp	2 ¾ tsp	1 tbsp
……	½ tsp	1 tsp	1 ½ tsp	2 tsp	2 ½ tsp	1 tbsp	3 ½ tsp	1 2/3 tbsp	1 7/8 tbsp	2 tbsp
¼ tsp	¾ tsp	1 ½ tsp	2 tsp	1 tbsp	3 ¾ tsp	1 1/3 tbsp	1 2/3 tbsp	2 1/3 tbsp	2 2/3 tbsp	3 tbsp
½ tsp	1 ½ tsp	1 tbsp	1 2/3 tbsp	2 tbsp	2 2/3 tbsp	3 tbsp	3 2/3 tbsp	5 tbsp	5 2/3 tbsp	6 tbsp
¾ tsp	2 ¼ tsp	1 2/3 tbsp	2 1/3 tbsp	3 tbsp	¼ cup	4 2/3 tbsp	5 tbsp	7 2/3 tbsp	½ cup	9 tbsp
1 tsp	1 tbsp	2 tbsp	3 tbsp	¼ cup	5 tbsp	6 tbsp	7 tbsp	10 tbsp	11 tbsp	¾ cup
1 ½ tsp	3 ¾ tsp	2 2/3 tbsp	4 tbsp	5 tbsp	6 tbsp	7 2/3 tbsp	9 tbsp	12 2/3 tbsp	14 tbsp	1 cup
1 ¾ tsp	4 ½ tsp	3 tbsp	4 2/3 tbsp	6 tbsp	7 2/3 tbsp	½ cup	10 2/3 tbsp	15 tbsp	1 cup	1 cup + 2 tbsp
2 tsp	5 ¼ tsp	3 2/3 tbsp	5 tbsp	7 tbsp	9 tbsp	10 2/3 tbsp	¾ cup	1 cup + 1 2/3 tbsp	1 cup + 3 tbsp	1 1/3 cups

A. GENERAL INFORMATION No. 16(1)

10 Portions	25 Portions	50 Portions	75 Portions	100 Portions	125 Portions	150 Portions	175 Portions	250 Portions	275 Portions	300 Portions
2 ¼ tsp	2 tbsp	4 tbsp	6 tbsp	½ cup	10 tbsp	¾ cup	14 tbsp	1 ¼ cups	1 cup + 6 tbsp	1 ½ cups
2 ½ tsp	2 tbsp	4 2/3 tbsp	7 tbsp	9 tbsp	11 tbsp	13 2/3 tbsp	1 cup	1 cup + 6 tbsp	1 ½ cups	1 ¾ cups
1 tbsp	2 2/3 tbsp	5 tbsp	7 2/3 tbsp	10 tbsp	¾ cup	1 cup	1 cup + 2 tbsp	1 ½ cups	1 ¾ cups	2 cups
3 ¼ tsp	3 tbsp	5 2/3 tbsp	8 tbsp	11 tbsp	14 tbsp	1 cup	1 cup + 3 tbsp	1 ¾ cups	2 cups	2 1/8 cups
3 ½ tsp	3 tbsp	6 tbsp	9 tbsp	¾ cup	1 cup	1 cup + 2 tbsp	1 ¼ cups	2 cups	2 cups + 2 tbsp	2 ¼ cups
3 ¾ tsp	3 tbsp	6 2/3 tbsp	10 tbsp	13 tbsp	1 cup	1 ¼ cups	1 ½ cups	2 cups	2 ¼ cups	2 ½ cups
1 1/3 tbsp	3 2/3 tbsp	7 tbsp	10 2/3 tbsp	14 tbsp	1 cup + 2 tbsp	1 1/3 cups	1 ½ cups	2 cups + 3 tbsp	2 1/3 cups	2 ½ cups
4 ½ tsp	3 ¾ tbsp	7 2/3 tbsp	11 tbsp	15 tbsp	1 ¼ cups	1 ½ cups	1 ¾ cups	2 1/3 cups	2 ¾ cups	2 7/8 cups
4 ¾ tsp	¼ cup	½ cup	¾ cup	1 cup	1 ¼ cups	1 ½ cups	1 ¾ cups	2 ½ cups	2 ¾ cups	3 cups
2 tbsp	5 tbsp	10 tbsp	1 cup	1 ¼ cups	1 ½ cups	2 cups	2 ¼ cups	3 cups	3 ½ cups	3 ¾ cups
7 tsp	6 tbsp	¾ cup	1 cup + 2 tbsp	1 ½ cups	2 cups	2 ¼ cups	2 ¾ cups	3 ¾ cups	1 qt	4 ½ cups

A. GENERAL INFORMATION No. 16(2)

CONVERSION OF QUANTITIES IN RECIPES
Measure Conversion Chart

10 Portions	25 Portions	50 Portions	75 Portions	100 Portions	125 Portions	150 Portions	175 Portions	250 Portions	275 Portions	300 Portions
8 ¼ tsp	7 tbsp	14 tbsp	1 1/3 cups	1 ¾ cups	2 ¼ cups	2 ¾ cups	3 cups	4 ½ cups	4 ¾ cups	5 ¼ cups
9 ½ tsp	½ cup	1 cup	1 ½ cups	2 cups	2 ½ cups	3 cups	3 ½ cups	5 cups	5 ½ cups	1 ½ qt
10 ¾ tsp	½ cup + 1 tbsp	1 cup + 2 tbsp	1 ¾ cups	2 ¼ cups	2 ¾ cups	3 ½ cups	1 qt	5 ¾ cups	1 ½ qt	6 ¾ cups
¼ cup	10 tbsp	1 ¼ cups	2 cups	2 ½ cups	3 cups + 2 tbsp	3 ¾ cups	4 ½ cups	6 ¼ cups	1 ¾ qt	7 ½ cups
4 ¾ tbsp	¾ cup	1 ½ cups	2 ¼ cups	3 cups	3 ¾ cups	4 ½ cups	5 ¼ cups	7 ½ cups	8 ¼ cups	2 ¼ qt
5 2/3 tbsp	14 tbsp	1 ¾ cups	2 ½ cups	3 ½ cups	4 ½ cups	1 ¼ qt	1 ½ qt	2 ¼ qt	9 ¾ cups	10 ½ cups
6 ¼ tbsp	1 cup	2 cups	3 cups	1 qt	1 ¼ qt	1 ½ qt	1 ¾ qt	2 ½ qt	2 ¾ qt	3 qt
½ cup	1 ¼ cups	2 ½ cups	3 ¾ cups	1 ¼ qt	6 ¼ cups	7 ½ cups	8 ¾ cups	12 ½ cups	3 ½ qt	3 ¾ qt
9 ¾ tbsp	1 ½ cups	3 cups	4 ½ cups	1 ½ qt	7 ½ cups	2 ¼ qt	10 ½ cups	3 ¾ qt	1 gal	4 ½ qt
11 tbsp	1 ¾ cups	3 ½ cups	5 ¼ cups	7 cups	8 ¾ cups	10 ½ cups	3 qt	1 gal + 1 ½ cups	1 gal + 3 ¼ cups	5 ¼ qt
12 ¾ tbsp	2 cups	1 qt	1 ½ qt	2 qt	2 ¼ qt	3 qt	3 ½ qt	1 ¼ gal	5 ½ qt	1 ½ gal
1 ¼ cups	3 cups	1 ½ qt	2 ¼ qt	3 qt	3 ¾ qt	4 ½ qt	5 ¼ qt	7 ½ qt	2 gal	2 ¼ gal
1 ½ cups	1 qt	2 qt	3 qt	1 gal	1 ¼ gal	1 ½ gal	1 ¾ gal	2 ½ gal	2 ¾ gal	3 gal
3 cups	2 qt	1 gal	1 ½ gal	2 gal	2 ¼ gal	3 gal	3 ½ gal	5 gal	5 ½ gal	6 gal
4 ½ cups	3 qt	1 ½ gal	2 ¼ gal	3 gal	3 ¾ gal	4 ½ gal	5 ¼ gal	7 ¼ gal	8 gal	9 gal
1 ½ qt	1 gal	2 gal	3 gal	4 gal	5 gal	6 gal	7 gal	10 gal	11 gal	12 gal
7 ½ cups	1 ¼ gal	2 ½ gal	3 ¾ gal	5 gal	6 ¼ gal	7 ½ gal	8 ¾ gal	12 ½ gal	13 ¾ gal	15 gal

A. GENERAL INFORMATION No. 19(1)

GUIDELINES FOR HANDLING FROZEN FOODS

Proper storage and thawing procedures for frozen foods are essential for keeping foods safe and palatable. Some foods, such as vegetables, do not need to be thawed before cooking. Many recipes require meat to be only partially thawed or tempered, to facilitate separation before cooking; this prevents excessive moisture loss. Unless otherwise indicated, preparation methods and cooking times are for thawed meat, fish and poultry.

Frozen foods should be stored at or below 0° F. and thawed at 36° F. DO NOT refreeze foods that have been thawed; cook and serve as soon as possible to promote maximum quality and safety.

FROZEN FRUITS: Thaw unopened under refrigeration (36° F. to 38° F.) or covered with cold water.

FROZEN FRUIT JUICES AND CONCENTRATES: These do not require thawing.

FROZEN VEGETABLES: These do not require thawing before cooking. For faster cooking, Brussels sprouts, broccoli, asparagus, cauliflower, and leafy greens may be partially thawed under refrigeration.

FROZEN MEATS: Improper thawing of meat encourages bacterial growth and also results in unnecessary loss of meat juices, poor quality and loss of yield and nutrients. To thaw meat, remove from shipping container, but leave inside wrappings (usually polyethylene bags) on meat. Thaw under refrigeration (36° F. to 38° F.) until almost completely thawed. Spread out large cuts, such as roasts, to allow air to circulate. The length of the thawing period will vary accordingly to the size of meat cut, the temperature and degree of air circulation in the chill space, and the quantity of meat being thawed in a given space. Boneless meats generally require 26 to 48 hours to thaw at 36° F. to 38° F.

A. GENERAL INFORMATION No. 19(1)

Meat may be cooked frozen or tempered except for a few cuts which require complete thawing (i.e., bulk ground beef, bulk beef patty mix, braising Swiss steak, bulk pork sausage and diced beef for stewing.)

Roasts, when cooked from the frozen state, will require one-third to one-half more cooking time than thawed roasts. The addition of seasonings, if required, must be delayed until the outside is somewhat thawed and the surface is sufficiently moist to retain the seasonings. The insertion of meat thermometers must also be delayed until roasts are partially thawed. Grill steaks, pork chops and liver should be tempered before cooking to ensure a moist, palatable product. (Temper - To remove from freezer and place under refrigeration for a period of time sufficient to facilitate separation and handling of frozen product. Internal temperature of the food should be approximately 26° F. to 28° F.). Pork sausage patties and pork and beef sausage links should be cooked frozen.

FROZEN SEAFOOD: Fish fillets and steaks may be cooked frozen or thawed. Any fish that is to be breaded or batter dipped should be thawed. Clams, crabmeat, oysters, scallops and shrimp should be kept wrapped while thawing. Fish and shellfish should be thawed under refrigeration (36° F. to 38° F.) and require 12 hours to thaw.

Frozen, whole lobster, king crab legs, spiny lobster tail, breaded fish portions or nuggets, batter-dipped fish portions, or breaded oysters and shrimp SHOULD NOT be thawed before cooking.

FROZEN POULTRY: Poultry must be thawed under refrigeration (36° F. to 38° F.). Proper thawing of poultry reduces bacterial growth, maintains quality and retains nutrients through less drip loss.

A. GENERAL INFORMATION No. 19(2)

GUIDELINES FOR HANDLING FROZEN FOODS

RAW CHICKEN: Remove whole chickens from shipping containers and thaw in individual wrappers (plastic bags). To thaw parts or quarters, remove intermediate containers from shipping containers; remove overwrapping from intermediate containers and open intermediate containers to expose inner wrapping. Length of thawing period under refrigeration (36° F. to 38° F.) will vary according to size of chicken and refrigeration conditions.
Approximate Thawing Times: Chicken, whole - 37 hours; Chicken, quarters - 52 hours; Chicken, cut-up - 52 hours
PRECOOKED BREADED CHICKEN, NUGGETS OR FILLETS: DO NOT THAW before cooking.
PRECOOKED UNBREADED CHICKEN FILLETS: Temper. DO NOT THAW before cooking.
PREPARED FROZEN CHILIES RELLENOS, BURRITOS, PIZZAS, ENCHILADAS, LASAGNA, TAMALES, MANICOTTI, CANNELLONI: DO NOT THAW before cooking.
TURKEY: Remove turkeys from shipping containers. Thaw in individual wrappers under refrigeration (36° F. to 38° F.)
Approximate Thawing Times: Turkey, whole (16 lbs or less - 2 days; Turkey, whole (over 16 lbs) - 3 to 4 days; Turkey, boneless - 12 to 16 hours; Turkey, ground – thaw; Turkey sausage patties and links - cook frozen
FROZEN EGGS: Thaw under refrigeration (36 F. to 38 F.) or covered with cold water. Thirty pound cans require at least 2 days to thaw, 10 lb cans or cartons require at least 1 day.
FROZEN PIZZA BLEND CHEESE: If pizza blend cheese is received and stored as a frozen product, it should be thawed under refrigeration (36° F. to 38° F.) to ensure retention of its characteristic flavor, texture, and appearance. Thawing at room temperature will encourage bacterial growth (inherent in the product) resulting in an undesirable flavor and swelling of the container.

A. GENERAL INFORMATION No. 20

GUIDELINES FOR USE OF ANTIBROWNING AGENT
(NON-SULFATING AGENTS)

The purpose of an antibrowning agent is to prevent browning and maintain color and crispness in fresh potatoes and fruits.

DIRECTIONS FOR USE

1. Dissolve 1-3/4 oz (3 tbsp) antibrowning agent per gallon of cold water in a clean stainless steel, glass or plastic container. DO NOT use galvanized metal containers.

2. Dip fresh white potatoes (peeled, whole, quarters, French fry cut, slices) or fruits (apples, avocados, bananas, peaches, pears) peeled, sliced and free from bruises in the antibrowning solution. Soak for 3 minutes.

3. Drain and refrigerate product until ready to use.

NOTE: 1. Keep antibrowning agent stored in its original container. Make the solution fresh daily. A plastic measuring spoon should be kept with the antibrowning agent for easy measuring.

2. Antibrowning agent is not required for lettuce, cauliflower, green peppers, cabbage, celery or pineapple.

A. GENERAL INFORMATION No. 21

GUIDELINES FOR USE OF STEAM COOKERS

Use of steamers in quantity food preparation can save cooking time, labor, help maintain appearance of food, and preserve nutrients normally lost by other cooking methods. Steamers are ideal for batch preparation. Foods may be steamed and served in the same pan, if steam table pans are used for preparation.

Steamers are either 5 lb pressure or 15 lb pressure (high speed) type. When food is steamed at 5 lb pressure, the internal temperature of the steamer is 225° F. to 228° F. At 15 lb pressure, the temperature is 245° F. to 250° F.

Most canned, fresh or frozen vegetables, in addition to other foods such as rice, pasta, poultry, meats, fish, and shellfish, can be cooked in steamers.

Foods may be steamed in perforated or solid pans. Perforated pans are usually used, particularly for vegetables, unless the cooking liquid is retained or manufacturer's directions specify solid type pans. Pans are normally filled no more than 2/3 full to allow steam to circulate for even cooking.

Cooking times will vary depending on the type steamer, food, and temperature and quantity of the product. For best results follow the manufacturer's cooking times and directions. Cooking time should be scheduled to include bringing food up to cooking temperature, as well as steaming time. <u>Timing begins</u> when the pressure gauge registers 3 lb on the 5 lb steamer and 9 lb on the 15 lb steamer. <u>Be sure to use timer, if available, to prevent overcooking.</u>

After cooking is completed, the steam should be exhausted slowly for safety and to preserve skins of vegetables such as peas. Leave steamer doors ajar for cooling and to preserve door gaskets.

A. GENERAL INFORMATION No. 23(1)

GUIDELINES FOR CONVECTION OVENS

A convection oven has a blower fan which circulates hot air throughout the oven, eliminating cold spots and promoting rapid cooking. Overall, cooking temperatures and times are shorter than in conventional ovens. The size, thickness, type of food, and amount loaded into the oven at one time will influence the cooking time.

TEMPERATURE SETTINGS: Follow the recommended temperature guide provided in the manufacturer's operating manual. If not available, follow the guidelines furnished on this card or check specific recipe for convection oven information. Note: At this time, not all AFRS oven recipes contain convection information. If food is cooked around the edges, but the center is still raw or not thoroughly cooked, or if there is much color variation, reduce the heat by 15° F. to 25° F. and return food to the oven. If necessary, continue to reduce the heat on successive loads until the desired results are achieved. Record most successful temperature on the recipe card for future reference.

TIME SETTING: Follow the recommended times provided in the manufacturer's operating manual. Should the manual not be available, follow the guidelines furnished on this card or check the specific recipe for convection oven information. Check progress halfway through the cooking cycle since time will vary with the quantity of food loaded, the temperature, and the type of pan used. NOTE: meat thermometers for roasting and visual examination of baked products are the most accurate methods of determining cooking times, both in convection ovens and in conventional ovens. Record most successful cooking time on the recipe card for future reference.

VENT DAMPER CONTROL SETTING: The vent damper control is located on or near the control panel. The damper should be kept closed for most foods of low moisture content such as roasts. If open during roasting, meats will be dry with excessive shrinkage.

A. GENERAL INFORMATION No. 23(1)

The damper should be kept open when baking high moisture content foods (cakes, muffins, yeast bread, etc.). Leaving the damper closed throughout a baking cycle will produce cakes which are too moist and will not rise. A "cloud" or water droplets on the window indicate excessive moisture which should be vented out of the oven through the open damper.

FAN SPEED SETTINGS: SEE GENERAL NOTES BELOW.

INTERIOR OVEN LIGHTS: Turn on lights only when loading, unloading, or checking product. Continual burning of lights will result in short bulb life.

TIMER: The oven timer will ring only as a reminder; it has no control over the functioning of the oven. To ensure proper operation, wind the timer to the maximum setting, then turn back to the desired setting for the product.

GENERAL OPERATION:

1. Select and make the proper rack arrangement for the product to be cooked.
2. Turn or push the main power switch "ON" (gas oven - turn burner valve "ON"). Set thermostat to the recommended temperature. The thermostat signal light will light. Adjust fan speed on two-speed blower, if available (see General Notes below).
3. PREHEAT oven until thermostat signal light goes out indicating that the oven has reached the desired temperature. The oven should preheat to 350° F. within 10 to 15 minutes. (Note: To conserve energy, DO NOT turn on the oven until absolutely necessary - about 15 minutes before actual cooking is to start.)
4. OPEN oven doors and load the oven quickly to prevent excessive loss of heat. Load the oven from the top, centering the pans on the rack toward the front of the oven. Place partial loads in the center of the oven. Allow 1 to 2 inches between pans and along oven sides to permit good air circulation. <u>Remember - overloading is the major cause of non-uniform baking and roasting</u>.

A. GENERAL INFORMATION No. 23(2)

GUIDELINES FOR CONVECTION OVENS

5. Close oven doors and set the timer for the desired cooking time. Check the baking/roasting progress periodically until product is ready.

CLEANING AND MAINTENANCE: Refer to the manufacturer's operating manual for cleaning and maintenance instructions.

GENERAL NOTES: Most convection ovens are equipped with an electric interlock which energizes/de-energizes both the heating elements and the fan motor when the doors are closed/open. Therefore, the heating elements and fan will not operate independently and will only operate with the doors closed.

(Only one known company manufactures an oven in which the fan can be controlled independently.) Some convection ovens are equipped with single-speed fan motors while others are equipped with two-speed fan motors. This information is particularly important to note when baking cakes, muffins or meringue pies, or similar products, and when oven-frying bacon. High speed air circulation may cause damage to the food (e.g., cakes slope to one side of the pan) or blow melted fat throughout the oven. Read the manufacturer's manuals and determine exactly what features you have and then, for the above products, proceed as follows.

A. GENERAL INFORMATION No. 23(2)

<u>Two-Speed Interlocked Fan Motor</u>: Set fan speed to "low."

<u>Single-Speed Interlocked Fan Motor</u>: Preheat oven 50° F. higher than the recommended cooking temperature. Load oven quickly, close doors, and reduce thermostat to recommended cooking temperature. (This action will allow the product to "set up" before the fan/heating elements come on again.)

<u>Single-Speed Independent Fan Motor</u>:

1. Preheat oven 25° F. above temperature specified in recipe.
2. Turn fan "OFF."
3. Reduce heat 25° F.
4. Load oven quickly and close doors.
5. Turn fan "ON" after 7 to 10 minutes and keep "ON" for remaining cooking time.

EXCEPTION: Leave fan "OFF" for bacon to prevent fat from blowing throughout the oven. READ AND UNDERSTAND THE MANUFACTURER'S MANUALS. THEY WILL MAKE YOUR JOB EASIER.

Note: Equipment is becoming more and more complex as the "state-of-the-art" progresses. It is absolutely essential that proper operating manuals be read and understood by everyone who either uses or maintains food service equipment. If you do not have the proper manuals available, proceed with extreme caution so as not to damage or misuse this equipment. Local food service equipment dealers, and/or your service's food service office should be contacted for assistance.

GUIDELINES FOR CONVECTION OVENS

FOOD	PAN SIZE (INCHES)	RECOMMENDED NO. OF SHELVES FOR ONE LOAD	RECOMMENDED TEMPERATURE (° F.)	TIME
BREADS				
Breads, yeast	10-1/2 by 5 by 3-1/2	3	375	30 min
Coffee cakes	18 by 26	4	325	15 min
Muffins	12-cup muffin pan	4	350	30 min
Rolls, yeast	18 by 26	4	350	10 to 15 min
Sweet rolls	18 by 26	4	325	15 min
CAKES				
Angel food	16 by 4-1/2 by 4-1/8	3	300	25 to 30 min
Layer	8 or 9	4	300	25 to 35 min
Loaf	16 by 4-1/2 by 4-1/8	3	325	65 min
Sheet	18 by 26	4	300 to 325	25 to 35 min
DESSERTS				
Brownies	18 by 26	4	325	25 to 30 min
Cookies, bar	18 by 26	5	325	15 min
Cookies, drop	18 by 26	5	325	12 min
Cookies, sliced	18 by 26	5	350	8 to 10 min
Pies, fruit	9	4	375	25 min

FOOD	PAN SIZE (INCHES)	RECOMMENDED NO. OF SHELVES FOR ONE LOAD	RECOMMENDED TEMPERATURE (° F.)	TIME
MEATS				
Bacon, oven fried	18 by 26	5	325	15 to 20 min
Chicken, quarters or pieces	18 by 26	5	350	30 min
Fish, baked or oven fried	18 by 26	4	325	15 to 20 min
Meatloaf	18 by 26	3	300	1 hr 15 min
Roasts, boneless,				
Beef	18 by 26	3	325	1 hr 45 min
Pork	18 by 26	3	325	1-1/2 hr to 2 hrs
Steak, grill (strip loin, ribeye roll, top sirloin butt)	18 by 26	7	400	See Recipe No. L00700
Turkey, boneless	18 by 26	3	325	3-1/2 to 4 hrs
MISCELLANEOUS				
Pizza	18 by 26	4	450	15 min
Potatoes, baked	18 by 26	5	400	35 to 40 min

A. GENERAL INFORMATION No. 24

GUIDELINES FOR USE OF TILTING FRY PANS

The tilting fry pan is a versatile piece of equipment. Although usually described as an oversized skillet because of its large flat cooking surface, this piece of equipment can perform almost any type of cooking except deep fat frying. The tilting fry pan can be used for braising, grilling, sautéing, pan frying, simmering, steaming, boiling, warming, and holding. The ability to tilt the pan allows for easy removal of food to the serving pans without heavy lifting. It can be used for successive cooking functions without having to move the food from one piece of equipment to another. The temperature dial is adjustable over a range of 200 F. to 400 F.

GENERAL OPERATION:

1. Turn or push main power switch to "on" position. The red light will signal that power is on.
2. Set thermostat to desired temperature. Yellow light will signal when heating unit has reached temperature. It will cycle on and off to maintain the temperature.
3. Preheat approximately 12 minutes before using as a griddle or fry pan.
4. To use as a steamer use 1 to 2 inches water with a rack for holding food above the water. Leave cover closed while steaming.
5. To use as a griddle, follow directions and temperature as shown on the recipe card.
6. For sautéing or pan frying, temperature should be between 300 F. and 365 F.
7. For simmering, temperature should be 200 F.

CLEANING AND MAINTENANCE: Refer to the manufacturer's operating manual for instructions.

A. GENERAL INFORMATION No. 25

GUIDELINES FOR CAPACITIES OF STEAM TABLE AND BAKING AND ROASTING PANS

PANS	DEPTH (Inches)	USABLE CAPACITY (Quarts)	USABLE CAPACITY (1/2 Cup Portions)
STEAM TABLE: 12 by 20 inch (full size)	2-1/2	7	56
	4	13	104
	6	18-1/2	148
	8	27	216
12 by 10 inch (1/2 size)	2-1/2	3-1/2	28
	4	6-1/2	52
	6	9	72
	8	12	96
6 by 12 inch (1/3 size)	2-1/2	2-1/2	20
	4	4	16
	6	6	24
6 by 10 inch (1/4 size)	2-1/2	1-2/3	13
	4	2-2/3	21
	6	4	32
BAKING AND ROASTING: 18 by 24 inch	4-1/2	24	192
16 by 16 inch	4	8	64

NOTE: Usable capacity: Pans are filled to about 1/2 inch from the brim. If pans are to be used for carrying liquids (i.e., soups, gravies), the capacity should be reduced to half full.

A. GENERAL INFORMATION No. 27(1)

METRIC CONVERSION

The metric system is an international language of measurement. Its symbols are based on the International System of Units (SI). Of these, food service preparation will be primarily involved with the following metric base units:

Weight (mass)	gram (g)
	kilogram (kg)
Volume	milliliter (mL)
	liter (L)
Length	centimeter (cm)
	meter (m)
Temperature	degree Celsius (°C.)

While the U. S. metric system is voluntary and the food service industry in the United States has not converted to metric system, except for a few soft conversions (e. g., labeling), military food service dining facilities/general messes outside CONUS may experience the metric system in food and equipment support provided by the host country. The information furnished in this guideline card is primarily for these food service personnel.

A. GENERAL INFORMATION No. 27(1)

CONVERSION OF U. S. CUSTOMARY TO METRIC UNITS

	U. S. Customary	**Metric**
Weight (or Mass)	1 ounce (oz) =	28.35 grams (g)
	1 pound (lb) =	453.6 grams (g) or .4536 kilograms
	2.2 pound (lb) =	1 kilogram (kg) or 1000 grams (g)
Volume	1 tsp =	4.93 milliliters (mL)
	1 tbsp =	14.79 milliliters (mL)
	1 cup =	236.59 milliliters (mL) or .237 liters (L)
	1 pint =	.473 liters (L)
	1 quart =	.946 liters (L)
	1 gallon =	3.785 liters (L)
	1.06 quarts =	1 liter (L) or 1000 milliliters (mL)
Length	1 inch =	2.54 centimeters (cm)
	1 foot =	.3048 meters (m)
	1 yard =	30.48 centimeters (cm) or .9144 meters (m)
	1.1 yards =	1 meter (m) or 100 centimeters (cm)

A. GENERAL INFORMATION No. 27(2)

GUIDELINES FOR METRIC CONVERSION - CONTINUED
Temperature Conversions

°F.	°C.	°F	°C.
0	-18	212	100
26	-3	225	107
28	-2	228	109
30	-1	245	118
32	0	250	121
36	2	275	135
38	3	300	149
40	4	325	163
70	21	350	177
90	32	360	182
140	60	365	185
160	71	375	191
170	77	400	204
175	79	425	218
180	82	450	232
185	85	500	260
		550	288

A. GENERAL INFORMATION No. 28(1)

GUIDELINES FOR CHEESES
USE OF DEHYDRATED CHEESES

Two types of dehydrated cheeses are used - dehydrated American cheese and dehydrated cottage cheese.

   a. Cheese, Cottage, Dehydrated

      (1) USE - Dehydrated cottage cheese may be substituted in any recipe using fresh cottage cheese.

      (2) PREPARATION - Measure 8-1/2 cups water (70° F.) into a shallow serving pan. Pour 1-No. 10 cn (1 lb 1 oz) canned dehydrated cottage cheese evenly over the water. Stir gently to wet all particles of cheese. Let stand 5 minutes, then stir gently. If more water is needed, sprinkle 1/2 to 1 cup water over cheese. Chill rehydrated cheese thoroughly before serving (3 to 4 hours).

      (3) SUBSTITUTION - Rehydration ratio - 1 pound dehydrated cottage cheese to 4 pounds (2 qt) water.

Dehydrated Cheese	Water Added	= Rehydrated Cheese	OR	Fresh Cheese Equivalent
1-No. 10 cn (1 lb 1 oz (2-3/4 qt))	8-1/2 cups	5 lb oz (3 qt)		6 lb (3qt)
2-No. 10 cn (2 lb 2 oz (5-1/2 qt))	4-1/4 qt	10 lb 2 oz (6-1/4 qt)		12 lb (1-1/2 gal)

A. GENERAL INFORMATION No. 28(1)

b. Cheese, American, Processed, Dehydrated

(1) USE - Dehydrated American processed cheese may be substituted in any recipe using processed American cheese. Rehydrate cheese before adding to any recipe to eliminate any un-rehydrated cheese in the end product. To store dehydrated cheese after being opened, place unused portion in a tightly covered container to prevent absorption of moisture. Refrigerate if possible.

(2) PREPARATION - Add water to cheese and mix until blended. For a moist semi-solid cheese, such as for an appetizer or omelet, use 1 lb (1 qt) dehydrated cheese and 1 cup water. For a semi-fluid cheese for sauces (better volume substitute), use 1 pound (1 qt) dehydrated cheese and 2 cups water.

(3) SUBSTITUTION:

Dehydrated Cheese +	WARM Water Added =	Rehydrated Cheese OR	Fresh Cheese Equivalent
Semi-solid 6 oz (1-1/2 cups)	3/8 cup	1-1/8 cups	1 lb
3 lb (3 qt) 1-No. 10 cn	3 cups	2-1/4 qt	8 lb
Fluid 6 oz (1-1/2 cups)	3/4 cup	1-1/2 cups	1 lb
3 lb (3 qt) 1-No. 10 cn	1-1/2 qt	3 qt	8 lb

A. GENERAL INFORMATION No. 30(1)

GUIDELINES FOR USING HERBS

The following information is provided as a guide in developing familiarity and creativity with using herbs. Start with a small amount, taste, then add more if necessary.

Herb	Appetizers Salad	Breads/Eggs Sauces/Cheese	Vegetables Pasta	Meat Poultry	Fish Shellfish
Basil	Green, Potato & Tomato Salads, Salad Dressing, Stewed Fruit	Breads, Fondue & Egg Dishes, Dips, Marinades, Sauces	Mushrooms, Tomatoes, Squash, Pasta, Bland Vegetables	Broiled, Roast Meat & Poultry Pies, Stews, Stuffing	Baked, Broiled & Poached Fish, Shellfish
Bay Leaf	Seafood Cocktail, Seafood Salad, Tomato Aspic, Stewed Fruit	Egg Dishes, Gravies, Marinades, Sauces	Dried Bean Dishes, Beets, Carrots, Onions, Potatoes, Rice, Squash	Corned Beef, Tongue Meat & Poultry Stews	Poached Fish, Shellfish Fish Stews

Guide to Cooking with Popular Herbs (continued)

Herb	Appetizers Salad	Breads/Eggs Sauces/Cheese	Vegetables Pasta	Meat Poultry	Fish Shellfish
Chives	Mixed Vegetables, Green, Potato & Tomato Salads, Salad Dressings	Egg & Cheese Dishes, Cream Cheese, Cottage Cheese, Gravies, Sauces	Hot Vegetables, Potatoes	Broiled Poultry, Rissoles, Poultry & Meat Pies, Stews, Casseroles	Baked Fish, Fish Casseroles, Fish Stews, Shellfish
Dill	Seafood Cocktail, Green, Potato & Tomato Salads, Salad Dressings	Breads, Egg & Cheese Dishes, Cream Cheese, Fish and Meat Sauces	Beans, Beets, Cabbage, Carrots, Cauliflower, Peas, Squash, Tomatoes	Beef, Veal Roasts, Lamb, Steaks, Chips, Stews, Roast & Creamed Poultry	Baked, Broiled, Poached & Stuffed Fish, Shellfish
Garlic	All Salads, Salad Dressings	Fondue Poultry Sauces, Fish and Meat Marinades	Beans, Eggplant, Potatoes, Rice, Tomatoes	Roast Meats, Meat & Poultry Pies, Hamburgers, Stews & Casseroles	Broiled Fish, Shellfish, Fish Stews, Casseroles
Marjoram	Seafood Cocktail, Green, Poultry & Seafood Salads	Breads, Cheese Spreads, Egg & Cheese Dishes, Gravies, Sauces	Carrots, Eggplant, Peas, Onions, Potatoes, Dried Bean Dishes, Spinach	Roast Meats & Poultry Meat & Poultry Pies, Stews & Casseroles	Baked, Broiled & Stuffed Fish, Shellfish

Guide to Cooking with Popular Herbs (continued)

Herb	Appetizers Salad	Breads/Eggs Sauces/Cheese	Vegetables Pasta	Meat Poultry	Fish Shellfish
Mustard	Fresh Green Salads, Prepared Meat, Macaroni & Potato Salads, Salad Dressing	Biscuits, Egg & Cheese Dishes, Sauces	Baked Beans, Cabbage, Eggplant, Squash, Dried Beans, Mushrooms, Pasta	Chops, Steaks, Ham, Pork, Poultry Cold Meats	Shellfish
Oregano	Green, Poultry & Seafood Salads	Breads, Egg & Cheese Dishes, Meat, Poultry & Vegetable Sauces	Artichokes, Cabbage, Eggplant, Squash, Dried Beans, Mushrooms, Pasta	Broiled, Roast Meats, Meat & Poultry Pies, Stews, Casseroles	Baked, Broiled & Poached Fish, Shellfish
Parsley	Green, Potato, Seafood & Vegetable Salads	Biscuits, Breads, Egg & Cheese Dishes, Gravies, Sauces	Asparagus, Beets, Eggplant, Squash, Dried Beans, Mushrooms, Pasta	Meat Loaf, Meat & Poultry Pies, Stews and Casseroles, Stuffing	Fish Stews, Stuffed Fish
Rosemary	Fruit Cocktail, Fruit & Green Salads	Biscuits, Egg Dishes, Herb Butter, Cream Cheese, Marinades, Sauces	Beans, Broccoli, Peas, Cauliflower, Mushrooms, Baked Potatoes, Parsnips	Roast Meat, Poultry & Meat Loaf, Meat & Poultry Pies, Stews & Casseroles, Stuffing	Stuffed Fish, Shellfish

Guide to Cooking with Popular Herbs (continued)

Herb	Appetizers / Salad	Breads/Eggs / Sauces/Cheese	Vegetables / Pasta	Meat / Poultry	Fish / Shellfish
Sage		Breads, Fondue, Egg & Cheese Dishes, Spreads, Gravies, Sauces	Beans, Beets, Onions, Peas, Spinach, Squash, Tomatoes	Roast Meat, Poultry, Meat Loaf, Stews, Stuffing	Baked, Poached, & Stuffed Fish
Tarragon	Seafood Cocktail, Avocado Salads (all), Salad Dressings	Cheese Spreads, Marinades, Sauces, Egg Dishes	Asparagus, Beans, Beets, Carrots, Mushrooms, Peas, Squash, Spinach	Steaks, Poultry, Roast Meats, Casseroles & Stews	Baked, Broiled & Poached Fish, Shellfish
Thyme	Seafood Cocktail, Green, Poultry, Seafood & Vegetable Salads	Biscuits, Breads, Egg & Cheese Dishes, Sauces, Spreads	Beets, Carrots, Mushrooms, Onions, Peas, Eggplant, Spinach, Potatoes	Roast Meat, Poultry & Meat Loaf, Meat & Poultry Pies, Stews & Casseroles	Baked, Broiled & Stuffed Fish, Shellfish, Fish Stews

HAZARD ANALYSIS CRITICAL CONTROL POINT
(HACCP)

HACCP System: A food safety system that identifies hazards and develops control points throughout the receiving, storage, preparation, service and holding of food. This system is designed to prevent foodborne illness.

- **Critical Control Point (CCP)**: A point in a specific food service process where loss of control may result in an unacceptable health risk. Implementing a control measure at this point may eliminate or prevent the food safety hazard.
- **Critical Limits:** Elements such as time and temperature that must be adhered to in order to keep food safe. The Temperature Danger Zone is defined by the Food and Drug Administration's Food Code as 41° F. to 140° F.
- **Foodborne Illness:** An illness transmitted to humans through food. Any food may cause a foodborne illness, however *potentially hazardous foods* are responsible for most foodborne illnesses. Symptoms may include abdominal pain/cramps, nausea and vomiting.
- **Potentially Hazardous Food:** A food that is used as an ingredient in recipes or served alone that is capable of supporting the growth of organisms responsible for foodborne illness. Typical foods include high protein foods such as meat, fish, poultry, eggs and dairy products.

A. GENERAL INFORMATION No. 32 (1)

COOKING TEMPERATURES *These temperatures represent the minimum required temperature. The time represents the minimum amount of time the temperature must be maintained.*	
Eggs, Raw shell eggs	155° F. for 15 seconds
Eggs, Egg products, pasteurized	145° F. for 15 seconds
Poultry	165° F. for 15 seconds
Pork	145° F. for 15 seconds
Whole Beef Roasts and Corned Beef Roasts	145° F. for 3 minutes
Fish	145° F. for 15 seconds
Stuffed meat, fish, poultry or pasta, OR stuffings containing meat, fish or poultry	165° F. for 15 seconds
Meat or fish that has been reduced in size by methods such as chopping (i.e., beef cubes), grinding (i.e., ground beef, sausage), restructuring (i.e., formed roast beef, gyro meat), or a mixture of two or more meats (i.e., sausage made from two or more meats)	155° F. for 15 seconds
CCP: SERVING AND HOLDING (hot foods)	140° F.
COOLING *FDA recommends a cooled product temperature of 41° F. In order to achieve a cooled internal product temperature of 34-38° F., the temperature of the refrigerator must be lower than 41° F.*	Cooling from 140° F. to 70° F. should take no longer than 2 hours. Cooling from 70° F. to 41° F. should take no longer than 4 hours.

A. GENERAL INFORMATION No. 32 (1)

GUIDELINES FOR COMBI-OVENS

A combi-oven is a versatile piece of equipment that combines three modes of cooking in one oven: steam, circulated hot air or a combination of both. The combi mode is used to re-heat foods and to roast, bake and "oven fry." The steam mode is ideal for rapid cooking of vegetables and shellfish. The hot air mode operates as a normal convection oven for baking cookies, cakes and pastries. The combi mode decreases overall cooking times, reduces product shrinkage and eliminates flavor transfer when multiple items are cooked simultaneously.

OVEN MODES

COMBI MODE: Use to roast and braise meats, bake poultry and fish and reheat prepared foods. The combination of steam and hot air will improve yield and reduce overall cooking times. To **OVEN FRY,** use food items that are labeled "ovenable" by the manufacturer. Refer to cooking guidelines for oven frying individual items. Place items on perforated sheet pan in a single layer. DO NOT place excess amount of product on pan. A solid sheet pan may be placed under perforated pan to catch excess oils and eliminate smoke.

HOT AIR MODE: Use to bake cakes, cookies and breads and to roast and bake meats and poultry. The hot air mode circulates air in the same manner as a convection oven.

GUIDELINES FOR COMBI-OVENS (continued)

STEAMING MODE: Use to steam fresh, frozen or canned vegetables and shellfish. Use of the Combi-oven to steam foods can save time, labor, and help maintain appearance, and preserve nutrients normally lost by other cooking methods. The oven is ideal for steaming more than one type of vegetable at the same time without flavor transfer. Foods may be steamed in perforated or solid pans. Perforated pans are generally used, particularly for vegetables, unless the cooking liquid is retained or manufacturer's directions specify solid pans. Pans are normally filled no more than 2/3 full to allow steam to circulate for even cooking.

Steam temperature is preset at 212° F. The cooking time will vary depending on the type of food and the number of pans in the oven. The cooking time should include the time it requires to heat food up to cooking temperature, as well as steaming.

TEMPERATURE SETTING: At this time the AFRS recipes do not contain combi-oven information. Refer to the attached cooking guidelines for individual items or begin by using the recommended convection oven temperature noted on individual recipes. If food is cooked around the edges, but the center is still raw or not thoroughly cooked, or if there is too much color variation (some is normal), turn pan or reduce the heat by 10° F. to 15° F. and return food to the oven and continue cooking until done.

TIME SETTING: Follow the recommended convection cooking times on recipe cards. Check progress halfway through the cooking cycle since times will vary in the Combi mode with the quantity of food being cooked, the temperature, and the type of pan used.

GUIDELINES FOR COMBI-OVENS (continued)

MEAT PROBE: The meat probe measures a product core temperature during the cooking process.

FAN SPEED SETTING: See general operations notes below.

GENERAL OPERATION NOTES:

1. **OVEN RACKS:** Position oven racks for the number of pans and product to be cooked.
2. **WATER SUPPLY:** Verify water supply is on.
3. **SELECT COOKING MODE AND TEMPERATURE:** Turn oven on; SELECT the cooking mode. To cook in the combi or hot air mode, set thermostat to desired temperature. To cook in the steam mode, set thermostat to 200° F. The thermostat light will come on indicating oven temperature is below set point.
4. **PREHEAT:** Heat oven until thermostat light goes out indicating that the oven has reached the set temperature. The oven should preheat to 350° F. within 10 to 15 minutes.
5. **FAN SPEED:** If two-speed fan is available, adjust the fan to recommended speed noted on individual recipe card. NOTE: The Combi-oven is equipped with electric interlock, which energizes/de-energizes both the heating element and fan motor when the doors are closed and open. Therefore, the heating elements and fan will not operate with the doors open, only when closed.

GUIDELINES FOR COMBI-OVENS (continued)

6. **MEAT PROBE:** Insert the meat probe in the thickest section of the product. NOTE: The tip of the probe should not be placed near bone or fat. This will result in inaccurate temperature readings. Turn the meat probe switch on and set the desired core temperature by using the up or down arrows. Press the set button to store the set point temperature. Set the timer to the STAY ON position. When the selected core temperature is reached the buzzer will sound and the oven automatically turns off.
7. **CLEANING AND MAINTENANCE:** Refer to the manufacturer's operating manual for cleaning and maintenance instructions. NOTE: Wipe out all spills as soon as they occur for ease of cleaning.

COMBI-OVEN COOKING GUIDELINES

Food	Cook Mode	Recommended Temperature	Time
MEATS			
Steak	Hot Air	400	See Recipe No. L 007 00
Bacon, oven fried	Hot Air	325	25-30 minutes
Roasts, boneless			
Beef	Combi	325	1 hr 45 minutes
Pork	Combi	325	2 to 2-1/2 hours
Spareribs	Combi	350	1 to 1-1/2 hours
Meatloaf	Combi	300	1 hour

GUIDELINES FOR COMBI-OVENS (continued)
COMBI-OVEN COOKING GUIDELINES

Food	Cook Mode	Recommended Temperature	Time
POULTRY			
Turkey, boneless	Combi	325	2 to 2-1/2 hours
Chicken, pieces (with bone)	Combi	350	20-30 minutes
FISH			
Fish, baked	Combi	325	10-20 minutes
Shrimp, raw, frozen	Steam	Preset	3-5 minutes
MISCELLANEOUS			
Casserole type dishes			
Macaroni & cheese	Combi	325	15-20 minutes
Lasagna	Combi	300	40-50 minutes
BREADS			
Breads, yeast	Hot Air	375	30 minutes
Coffee cakes	Hot Air	325	15 minutes
Muffins	Hot Air	350	30 minutes
Rolls Yeast	Hot Air	350	10-15 minutes
Sweet rolls	Hot Air	325	15 minutes

A. GENERAL INFORMATION No. 33 (3)

GUIDELINES FOR COMBI-OVENS (continued)

COMBI-OVEN COOKING GUIDELINES

Food	Cook Mode	Recommended Temperature	Time
EGGS			
Hard Cooked Eggs	Steam	Preset	12 minutes
CAKES			
Angel Food	Hot Air	300	30-35 minutes
Layer	Hot Air	300	25-35 minutes
Loaf	Hot Air	325	65-75 minutes
Sheet	Hot Air	300-325	25-35 minutes
DESSERTS			
Brownies	Hot Air	325	25-30 minutes
Cookies	Hot air	325	12-15 minutes
Pies, Fruit	Hot air	375	25 minutes
VEGETABLES			
Frozen	Steam	Preset	12-15 minutes
Canned	Steam	Preset	10-12 minutes
Fresh*	Steam	Preset	*See individual recipe cards

GUIDELINES FOR COMBI-OVENS (continued)

COMBI-OVEN COOKING GUIDELINES

Food	Cook Mode	Recommended Temperature	Time
EGGS			
Hard Cooked Eggs	Steam	Preset	12 minutes
CAKES			
Angel Food	Hot Air	300	30-35 minutes
Layer	Hot Air	300	25-35 minutes
Loaf	Hot Air	325	65-75 minutes
Sheet	Hot Air	300-325	25-35 minutes
DESSERTS			
Brownies	Hot Air	325	25-30 minutes
Cookies	Hot air	325	12-15 minutes
Pies, Fruit	Hot air	375	25 minutes
VEGETABLES			
Frozen	Steam	Preset	12-15 minutes
Canned	Steam	Preset	10-12 minutes
Fresh*	Steam	Preset	*See individual recipe cards

GUIDELINES FOR COMBI-OVENS (continued)

COMBI-OVEN COOKING GUIDELINES

Food	Cook Mode	Recommended Temperature	Time
OVEN FRYING			
French Fries	Combi	400	7-9 minutes
Fish Portions	Combi	400	10-12 minutes
Shrimp, Battered	Combi	400	7-8 minutes
Chicken Pieces	Combi	400	20 minutes
Chicken Nuggets	Combi	400	8-14 minutes
Onion Rings	Combi	400	6-8 minutes
Jalapeno Popper	Combi	400	9-12 minutes
Egg rolls	Combi	400	12-18 minutes

GUIDELINES FOR COMBI-OVENS (continued)

COMBI-OVEN COOKING GUIDELINES

Food	Cook Mode	Recommended Temperature	Time
OVEN FRYING			
French Fries	Combi	400	7-9 minutes
Fish Portions	Combi	400	10-12 minutes
Shrimp, Battered	Combi	400	7-8 minutes
Chicken Pieces	Combi	400	20 minutes
Chicken Nuggets	Combi	400	8-14 minutes
Onion Rings	Combi	400	6-8 minutes
Jalapeno Popper	Combi	400	9-12 minutes
Egg rolls	Combi	400	12-18 minutes

A. GENERAL INFORMATION No. 34 (1)

GUIDELINES FOR SKITTLE

A skittle is a multipurpose piece of equipment that can be used as a pressureless steamer, braising pan or griddle. The griddle mode is ideal for cooking steaks, sandwiches, eggs, pancakes, breakfast meats and potatoes. The steam mode may be used to cook vegetables, seafood, rice and pasta. The braising mode is used for slow moist-heat cooking of meats, poultry and vegetables.

TO OPERATE AS A STEAMER:

1. Add 5 gallons (2"- 3") of water to the skittle using the spray hose.
2. Position steaming racks for the number of pans and product to be cooked.
3. Close the lid and the steam vent.
4. Set the thermostat at 350° Fahrenheit and allow 6-8 minutes to preheat. The skittle is ready when the heater power light goes out.
5. When the skittle is preheated, raise the lid to the top of the steamer racks and place food pans in the racks and close the lid. **(NOTE: To retain maximum steam, do not raise the lid beyond steamer racks. The lid should be kept in a horizontal position)**
6. If steam escapes from the closed lid, open the rear vent until excess is released.

The skittle is ideal for steaming more than one type of vegetable at the same time without flavor transfer. Foods may be steamed in perforated or solid pans. Perforated pans are normally used, particularly for vegetables, unless the cooking liquid is retained or manufacturer's directions specify solid pans. Pans should not be filled more than 2/3 to the top to allow steam to circulate for even cooking.

A. GENERAL INFORMATION No. 34 (1)

Cooking times will vary depending on the type of food and the number of pans used. The cooking time should include the time it requires to heat food up to cook temperature, as well as steaming. Be sure to record the most successful steaming times on individual recipe cards for future reference.

TO OPERATE AS A BRAISING PAN:
    1. Set the thermostat at 375° Fahrenheit and allow 6-8 minutes to preheat. The skittle is ready when the heater power light goes out. Brown food according to individual AFRS recipe card instructions.
    2. Lower temperature to 325° Fahrenheit and add cooking liquid. Lower hood and cook according to individual recipe card instructions.
    3. To remove liquid, tilt the pan 10° using the tilt handle and drain the liquid through the drain valve into a food pan.

The Skittle may be used for braising pot roast, Swiss steaks, spareribs, stews and for preparing gravy, soups and sauces. Cooking times will vary according to individual foods and amount prepared.

TO OPERATE AS A GRIDDLE:
    1. Set the thermostat to 350° Fahrenheit and allow 6-8 minutes to preheat. The griddle is ready when the heater power light goes out.
    2. Raise the lid and cook foods according to individual AFRS guideline cards.
    3. To drain any accumulated grease, place a #10 can into the can holder attached to the drain valve. Tilt the pan 10° using the tilt handle and allow grease to drain into the can. The griddle can be used to cook hamburgers, steak, sandwiches, eggs, pancakes, breakfast meats and potatoes. Heat is distributed evenly over the entire pan surface ensuring food products cook uniformly.

A. GENERAL INFORMATION No. 34 (2)

GUIDELINES FOR SKITTLE (continued)

GENERAL OPERATION NOTES:

1. **STEAMING MODE:** The recommended thermostat temperature for steaming is 350° Fahrenheit. Higher temperatures may be used but water will evaporate quickly and cooking time will not be decreased.

2. **WATER SUPPLY:** The easiest way to fill the skittle with water is with the attached flexible spray hose.

3. **SELECT COOKING TEMPERATURE**: SELECT desired cooking temperature according to cook mode or individual recipe cards. The thermostat light will come on indicating oven temperature is below set point.

4. **PREHEAT:** Heat Skittle until thermostat light goes out indicating that the unit has reached the set temperature. The Skittle should preheat to 350° F. within 6 to 8 minutes. (Note: Lower the lid for faster preheating.)

5. **CLEANING AND MAINTENANCE:** Remove food waste. Fill the pan with warm water using the spray hose. Add mild detergent and scrub with a nylon scrub pad if necessary. Tilt the pan 10° using the tilt handle and allow water to drain into container placed directly under the drain valve. Rinse with clean water and drain again. Refer to the manufacturer's operating manual for cleaning and maintenance instructions.

A. GENERAL INFORMATION No. 35

GUIDELINES FOR USE OF CONVENIENCE PREPARED FOODS

Convenience prepared foods reduce labor since they only require heating. Specific cooking instructions should be located on each advanced foods package. Items to be considered when using convenience prepared foods are cooking times, nutrient content and serving size. Cooking times, nutrient content and serving size will vary among manufacturers for identical food items, therefore, in order to maintain the quality of these convenience prepared foods, instructions must be read and followed every time a convenience prepared food is utilized.

GUIDELINES FOR RELISH TRAYS OR SALAD BARS

Crisp colorful relishes may be served on relish trays or salad bars. Raw vegetable relishes (celery sticks, carrot sticks, or radishes) and pickles, pickled peppers, or olives may be used. Salad greens along with the other foods may be added for "make-your-own" salads from the Salad Bar. Place prepared relishes in covered containers. Refrigerate until served.

ITEM	APPROXIMATE PORTION SIZE	100 PORTIONS	
		A.P. or Container	E.P.
Alfalfa sprouts, fresh	2 tbsp	1 lb 9 oz	1 lb 9 oz
Apple rings, spiced	1 ring	1-1/6-No. 10 cn (7 lb 14 oz)	3 lb 15 oz
Bacon bits, imitation	1 tbsp	1-1/8-22 oz cn (1 lb 9 oz)	
Beans, kidney	2 tbsp	1-1/6-No. 10 cn (8 lb)	5 lb
Bean sprouts, canned	1/4 cup	4-No. 10 cn (25 lb 8 oz)	12 lb 12 oz
Bean sprouts, fresh	2 tbsp	3 lb 2 oz	3 lb 2 oz
Beets, sliced	4 slices	3-No. 10 cn (19 lb 8 oz)	12 lb 3 oz
Broccoli, fresh	2 to 3 stalks	25 lb 10 oz	20 lb
Broccoli, fresh, flowerets (2 to 2-1/2 inch)	2 flowerets	9 lb 14 oz	4 lb 8 oz
Cabbage, fresh, shredded	2 tbsp	2 lb 7 oz	1 lb 15 oz
Carrots, fresh, slices (1/4 inch)	2 tbsp	5 lb 2 oz	4 lb 3 oz
Carrots, fresh, strips (4 by 1/2 inch)	6 strips	8 lb	6 lb 9 oz
Cauliflower, fresh	2 flowerets	12 lb	10 lb
Celery, fresh, diced (1/4 inch)	2 tbsp	4 lb 5 oz	3 lb 2 oz
Celery, fresh, sticks or strips (1/2 inch)	4 strips	9 lb	6 lb 9 oz

ITEM	APPROXIMATE PORTION SIZE	100 PORTIONS A.P. or Container	E.P.
Cheese, Cheddar, American, Monterey Jack or Mozzarella, shredded	2 tbsp	2 lb 1 oz	
Cheese, cottage	2 tbsp	6 lb 4 oz	
Chow mein noodles	1/3 cup	3-No. 10 cn (4 lb 8 oz)	
Crabapples, spiced, whole	1 crabapple	2-No. 10 cn (13 lb 4 oz)	7 lb
Croutons	8 croutons		1 gal
Cucumbers, fresh, pared, sliced	4 slices	9 lb	7 lb 9 oz
Endive or escarole, fresh	variable	5 lb	4 lb 8 oz
Lettuce, fresh, trimmed, separated	variable	4 lb 5 oz	4 lb
Mushrooms, sliced, fresh, trimmed	2 tbsp	3 lb 7 oz	3 lb 2 oz
Mushrooms, sliced, canned	1 tbsp	2-1/4 jumbo cn (4 lb)	2 lb 4 oz
Olives, green, unpitted	3 olives	3-3/4 1 qt jr (6 lb 12 oz)	4 lb 14 oz
Olives, ripe, whole, unpitted or pitted	3 olives	6-1/4 No. 300 cn (5 lb 13 oz)	2 lb 9 oz
Onions, dry, chopped	2 tbsp	4 lb 10 oz	4 lb 3 oz
Onions, dry, sliced	3 to 4 slices	6 lb 11 oz	6 lb
Onions, green, whole	1 green onion	3 lb	2 lb 8 oz
Onions, green, chopped	2 tbsp	3 lb 5 oz	2 lb 12 oz
Peas, chick (garbanzo beans)	2 tbsp	7-15 to 16 oz cn (7 lb)	4 lb 11 oz
Peppers, pickled, cherry, whole	1 to 2 peppers	6-1 qt jr (9 lb 9 oz)	6 lb 12 oz

GUIDELINES FOR RELISH TRAYS OR SALAD BARS

ITEM	APPROXIMATE PORTION SIZE	100 PORTIONS A.P. or Container	E.P.
Peppers, pickled, jalapeno	1 to 2 peppers	2-No. 10 cn (12 lb 12 oz) or 6-1/3-1 qt jr	8 lb
Peppers, sweet, fresh, diced, 1/2 inch	2 tbsp	3 lb 13 oz	3 lb 2 oz
Peppers, sweet, fresh, strips	variable	6 lb 8 oz	5 lb 5 oz
Pickles, cucumber, dill, whole (cut in sticks, 6 per pickle)	4 sticks	2-1 gal jr (17 lb 7 oz) or 2-1/2-No. 10 cn (17 lb 2 oz)	10 lb 7 oz 10 lb 7 oz
Pickles, cucumber, sweet, whole	1 to 2 pickles	2-No. 10 cn (15 lb 13 oz)	9 lb 8 oz
Pickles, mixed, sweet	3 to 4 pickles	1 gal jar (9 lb 14 oz) or 1-1/4-No. 10 cn (10 lb 2 oz)	5 lb 15 oz 6 lb 1 oz
Radishes, fresh	3 radishes	7 lb 12 oz	7 lb 2 oz
Radishes, fresh, slices, 1/8 inch	2 tbsp	3 lb 8 oz	3 lb 3 oz
Romaine, fresh	variable	4 lb 12 oz	4 lb 8 oz
Rutabagas, fresh, pared, 3-1/2 by 1/2 by 1/4 inch	3 strips	4 lb 14 oz	4 lb 2 oz
Spinach, fresh	variable	4 lb 8 oz	4 lb 2 oz
Tomatoes, fresh, cherry	2 to 3 tomatoes	7 lb	
Tomatoes, fresh, wedges (8 wedges)	2 wedges	8 lb 5 oz	8 lb 2 oz
Tomatoes, fresh, sliced (6 slices)	2 slices	11 lb 5 oz	11 lb 2 oz
Turnips, fresh, pared, 2-1/2 by 1/2 by 1/4	3 strips	3 lb 11 oz	3 lb

M. SALADS, SALAD DRESSINGS, AND RELISHES No. 0(1)

INDEX

Card No.		Card No.	
M 001 00	Apple, Celery, and Pineapple Salad	M 009 02	Cabbage and Carrot Slaw with Creamy Dressing
M 001 01	Cabbage, Apple, and Celery Salad	M 009 03	Pineapple Cole Slaw
M 001 02	Cabbage, Apple, and Raisin Salad	M 009 04	Pineapple Marshmallow Cole Slaw
M 002 00	Spinach Salad	M 009 05	Vegetable Slaw with Creamy Dressing
M 002 01	Spinach and Apple Salad	M 010 00	Honey Mustard Dressing
M 002 02	Spinach and Mushroom Salad	M 011 00	Low Calorie Yogurt Dressing
M 003 00	Red Wine Vinaigrette Dressing	M 012 00	Cottage Cheese Salad
M 004 00	Frijole Salad	M 013 00	Cottage Cheese and Peach Salad
M 005 00	Carrot Salad	M 013 01	Cottage Cheese and Apricot Salad
M 005 01	Carrot and Pineapple Salad	M 013 02	Cottage Cheese and Pear Salad
M 005 02	Carrot, Celery, and Apple Salad	M 013 03	Cottage Cheese and Pineapple Salad
M 006 00	Pimiento Cheese Stuffed Celery	M 014 00	Cottage Cheese and Tomato Salad
M 006 01	Cottage Cheese Stuffed Celery	M 015 00	Cucumber and Onion Salad
M 006 02	Peanut Butter Stuffed Celery	M 016 00	Low Calorie Thousand Island Dressing
M 006 03	Cream Cheese Stuffed Celery	M 017 00	Fruit Salad
M 007 00	Chef's Salad	M 018 00	Garden Cottage Cheese Salad
M 007 01	Chef's Salad (Entree)	M 019 00	Garden Vegetable Salad
M 008 00	Cole Slaw	M 020 00	Marinated Carrots
M 008 01	Mexican Cole Slaw	M 021 00	Low Calorie Tangy Tarragon Dressing
M 009 00	Cole Slaw with Creamy Dressing	M 021 01	Low Calorie Basil Dressing
M 009 01	Cole Slaw with Vinegar Dressing	M 022 00	Tangy Yogurt Salad Dressing

M. SALADS, SALAD DRESSINGS, AND RELISHES No. 0(1)

Card No.		Card No.	
M 023 00	Jellied Cranberry and Orange Salad	M 038 00	Pasta Salad
M 023 01	Jellied Cranberry and Orange Salad (Canned)	M 039 00	Corn Relish
M 024 00	Jellied Cranberry and Pineapple Salad	M 040 00	Potato Salad
M 025 00	Jellied Fruit Salad	M 040 01	Deviled Potato Salad
M 025 01	Jellied Orange Salad	M 040 02	Potato Salad with Vinegar Dressing
M 025 02	Jellied Pear Salad	M 041 00	Potato Salad (Dehydrated Sliced Potatoes)
M 025 03	Jellied Pineapple, Pear, and Banana Salad	M 042 00	Hot Potato Salad
M 025 04	Jellied Strawberry Salad	M 043 00	Hot Potato Salad (Dehydrated Sliced Potatoes)
M 025 05	Jellied Banana Salad	M 044 00	Spring Salad
M 026 00	Jellied Fruit Cocktail Salad	M 045 00	Three Bean Salad
M 027 00	German Cole Slaw	M 045 01	Pickled Green Bean Salad
M 028 00	Taco Salad	M 046 00	Tossed Lettuce, Cucumber and Tomato Salad
M 029 00	Italian Style Pasta Salad	M 046 01	Tossed Garden Salad
M 030 00	Cobb Salad	M 046 02	Tossed Calico Garden Salad
M 031 00	Kidney Bean Salad	M 046 03	Tossed Romaine, Cucumber and Tomato Salad
M 032 00	Fruit Medley Salad	M 046 04	Tossed Red Leaf Lettuce, Cucumber and Tomato Salad
M 033 00	Lettuce and Tomato Salad		
M 034 00	Macaroni Salad	M 046 05	Green Leaf Lettuce, Cucumber and Tomato Salad
M 035 00	Mixed Fruit Salad		
M 036 00	Perfection Salad	M 047 00	Tossed Green Salad
M 036 01	Golden Glow Salad	M 048 00	Tossed Vegetable Salad
M 036 02	Jellied Spring Salad	M 048 01	Tossed Calico Vegetable Salad
M 037 00	Pickled Beet and Onion Salad	M 049 00	Vegetable Salad

M. SALADS, SALAD DRESSINGS, AND RELISHES

INDEX

Card No.		Card No.	
M 050 00	Waldorf Salad	M 068 01	Blue Cheese and Sour Cream Dressing
M 050 01	Apple, Celery, and Raisin Salad	M 069 00	Vinegar and Oil Dressing
M 051 00	Cranberry Orange Relish	M 070 00	Zesty Rotini Pasta Salad
M 052 00	Guacamole	M 071 00	Salsa Pasta Salad
M 053 00	German Style Tomato Salad	M 072 00	Confetti Rice Salad
M 053 01	Country Style Tomato Salad	M 072 01	Creamy Cucumber Rice Salad
M 054 00	Tomato French Dressing	M 073 00	Kiwi Fruit Salad
M 055 00	Vinaigrette Dressing	M 074 00	Marinated Black Bean Salad
M 056 00	Quick Fruit Dressing	M 504 00	Broccoli Salad
M 057 00	Zero Salad Dressing	M 801 00	Salad Bar
M 058 00	French Dressing		
M 058 01	Low Calorie French Dressing		
M 059 00	Blue Cheese Dressing		
M 060 00	Garlic French Dressing		
M 061 00	Tangy Salad Dressing		
M 062 00	Mexican Potato Salad		
M 063 00	Thousand Island Dressing		
M 064 00	Creamy Italian Dressing		
M 065 00	Creamy Horseradish Dressing		
M 066 00	Low Calorie Tomato Dressing		
M 067 00	Russian Dressing		
M 068 00	Sour Cream Dressing		

SALADS, SALAD DRESSINGS, AND RELISHES   No.M 001 00

APPLE, CELERY, AND PINEAPPLE SALAD

Yield 100   Portion   1/2 Cup

Calories	Carbohydrates	Protein	Fat	Cholesterol	Sodium	Calcium
86 cal	12 g	0 g	5 g	3 mg	68 mg	24 mg

Ingredient	Weight	Measure	Issue
SALAD DRESSING,MAYONNAISE TYPE	2 lbs	1 qts	
APPLES,FRESH,MEDIUM,UNPEELED,DICED	9 lbs	2 gal 1/8 qts	10-5/8 lbs
PINEAPPLE,CANNED,CHUNKS,JUICE PACK,DRAINED	4-1/4 lbs	1 qts 3-3/4 cup	
CELERY,FRESH,CHOPPED	3 lbs	2 qts 3-3/8 cup	4-1/8 lbs
LETTUCE,LEAF,FRESH,HEAD	4 lbs		6-1/4 lbs

Method

1. Combine Salad Dressing or fat free Salad Dressing and apples.
2. Drain pineapple.
3. Add pineapple and celery to apple mixture; mix lightly.
4. Place 1 lettuce leaf on each serving dish; add 1/2 cup salad mixture. CCP: Cover and refrigerate until ready to serve. Hold for service at 41 F. or lower.

SALADS, SALAD DRESSINGS, AND RELISHES   No.M 001 01
CABBAGE, APPLE, AND CELERY SALAD

Yield 100　　　　　　　　　　　　　　　　　　　　　　**Portion**　1/2 Cup

Calories	Carbohydrates	Protein	Fat	Cholesterol	Sodium	Calcium
73 cal	8 g	1 g	5 g	3 mg	77 mg	35 mg

Ingredient	Weight	Measure	Issue
SALAD DRESSING,MAYONNAISE TYPE	2 lbs	1 qts	
APPLES,FRESH,MEDIUM,UNPEELED,DICED	4-3/8 lbs	0 gal 4 qts	5-1/8 lbs
CELERY,FRESH,CHOPPED	3-1/8 lbs	2 qts 3-3/4 cup	4-1/4 lbs
CABBAGE,GREEN,FRESH,SHREDDED	13 lbs	5 gal 1 qts	16-1/4 lbs

Method

1 Combine Salad dressing or fat free Salad Dressing and apples.
2 Combine apple mixture, cabbage, and celery.
3 Cover; refrigerate at least 2 to 3 hours. Refrigerate until ready to serve. CCP: Hold for service at 41 F. or lower.

SALADS. SALAD DRESSINGS. AND RELISHES   No.M 001 02

CABBAGE, APPLE, AND RAISIN SALAD

Yield 100   Portion   1/2 Cup

Calories	Carbohydrates	Protein	Fat	Cholesterol	Sodium	Calcium
105 cal	17 g	1 g	5 g	3 mg	66 mg	35 mg

Ingredient	Weight	Measure	Issue
SALAD DRESSING,MAYONNAISE TYPE	2 lbs	1 qts	
APPLES,FRESH,MEDIUM,UNPEELED,DICED	4-3/8 lbs	0 gal 4 qts	5-1/8 lbs
RAISINS	2-1/2 lbs	2 qts	
CABBAGE,GREEN,FRESH,SHREDDED	13 lbs	5 gal 1 qts	16-1/4 lbs

Method

1. Combine Salad Dressing or fat free Salad Dressing and apples.
2. Combine apple mixture, cabbage, and raisins.
3. Cover; refrigerate at least 2 to 3 hours. Refrigerate until ready to serve. CCP: Hold for service at 41 F. or lower.

SALADS, SALAD DRESSINGS, AND RELISHES No.M 002 00

SPINACH SALAD

Yield 100   **Portion** 1 Cup

Calories	Carbohydrates	Protein	Fat	Cholesterol	Sodium	Calcium
49 cal	3 g	4 g	3 g	32 mg	99 mg	43 mg

Ingredient	Weight	Measure	Issue
EGG,HARD COOKED,CHOPPED	1-1/2 lbs	1 qts 1 cup	
ONIONS,FRESH,SLICED	2-3/4 lbs	2 qts 3 cup	3-1/8 lbs
MUSHROOMS,FRESH,WHOLE,SLICED	2-7/8 lbs	1 gal 3/4 qts	3-1/4 lbs
SPINACH,FRESH,BUNCH	8 lbs	7 gal 2-1/4 qts	8-2/3 lbs
BACON,SLICED,RAW	3 lbs		

Method

1. Combine eggs, onions, mushrooms, and spinach. Toss lightly to mix ingredients.
2. Cover; refrigerate until ready to serve. CCP: Hold for service at 41 F. or lower.
3. Cook bacon until crisp; drain, chop.
4. Add bacon just before serving. Toss lightly.

Notes

1. In Step 3, 7 ounces or 1-3/4 cups imitation bacon bits may be used per 100 servings.
2. In Step 4, for 100 portions: Serve with 12-1/2 cups prepared fat free Red Wine Vinaigrette Dressing or 1 recipe Vinaigrette Dressing, Recipe No. M 055 00 or 1 recipe Red Wine Vinaigrette Dressing, Recipe No. M 003 00.

SALADS. SALAD DRESSINGS. AND RELISHES No.M 002 01
SPINACH AND APPLE SALAD

Yield 100   Portion   1-1/3 Cups

Calories	Carbohydrates	Protein	Fat	Cholesterol	Sodium	Calcium
60 cal	15 g	2 g	0 g	0 mg	30 mg	45 mg

Ingredient

Ingredient	Weight	Measure	Issue
SPINACH,FRESH,BUNCH	8 lbs	7 gal 2-1/4 qts	8-2/3 lbs
APPLES,FRESH,MEDIUM,UNPEELED,DICED	7-3/4 lbs	1 gal 3 qts	9-1/8 lbs
ONIONS,FRESH,CHOPPED	2-7/8 lbs	2 qts 1/8 cup	3-1/4 lbs
RAISINS	1-7/8 lbs	1 qts 2 cup	

Method

1 Combine apples, onions, and raisins. Toss lightly with spinach to mix ingredients.
2 Cover; refrigerate until ready to serve. CCP: Hold for service at 41 F. or lower.

SALADS, SALAD DRESSINGS, AND RELISHES No.M 002 02

SPINACH AND MUSHROOM SALAD

Yield 100   Portion   1-3/4 Cups

Calories	Carbohydrates	Protein	Fat	Cholesterol	Sodium	Calcium
18 cal	3 g	2 g	0 g	0 mg	30 mg	39 mg

Ingredient

Ingredient	Weight	Measure	Issue
SPINACH,FRESH,BUNCH	8 lbs	7 gal 2-1/4 qts	8-2/3 lbs
ONIONS,FRESH,SLICED	2-3/4 lbs	2 qts 3 cup	3-1/8 lbs
MUSHROOMS,FRESH,WHOLE,SLICED	4-1/2 lbs	1 gal 3-1/4 qts	4-7/8 lbs

Method

1. Combine onions, mushrooms, and spinach. Toss lightly to mix ingredients.
2. Cover; refrigerate until ready to serve. CCP: Hold for service at 41 F. or lower.

SALADS, SALAD DRESSINGS, AND RELISHES   No. M 003 00
RED WINE VINAIGRETTE DRESSING

Yield 100   Portion 1 Tablespoon

Calories	Carbohydrates	Protein	Fat	Cholesterol	Sodium	Calcium
50 cal	1 g	0 g	5 g	0 mg	0 mg	3 mg

Ingredient	Weight	Measure	Issue
SUGAR,GRANULATED	2-1/4 oz	1/4 cup 1-1/3 tbsp	
MUSTARD,DRY	1 oz	2-2/3 tbsp	
PAPRIKA,GROUND	1/2 oz	2 tbsp	
PEPPER,BLACK,GROUND	1/2 oz	2 tbsp	
VINEGAR,RED WINE	1-1/3 lbs	2-1/2 cup	
WATER	1-1/3 lbs	2-1/2 cup	
OIL,SALAD	1-1/8 lbs	2-3/8 cup	

Method

1 Combine sugar, mustard, paprika, pepper, vinegar, and water; blend well. Mix at medium speed 2 minutes using a wire whip.
2 Add oil gradually while mixing at low speed 3 minutes; scrape down bowl.
3 Mix at medium speed 2 minutes or until well blended.
4 Cover; refrigerate until ready to serve. CCP: Hold for service at 41 F. or lower.
5 Whip or stir well before using.

SALADS, SALAD DRESSINGS, AND RELISHES   No.M 004 00
FRIJOLE SALAD

Yield 100   Portion   3/4 Cup

Calories	Carbohydrates	Protein	Fat	Cholesterol	Sodium	Calcium
63 cal	12 g	2 g	2 g	0 mg	247 mg	30 mg

Ingredient	Weight	Measure	Issue
BEANS,KIDNEY,DARK RED,CANNED,DRAINED	4-2/3 lbs	3 qts	
SALAD DRESSING,FRENCH,PREPARED,L/C	4-5/8 lbs	2 qts	
CABBAGE,GREEN,FRESH,SHREDDED	8-5/8 lbs	3 gal 2 qts	10-3/4 lbs
TOMATOES,FRESH,CHOPPED	6-1/3 lbs	0 gal 4 qts	6-1/2 lbs
CUCUMBERS,FRESH,PEELED,SLICED	3-2/3 lbs	3 qts 2 cup	7-1/4 each

Method

1 Drain beans; rinse well; drain.
2 Combine beans and French Dressing.
3 Cover; refrigerate at least 6 hours. CCP: Hold for service at 41 F. or lower.
4 Add cabbage, tomatoes, and cucumbers just before serving. Mix lightly.

SALADS, SALAD DRESSINGS, AND RELISHES No. M 005 00
CARROT SALAD

Yield 100   Portion 1/2 Cup

Calories	Carbohydrates	Protein	Fat	Cholesterol	Sodium	Calcium
109 cal	15 g	1 g	6 g	4 mg	214 mg	38 mg

Ingredient	Weight	Measure	Issue
CELERY,FRESH,CHOPPED	2-3/8 lbs	2 qts 1 cup	3-1/4 lbs
RAISINS	1-7/8 lbs	1 qts 2 cup	
CARROTS,FRESH,SHREDDED	9-2/3 lbs	2 gal 2 qts	11-3/4 lbs
MILK,NONFAT,DRY	1-1/3 oz	1/2 cup 1 tbsp	
WATER,WARM	10-1/2 oz	1-1/4 cup	
SALAD DRESSING,MAYONNAISE TYPE	2-1/2 lbs	1 qts 1 cup	
SALT	1 oz	1 tbsp	
SUGAR,GRANULATED	1-3/4 oz	1/4 cup 1/3 tbsp	
JUICE,LEMON	2-1/8 oz	1/4 cup 1/3 tbsp	
LETTUCE,LEAF,FRESH,HEAD	4 lbs		6-1/4 lbs

Method

1. Combine carrots, celery, and raisins.
2. Reconstitute milk; combine with Regular Salad Dressing or Fat Free Salad Dressing, salt, sugar, and lemon juice. Blend well.
3. Add to vegetables; toss together lightly.
4. Place 1 lettuce leaf on each serving dish; add salad mixture. Cover; refrigerate at least 2 to 3 hours. CCP: Hold for service at 41 F. or lower.

SALADS, SALAD DRESSINGS, AND RELISHES No.M 005 01
CARROT AND PINEAPPLE SALAD

Yield 100   Portion 1/2 Cup

Calories	Carbohydrates	Protein	Fat	Cholesterol	Sodium	Calcium
96 cal	11 g	1 g	6 g	4 mg	204 mg	33 mg

Ingredient	Weight	Measure	Issue
CARROTS,FRESH,SHREDDED	9-2/3 lbs	2 gal 2 qts	11-3/4 lbs
PINEAPPLE,CANNED,CHUNKS,JUICE PACK,DRAINED	5-1/2 lbs	2 qts 2 cup	
MILK,NONFAT,DRY	1-1/3 oz	1/2 cup 1 tbsp	
WATER,WARM	10-1/2 oz	1-1/4 cup	
SALAD DRESSING,MAYONNAISE TYPE	2-1/2 lbs	1 qts 1 cup	
SALT	1 oz	1 tbsp	
SUGAR,GRANULATED	1-3/4 oz	1/4 cup 1/3 tbsp	
JUICE,LEMON	2-1/8 oz	1/4 cup 1/3 tbsp	
LETTUCE,LEAF,FRESH,HEAD	4 lbs		6-1/4 lbs

Method

1. Combine carrots and pineapple.
2. Reconstitute milk; combine with Regular or Fat Free Salad Dressing, salt, sugar, and lemon juice. Blend well.
3. Add to vegetables; toss together lightly.
4. Place 1 lettuce leaf on each serving dish; add salad mixture. Cover; refrigerate at least 2 to 3 hours. CCP: Hold for service at 41 F. or lower.

SALADS, SALAD DRESSINGS, AND RELISHES No. M 005 02

CARROT, CELERY, AND APPLE SALAD

Yield 100   **Portion** 1/2 Cup

Calories	Carbohydrates	Protein	Fat	Cholesterol	Sodium	Calcium
95 cal	11 g	1 g	6 g	4 mg	210 mg	33 mg

Ingredient	Weight	Measure	Issue
CARROTS,FRESH,SHREDDED	8 lbs	2 gal 1/4 qts	9-3/4 lbs
CELERY,FRESH,CHOPPED	2-3/8 lbs	2 qts 1 cup	3-1/4 lbs
APPLES,FRESH,MEDIUM,UNPEELED,DICED	5-1/2 lbs	1 gal 1 qts	6-1/2 lbs
MILK,NONFAT,DRY	1-1/3 oz	1/2 cup 1 tbsp	
WATER,WARM	10-1/2 oz	1-1/4 cup	
SALAD DRESSING,MAYONNAISE TYPE	2-1/2 lbs	1 qts 1 cup	
SALT	1 oz	1 tbsp	
SUGAR,GRANULATED	1-3/4 oz	1/4 cup 1/3 tbsp	
JUICE,LEMON	2-1/8 oz	1/4 cup 1/3 tbsp	
LETTUCE,LEAF,FRESH,HEAD	4 lbs		6-1/4 lbs

Method

1. Combine carrots, diced celery, and diced unpared apples.
2. Reconstitute milk; combine with Regular or Fat Free Salad Dressing, salt, sugar, and lemon juice. Blend well.
3. Add to vegetables; toss together lightly.
4. Place 1 lettuce leaf on each serving dish; add salad mixture. Cover; refrigerate at least 2 to 3 hours. CCP: Hold for service at 41 F. or lower.

SALADS, SALAD DRESSINGS, AND RELISHES No.M 006 00

PIMIENTO CHEESE STUFFED CELERY

Yield 100   Portion   2 Pieces

Calories	Carbohydrates	Protein	Fat	Cholesterol	Sodium	Calcium
57 cal	2 g	3 g	4 g	11 mg	101 mg	84 mg

Ingredient	Weight	Measure	Issue
CELERY,FRESH,BUNCH	6 lbs	1 gal 1-2/3 qts	8-1/4 lbs
CHEESE,CHEDDAR,SHREDDED	2-1/4 lbs	2 qts 1 cup	
PIMIENTO,CANNED,DRAINED,CHOPPED	10-1/8 oz	1-1/2 cup	
SALAD DRESSING,MAYONNAISE TYPE	7-7/8 oz	1 cup	
PEPPER,RED,GROUND	<1/16th oz	1/8 tsp	
WORCESTERSHIRE SAUCE	1/2 oz	1 tbsp	

Method

1. Cut celery into 2 to 3 inch pieces. Place celery in ice water 1 hour or until crisp. Drain. Place on sheet pans lined with waxed paper. Refrigerate for use in Step 3.
2. Combine cheese, pimientos, Salad Dressing, red pepper, and Worcestershire sauce; blend well.
3. Fill hollow section of each celery piece with mixture.
4. Cover; refrigerate until ready to serve. CCP: Hold for service at 41 F. or lower.

SALADS. SALAD DRESSINGS. AND RELISHES No.M 006 01
COTTAGE CHEESE STUFFED CELERY

Yield 100   Portion 2 Pieces

Calories	Carbohydrates	Protein	Fat	Cholesterol	Sodium	Calcium
14 cal	1 g	1 g	0 g	1 mg	68 mg	17 mg

Ingredient	Weight	Measure	Issue
CELERY,FRESH,BUNCH	6 lbs	1 gal 1-2/3 qts	8-1/4 lbs
CHEESE,COTTAGE	2 lbs	1 qts	
CATSUP	2-1/8 oz	1/4 cup 1/3 tbsp	
HORSERADISH,PREPARED	1 oz	2 tbsp	
ONIONS,FRESH,GRATED	1/4 oz	1/3 tsp	1/4 oz

Method

1. Cut celery into 2 to 3 inch pieces. Place celery in ice water 1 hour or until crisp. Drain. Place on sheet pans lined with waxed paper. Refrigerate for use in Step 3.
2. Combine cottage cheese, tomato catsup, prepared horseradish, and grated onions; blend thoroughly.
3. Fill hollow section of each celery piece with mixture.
4. Cover; refrigerate until ready to serve. CCP: Hold for service at 41 F. or lower.

SALADS, SALAD DRESSINGS, AND RELISHES   No.M 006 02

PEANUT BUTTER STUFFED CELERY

Yield 100   Portion   2 Pieces

Calories	Carbohydrates	Protein	Fat	Cholesterol	Sodium	Calcium
50 cal	6 g	2 g	3 g	0 mg	48 mg	13 mg

Ingredient	Weight	Measure	Issue
CELERY,FRESH,BUNCH	6 lbs	1 gal 1-2/3 qts	8-1/4 lbs
PEANUT BUTTER	1-1/8 lbs	2 cup	
HONEY	1-1/8 lbs	1-1/2 cup	

Method

1. Cut celery into 2 to 3 inch pieces. Place celery in ice water 1 hour or until crisp. Drain. Place on sheet pans lined with waxed paper.
2. Combine peanut butter with honey; blend thoroughly.
3. Fill hollow section of each celery piece with mixture.
4. Cover; refrigerate until ready to serve. CCP: Hold for service at 41 F. or lower.

SALADS, SALAD DRESSINGS, AND RELISHES   No.M 006 03

CREAM CHEESE STUFFED CELERY

Yield 100   Portion   2 Pieces

Calories	Carbohydrates	Protein	Fat	Cholesterol	Sodium	Calcium
37 cal	1 g	1 g	3 g	10 mg	51 mg	18 mg

Ingredient

Ingredient	Weight	Measure	Issue
CELERY,FRESH,BUNCH	6 lbs	1 gal 1-2/3 qts	8-1/4 lbs
CHEESE,CREAM	2 lbs	1 qts	

Method

1. Cut celery into 2 to 3 inch pieces. Place celery in ice water 1 hour or until crisp. Drain. Place on sheet pans lined with waxed paper. Refrigerate for use in Step 2.
2. Fill hollow section of each celery piece with softened cream cheese.
3. Cover; refrigerate until ready to serve. CCP: Hold for service at 41 F. or lower.

SALADS, SALAD DRESSINGS, AND RELISHES   No.M 007 00
CHEF'S SALAD

Yield 100   Portion 1 Cup

Calories	Carbohydrates	Protein	Fat	Cholesterol	Sodium	Calcium
102 cal	5 g	9 g	5 g	71 mg	228 mg	130 mg

Ingredient	**Weight**	**Measure**	**Issue**
LETTUCE,LEAF,FRESH,HEAD	7 lbs		11 lbs
CABBAGE,GREEN,FRESH,SHREDDED	1 lbs	1 qts 2-1/2 cup	1-1/4 lbs
PEPPERS,GREEN,FRESH,CHOPPED	2-5/8 lbs	2 qts	3-1/4 lbs
CELERY,FRESH,CHOPPED	3-1/8 lbs	2 qts 3-3/4 cup	4-1/4 lbs
CUCUMBERS,FRESH,PEELED,SLICED	1-5/8 lbs	1 qts 2-1/4 cup	3-1/4 each
TURKEY,BNLS,WHITE AND DARK MEAT	2 lbs		
CHEESE,SWISS,CUBED	2 lbs	1 qts 2-7/8 cup	
HAM,COOKED,BONELESS,SLICED	2 lbs		
EGG,HARD COOKED,CHOPPED	2-3/4 lbs	25 Eggs	
TOMATOES,FRESH,THIN WEDGES	8-1/8 lbs	1 gal 1-1/8 qts	8-1/4 lbs

Method

1 Wash lettuce. Tear or cut into large pieces.
2 Wash vegetables. Combine lettuce with cabbage, peppers, celery, and cucumbers; toss lightly.
3 Cut turkey, ham and cheese into 1/2 inch strips and eggs and tomatoes into 8 wedges each.
4 Place 1 cup salad vegetables in salad bowls. Add 2 thin strips meat, 4 thin strips cheese, 2 egg wedges, and 2 tomato wedges.
5 Cover; refrigerate until ready to serve. CCP: Hold for service at 41 F. or lower. If desired, 3/4 Garlic Croutons (Recipe D 016 01) may be prepared.

SALADS. SALAD DRESSINGS. AND RELISHES No.M 007 01

CHEF'S SALAD (ENTREE)

Yield 100 **Portion** 1-1/2 Cups

Calories	Carbohydrates	Protein	Fat	Cholesterol	Sodium	Calcium
193 cal	8 g	16 g	11 g	92 mg	363 mg	320 mg

Ingredient	Weight	Measure	Issue
LETTUCE,LEAF,FRESH,HEAD	10-1/3 lbs		16-1/8 lbs
CABBAGE,GREEN,FRESH,SHREDDED	1-2/3 lbs	2 qts 2-3/4 cup	2-1/8 lbs
PEPPERS,GREEN,FRESH,CHOPPED	4 lbs	3 qts 1/8 cup	4-7/8 lbs
CUCUMBERS,FRESH,PEELED,SLICED	2-3/8 lbs	2 qts 1 cup	4-2/3 each
CELERY,FRESH,CHOPPED	3-1/8 lbs	2 qts 3-3/4 cup	4-1/4 lbs
HAM,COOKED,BONELESS,SLICED	3 lbs		
TURKEY,BNLS,WHITE AND DARK MEAT	3 lbs		
CHEESE,SWISS,CUBED	6 lbs	1 gal 1-1/8 qts	
EGG,HARD COOKED,CHOPPED	2-3/4 lbs	25 Eggs	
TOMATOES,FRESH,THIN WEDGES	12-1/4 lbs	1 gal 3-2/3 qts	12-1/2 lbs

Method

1. Wash lettuce. Tear or cut lettuce into large pieces.
2. Wash vegetables. Combine lettuce with cabbage, peppers, celery, and cucumbers; toss lightly.
3. Cut ham, turkey and cheese into thin strips and eggs and tomatoes into 8 wedges each.
4. Place about 1-1/2 cups of salad vegetables in salad bowls. Add 6 thin strips meat, 12 thin strips cheese, 3 egg wedges, and 2 tomato wedges.
5. Cover; refrigerate until ready to serve. CCP: Hold for service at 41 F. or lower. If desired, 3/4 Garlic Croutons (Recipe D 016 01) may be prepared.

SALADS, SALAD DRESSINGS, AND RELISHES No.M 008 00

COLE SLAW

Yield 100   **Portion** 1/2 Cup

Calories	Carbohydrates	Protein	Fat	Cholesterol	Sodium	Calcium
115 cal	9 g	1 g	9 g	6 mg	258 mg	26 mg

Ingredient	**Weight**	**Measure**	**Issue**
CABBAGE,GREEN,FRESH,SHREDDED	12 lbs	4 gal 3-1/2 qts	15 lbs
SALAD DRESSING,MAYONNAISE TYPE	4 lbs	2 qts	
SALT	1-1/4 oz	2 tbsp	
SUGAR,GRANULATED	10-5/8 oz	1-1/2 cup	
VINEGAR,DISTILLED	4-1/8 oz	1/2 cup	
PAPRIKA,GROUND	1/2 oz	2 tbsp	

Method

1 Chill cabbage in covered container until crisp.
2 Combine Salad Dressing, salt, sugar, and vinegar.
3 Add to cabbage; mix well.
4 Cover; refrigerate until ready to serve. Just before serving, sprinkle lightly with paprika to garnish. CCP: Hold for service at 41 F. or lower.

SALADS. SALAD DRESSINGS. AND RELISHES   No. M 008 01

MEXICAN COLE SLAW

Yield 100   Portion   1/2 Cup

Calories	Carbohydrates	Protein	Fat	Cholesterol	Sodium	Calcium
121 cal	10 g	1 g	9 g	6 mg	268 mg	23 mg

Ingredient	Weight	Measure	Issue
CABBAGE,GREEN,FRESH,SHREDDED	6-3/4 lbs	2 gal 2-7/8 qts	8-1/2 lbs
SALAD DRESSING,MAYONNAISE TYPE	4 lbs	2 qts	
SALT	1-1/4 oz	2 tbsp	
SUGAR,GRANULATED	10-5/8 oz	1-1/2 cup	
CELERY,FRESH,CHOPPED	3-1/8 lbs	2 qts 3-3/4 cup	4-1/4 lbs
TOMATOES,FRESH,CHOPPED	4-3/8 lbs	2 qts 3 cup	4-1/2 lbs
ONIONS,FRESH,CHOPPED	7-1/2 oz	1-3/8 cup	8-1/3 oz
PEPPERS,GREEN,FRESH,MEDIUM,SHREDDED	4 lbs	3 qts 1/8 cup	4-7/8 lbs
VINEGAR,DISTILLED	4-1/8 oz	1/2 cup	

Method

1 Chill cabbage in covered container until crisp.
2 Combine Salad Dressing, salt, sugar, celery, tomatoes, onions, peppers, and vinegar.
3 Add to cabbage; mix well.
4 Cover; refrigerate until ready to serve. CCP: Hold for service at 41 F. or lower.

SALADS, SALAD DRESSINGS, AND RELISHES   No.M 009 00

COLE SLAW WITH CREAMY DRESSING

Yield 100   Portion   1/2 Cup

Calories	Carbohydrates	Protein	Fat	Cholesterol	Sodium	Calcium
75 cal	8 g	1 g	5 g	3 mg	188 mg	35 mg

Ingredient | Weight | Measure | Issue

Ingredient	Weight	Measure	Issue
MILK,NONFAT,DRY	1-3/4 oz	3/4 cup	
WATER,WARM	14-5/8 oz	1-3/4 cup	
SALAD DRESSING,MAYONNAISE TYPE	2 lbs	1 qts	
PEPPER,BLACK,GROUND	1/8 oz	1/3 tsp	
MUSTARD,PREPARED	1-1/8 oz	2 tbsp	
SALT	1 oz	1 tbsp	
SUGAR,GRANULATED	12-1/3 oz	1-3/4 cup	
VINEGAR,DISTILLED	8-1/3 oz	1 cup	
CABBAGE,GREEN,FRESH,SHREDDED	13 lbs	5 gal 1 qts	16-1/4 lbs

Method

1 Reconstitute milk; add Salad Dressing, pepper, mustard, salt, and sugar; mix well.
2 Add vinegar gradually; blend well.
3 Pour dressing over cabbage; toss lightly until well mixed.
4 Cover; refrigerate until ready to serve. CCP: Hold for service at 41 F. or lower.

SALADS. SALAD DRESSINGS. AND RELISHES No.M 009 01

COLE SLAW WITH VINEGAR DRESSING

Yield 100   Portion   1/2 Cup

Calories	Carbohydrates	Protein	Fat	Cholesterol	Sodium	Calcium
47 cal	12 g	1 g	0 g	0 mg	127 mg	29 mg

Ingredient	Weight	Measure	Issue
PEPPER,BLACK,GROUND	1/8 oz	1/3 tsp	
SALT	1 oz	1 tbsp	
SUGAR,GRANULATED	1-3/4 lbs	1 qts	
VINEGAR,DISTILLED	2-1/8 lbs	1 qts	
WATER	8-1/3 oz	1 cup	
CABBAGE,GREEN,FRESH,SHREDDED	13 lbs	5 gal 1 qts	16-1/4 lbs

Method

1 Combine black pepper, salt, granulated sugar, vinegar, and water; mix well.
2 Pour dressing over cabbage; toss lightly until well mixed.
3 Cover; refrigerate until ready to serve. CCP: Hold for service at 41 F. or lower.

SALADS, SALAD DRESSINGS, AND RELISHES   No.M 009 02

CABBAGE AND CARROT SLAW WITH CREAMY DRESSING

Yield 100   Portion   1/2 Cup

Calories	Carbohydrates	Protein	Fat	Cholesterol	Sodium	Calcium
78 cal	9 g	1 g	5 g	3 mg	190 mg	33 mg

Ingredient	Weight	Measure	Issue
MILK,NONFAT,DRY	1-3/4 oz	3/4 cup	
WATER,WARM	14-5/8 oz	1-3/4 cup	
SALAD DRESSING,MAYONNAISE TYPE	2 lbs	1 qts	
PEPPER,BLACK,GROUND	1/8 oz	1/3 tsp	
MUSTARD,PREPARED	1-1/8 oz	2 tbsp	
SALT	1 oz	1 tbsp	
SUGAR,GRANULATED	12-1/3 oz	1-3/4 cup	
VINEGAR,DISTILLED	8-1/3 oz	1 cup	
CABBAGE,GREEN,FRESH,SHREDDED	10-1/2 lbs	4 gal 1 qts	13-1/8 lbs
CARROTS,FRESH,SHREDDED	2-7/8 lbs	2 qts 3-7/8 cup	3-1/2 lbs

Method

1 Reconstitute milk; add Salad Dressing, pepper, mustard, salt and sugar; mix well.
2 Add vinegar gradually; blend well.
3 Use finely shredded cabbage and finely shredded carrots. Pour dressing over cabbage; toss lightly until well mixed.
4 Cover; refrigerate until ready to serve. CCP: Hold for service at 41 F. or lower.

SALADS, SALAD DRESSINGS, AND RELISHES No. M 009 03
PINEAPPLE COLE SLAW

Yield 100   Portion   1/2 Cup

Calories	Carbohydrates	Protein	Fat	Cholesterol	Sodium	Calcium
87 cal	12 g	1 g	5 g	3 mg	184 mg	33 mg

Ingredient	Weight	Measure	Issue
MILK,NONFAT,DRY	1-3/4 oz	3/4 cup	
WATER,WARM	14-5/8 oz	1-3/4 cup	
SALAD DRESSING,MAYONNAISE TYPE	2 lbs	1 qts	
MUSTARD,PREPARED	3/8 oz	1/3 tsp	
SALT	1 oz	1 tbsp	
SUGAR,GRANULATED	12-1/3 oz	1-3/4 cup	
VINEGAR,DISTILLED	8-1/3 oz	1 cup	
CABBAGE,GREEN,FRESH,SHREDDED	10-1/2 lbs	4 gal 1 qts	13-1/8 lbs
PINEAPPLE,CANNED,CHUNKS,JUICE PACK,DRAINED	6-3/4 lbs	2 qts 1-7/8 cup	

Method

1. Reconstitute milk; add Salad Dressing, mustard, salt, and sugar; mix well.
2. Add vinegar gradually; blend well.
3. Use finely shredded cabbage and pineapple chunks or tidbits. Pour dressing over cabbage; toss lightly until well mixed.
4. Cover; refrigerate until ready to serve. CCP: Hold for service at 41 F. or lower.

SALADS, SALAD DRESSINGS, AND RELISHES No.M 009 04

PINEAPPLE MARSHMALLOW COLE SLAW

Yield 100   **Portion** 2/3 Cup

Calories	Carbohydrates	Protein	Fat	Cholesterol	Sodium	Calcium
106 cal	16 g	1 g	5 g	3 mg	186 mg	33 mg

Ingredient	Weight	Measure	Issue
MILK,NONFAT,DRY	1-3/4 oz	3/4 cup	
WATER,WARM	14-5/8 oz	1-3/4 cup	
SALAD DRESSING,MAYONNAISE TYPE	2 lbs	1 qts	
MUSTARD,PREPARED	3/8 oz	1/3 tsp	
SALT	1 oz	1 tbsp	
SUGAR,GRANULATED	12-1/3 oz	1-3/4 cup	
VINEGAR,DISTILLED	8-1/3 oz	1 cup	
CABBAGE,GREEN,FRESH,SHREDDED	10-1/2 lbs	4 gal 1 qts	13-1/8 lbs
PINEAPPLE,CANNED,CHUNKS,JUICE PACK,DRAINED	6-3/4 lbs	2 qts 1-7/8 cup	
MARSHMALLOWS,MINIATURE	1-1/3 lbs	3 qts	

Method

1 Reconstitute milk; add Salad Dressing, mustard, salt, and sugar; mix well.
2 Add vinegar gradually; blend well.
3 Use finely shredded cabbage and pineapple chunks or tidbits. Pour dressing over cabbage; toss lightly until well mixed.
4 Cover; refrigerate until ready to serve. CCP: Hold for service at 41 F. or lower. Just before serving, add miniature marshmallows.

SALADS, SALAD DRESSINGS, AND RELISHES No. M 009 05

VEGETABLE SLAW WITH CREAMY DRESSING

Yield 100   **Portion** 1/2 Cup

Calories	Carbohydrates	Protein	Fat	Cholesterol	Sodium	Calcium
79 cal	9 g	1 g	5 g	3 mg	189 mg	33 mg

Ingredient	**Weight**	**Measure**	**Issue**
MILK,NONFAT,DRY	1-3/4 oz	3/4 cup	
WATER,WARM	14-5/8 oz	1-3/4 cup	
SALAD DRESSING,MAYONNAISE TYPE	2 lbs	1 qts	
PEPPER,BLACK,GROUND	1/8 oz	1/3 tsp	
MUSTARD,PREPARED	1-1/8 oz	2 tbsp	
SALT	1 oz	1 tbsp	
SUGAR,GRANULATED	12-1/3 oz	1-3/4 cup	
VINEGAR,DISTILLED	8-1/3 oz	1 cup	
CABBAGE,GREEN,FRESH,SHREDDED	10-1/2 lbs	4 gal 1 qts	13-1/8 lbs
CARROTS,FRESH,SHREDDED	2 lbs	2 qts 1/4 cup	2-1/2 lbs
ONIONS,FRESH,CHOPPED	5-5/8 oz	3/4 cup	6-1/4 oz
PEPPERS,GREEN,FRESH,CHOPPED	1-1/2 lbs	1 qts 1/2 cup	1-7/8 lbs

Method

1. Reconstitute milk; add Salad Dressing, pepper, mustard, salt, and sugar; mix well.
2. Add vinegar gradually; blend well.
3. Combine finely shredded cabbage, finely shredded carrots, fresh onions, and sweet peppers. Pour dressing over vegetables; toss lightly until well mixed.
4. Cover; refrigerate until ready to serve. CCP: Hold for service at 41 F. or lower.

SALADS, SALAD DRESSINGS, AND RELISHES No.M 010 00

HONEY MUSTARD DRESSING

Yield 100   **Portion** 1 Tablespoon

Calories	Carbohydrates	Protein	Fat	Cholesterol	Sodium	Calcium
27 cal	7 g	0 g	0 g	0 mg	36 mg	4 mg

Ingredient	**Weight**	**Measure**	**Issue**
WATER	1-7/8 lbs	3-1/2 cup	
HONEY	1-2/3 lbs	2-1/4 cup	
VINEGAR, DISTILLED	6-1/4 oz	3/4 cup	
GARLIC POWDER	1/3 oz	1 tbsp	
MUSTARD, DRY	2 oz	1/4 cup 1-1/3 tbsp	
ONION POWDER	1/2 oz	2 tbsp	
SALT	1/3 oz	1/4 tsp	

Method

1. Combine water, honey, and vinegar in mixer bowl.
2. Add garlic powder, mustard, onion powder, and salt to mixture.
3. Using a wire whip, mix at medium speed 3 minutes or until well blended.
4. Cover; refrigerate at 41 F. or lower until ready to serve.
5. Whip or stir well before serving.

SALADS. SALAD DRESSINGS. AND RELISHES No.M 011 00
LOW CALORIE YOGURT DRESSING

Yield 100   Portion   2 Tablespoons

Calories	Carbohydrates	Protein	Fat	Cholesterol	Sodium	Calcium
25 cal	4 g	2 g	0 g	2 mg	93 mg	57 mg

Ingredient	Weight	Measure	Issue
YOGURT,PLAIN,LOWFAT	6-1/2 lbs	3 qts	
ONIONS,FRESH,CHOPPED	1 lbs	2-7/8 cup	1-1/8 lbs
PARSLEY,FRESH,BUNCH,CHOPPED	3-1/8 oz	1-1/2 cup	3-1/4 oz
CELERY,FRESH,CHOPPED	6-1/3 oz	1-1/2 cup	8-2/3 oz
SUGAR,GRANULATED	3-1/2 oz	1/2 cup	
VINEGAR,DISTILLED	6-1/4 oz	3/4 cup	
SALT	5/8 oz	1 tbsp	
GARLIC POWDER	1/4 oz	1/3 tsp	

Method

1 Mix together yogurt, onions, parsley, celery leaves, sugar, vinegar, salt, and garlic. Stir well to blend.
2 Cover; refrigerate until ready to serve. CCP: Hold for service at 41 F. or lower.

SALADS, SALAD DRESSINGS, AND RELISHES No.M 012 00

COTTAGE CHEESE SALAD

Yield 100 **Portion** 1/4 Cup

Calories	Carbohydrates	Protein	Fat	Cholesterol	Sodium	Calcium
62 cal	2 g	7 g	3 g	8 mg	229 mg	46 mg

Ingredient

Ingredient	Weight	Measure	Issue
LETTUCE,LEAF,FRESH,HEAD	4 lbs		6-1/4 lbs
CHEESE,COTTAGE	12-3/8 lbs	1 gal 2-1/4 qts	
PAPRIKA,GROUND	1/4 oz	1 tbsp	

Method

1 Place 1 lettuce leaf on each serving dish; add 1/4 cup cottage cheese.
2 Garnish with paprika. Cover; refrigerate until ready to serve. CCP: Hold for service at 41 F. or lower.

SALADS. SALAD DRESSINGS. AND RELISHES No.M 013 00
COTTAGE CHEESE AND PEACH SALAD

Yield 100   Portion   1/4 Cup

Calories	Carbohydrates	Protein	Fat	Cholesterol	Sodium	Calcium
102 cal	12 g	8 g	3 g	8 mg	233 mg	52 mg

Ingredient — Weight — Measure — Issue

Ingredient	Weight	Measure	Issue
LETTUCE,LEAF,FRESH,HEAD	4 lbs		6-1/4 lbs
PEACHES,CANNED,HALVES,JUICE PACK,DRAINED	19-2/3 lbs	2 gal 1 qts	
CHEESE,COTTAGE	12-3/8 lbs	1 gal 2-1/4 qts	
PAPRIKA,GROUND	1/4 oz	1 tbsp	

Method

1. Separate leaves. Place 1 lettuce leaf on each serving dish.
2. Drain peach halves. Place one peach half, hollow side up, on each lettuce leaf.
3. Place 1/4 cup cottage cheese on each peach hollow.
4. Garnish with paprika. Cover; refrigerate until ready to serve. CCP: Hold for service at 41 F. or lower.

SALADS, SALAD DRESSINGS, AND RELISHES   No.M 013 01

COTTAGE CHEESE AND APRICOT SALAD

Yield 100   Portion   1/4 Cup

Calories	Carbohydrates	Protein	Fat	Cholesterol	Sodium	Calcium
91 cal	10 g	8 g	3 g	8 mg	232 mg	54 mg

Ingredient

Ingredient	Weight	Measure	Issue
LETTUCE,LEAF,FRESH,HEAD	4 lbs		6-1/4 lbs
APRICOTS,CANNED,JUICE PACK,DRAINED	13-1/2 lbs	1 gal 2-1/4 qts	
CHEESE,COTTAGE	12-3/8 lbs	1 gal 2-1/4 qts	
PAPRIKA,GROUND	1/4 oz	1 tbsp	

Method

1 Separate leaves. Place 1 lettuce leaf on each serving dish.
2 Drain apricots. Add 2 apricot halves per leaf.
3 Place 2 tablespoons cottage cheese on each apricot half.
4 Garnish with paprika. Cover; refrigerate until ready to serve. CCP: Hold for service at 41 F. or lower.

SALADS. SALAD DRESSINGS. AND RELISHES No.M 013 02

COTTAGE CHEESE AND PEAR SALAD

Yield 100   Portion   1/4 Cup

Calories	Carbohydrates	Protein	Fat	Cholesterol	Sodium	Calcium
107 cal	14 g	8 g	3 g	8 mg	233 mg	54 mg

Ingredient	Weight	Measure	Issue
LETTUCE,LEAF,FRESH,HEAD	4 lbs		6-1/4 lbs
PEARS,CANNED,HALVES,DRAINED	19-7/8 lbs	2 gal 1-1/8 qts	
CHEESE,COTTAGE	12-3/8 lbs	1 gal 2-1/4 qts	
PAPRIKA,GROUND	1/4 oz	1 tbsp	

Method

1 Separate leaves. Place 1 lettuce leaf on each serving dish.
2 Drain pear halves. Place 1 pear half on each lettuce leaf, hollow side up.
3 Place 1/4 cup cottage cheese in each pear hollow.
4 Garnish with paprika. Cover; refrigerate until ready to serve. CCP: Hold for service at 41 F. or lower.

SALADS, SALAD DRESSINGS, AND RELISHES   No.M 013 03

COTTAGE CHEESE AND PINEAPPLE SALAD

Yield 100   Portion   1/4 Cup

Calories	Carbohydrates	Protein	Fat	Cholesterol	Sodium	Calcium
98 cal	12 g	8 g	3 g	8 mg	230 mg	56 mg

Ingredient	Weight	Measure	Issue
LETTUCE,LEAF,FRESH,HEAD	4 lbs		6-1/4 lbs
PINEAPPLE,CANNED,SLICED,DRAINED	13-1/2 lbs	2 gal 1-1/2 qts	
CHEESE,COTTAGE	12-3/8 lbs	1 gal 2-1/4 qts	
PAPRIKA,GROUND	1/4 oz	1 tbsp	

Method

1 Separate leaves. Place 1 lettuce leaf on each serving dish.
2 Drain pineapple slices. Place 1 slice pineapple on lettuce leaf.
3 Place 1/4 cup cottage cheese on each pineapple slice.
4 Garnish with paprika. Cover; refrigerate until ready to serve. CCP: Hold for service at 41 F. or lower.

SALADS, SALAD DRESSINGS, AND RELISHES   No.M 014 00

COTTAGE CHEESE AND TOMATO SALAD

Yield 100   Portion   1/4 Cup

Calories	Carbohydrates	Protein	Fat	Cholesterol	Sodium	Calcium
74 cal	5 g	8 g	3 g	8 mg	234 mg	49 mg

Ingredient	Weight	Measure	Issue
TOMATOES,FRESH	12-1/2 lbs		12-3/4 lbs
LETTUCE,LEAF,FRESH,HEAD	4 lbs		6-1/4 lbs
CHEESE,COTTAGE	12-3/8 lbs	1 gal 2-1/4 qts	
PAPRIKA,GROUND	1/4 oz	1 tbsp	

Method

1. Cut each tomato into 8 wedges; set aside for use in Step 4.
2. Place 1 lettuce leaf on each serving dish.
3. Place 1/4 cup cottage cheese in center of each lettuce leaf.
4. Arrange 3 tomato wedges around cottage cheese.
5. Sprinkle with paprika. Cover; refrigerate until ready to serve. CCP: Hold for service at 41 F. or lower.

SALADS. SALAD DRESSINGS. AND RELISHES No.M 015 00

CUCUMBER AND ONION SALAD

Yield 100   Portion   1/2 Cup

Calories	Carbohydrates	Protein	Fat	Cholesterol	Sodium	Calcium
30 cal	8 g	0 g	0 g	0 mg	118 mg	11 mg

Ingredient	Weight	Measure	Issue
CUCUMBERS,FRESH,PEELED,SLICED	11-1/2 lbs	2 gal 3 qts	22-5/8 each
ONIONS,FRESH,SLICED	2-1/4 lbs	2 qts 7/8 cup	2-1/2 lbs
SALT	1 oz	1 tbsp	
PEPPER,BLACK,GROUND	1/8 oz	1/8 tsp	
SUGAR,GRANULATED	1 lbs	2-1/4 cup	
VINEGAR,DISTILLED	3-1/8 lbs	1 qts 2 cup	
WATER	1 lbs	2 cup	

Method

1 Combine cucumbers and onions.
2 Combine salt, pepper, sugar, vinegar, and water; blend well.
3 Pour over cucumbers and onions.
4 Cover and refrigerate for at least 1-1/2 hours. Keep refrigerated until ready to serve. CCP: Hold for service at 41 F. or lower.

SALADS. SALAD DRESSINGS. AND RELISHES No.M 016 00
LOW CALORIE THOUSAND ISLAND DRESSING

Yield 100   Portion   2 Tablespoons

Calories	Carbohydrates	Protein	Fat	Cholesterol	Sodium	Calcium
29 cal	4 g	2 g	1 g	2 mg	242 mg	57 mg

Ingredient	Weight	Measure	Issue
SAUCE,CHILI	1-3/4 lbs	3-1/4 cup	
MUSTARD,PREPARED	2-1/4 oz	1/4 cup 1/3 tbsp	
ONIONS,FRESH,CHOPPED	9-7/8 oz	1-3/4 cup	11 oz
VINEGAR,DISTILLED	2-1/8 oz	1/4 cup 1/3 tbsp	
SALT	1-1/2 oz	2-1/3 tbsp	
YOGURT,PLAIN,LOWFAT	6-1/2 lbs	3 qts	

Method

1 Combine chili sauce, mustard, onions, vinegar, and salt; blend well.
2 Add yogurt, stir until well blended.
3 Cover; refrigerate until ready to serve. CCP: Hold for service at 41 F. or lower.

SALADS, SALAD DRESSINGS, AND RELISHES No.M 017 00

FRUIT SALAD

Yield 100   Portion   1/2 Cup

Calories	Carbohydrates	Protein	Fat	Cholesterol	Sodium	Calcium
79 cal	19 g	1 g	0 g	1 mg	10 mg	53 mg

Ingredient	Weight	Measure	Issue
PINEAPPLE,CANNED,CHUNKS,JUICE PACK,DRAINED	5-1/4 lbs	3 qts	
ORANGE,FRESH,SECTIONS,PEELED,DICED	6-1/3 lbs	0 gal 4 qts	21-7/8 each
GRAPEFRUIT,FRESH,PEELED,CHUNKS	1-3/8 lbs	2-3/4 cup	2-5/8 lbs
APPLES,FRESH,MEDIUM,PEELED,CORED,CHOPPED	7-1/8 lbs	1 gal 2-1/2 qts	9-1/8 lbs
BANANA,FRESH,CHOPPED	2-5/8 lbs	1 qts 3-7/8 cup	4 lbs
QUICK FRUIT DRESSING		2 qts	
LETTUCE,LEAF,FRESH,HEAD	4 lbs		6-1/4 lbs

Method

1 Drain pineapple. Reserve about 3 cups juice for use in Step 4.
2 Add oranges and grapefruit.
3 Add apples and bananas. Toss lightly.
4 Prepare 1 recipe Quick Fruit Dressing, Recipe No. M 056 00 using reserved pineapple juice. Fold dressing into fruit salad. Toss lightly. Cover; refrigerate.
5 Separate leaves. Place 1 lettuce leaf on each serving dish; add 1/2 cup salad mixture. Cover; refrigerate until ready to serve. CCP: Hold for service at 41 F. or lower.

SALADS, SALAD DRESSINGS, AND RELISHES No.M 018 00

GARDEN COTTAGE CHEESE SALAD

Yield 100 **Portion** 1/2 Cup

Calories	Carbohydrates	Protein	Fat	Cholesterol	Sodium	Calcium
56 cal	3 g	6 g	2 g	7 mg	191 mg	46 mg

Ingredient	Weight	Measure	Issue
CUCUMBER,FRESH,CHOPPED	4 lbs	3 qts 3-1/4 cup	4-3/4 lbs
RADISH,FRESH,CHOPPED	1 lbs	3-7/8 cup	1-1/8 lbs
ONIONS,GREEN,FRESH,CHOPPED	8 oz	2-1/4 cup	8-7/8 oz
CELERY,FRESH,CHOPPED	1-1/2 lbs	1 qts 1-5/8 cup	2 lbs
PEPPERS,GREEN,FRESH,CHOPPED	1 lbs	3 cup	1-1/4 lbs
CHEESE,COTTAGE	9-7/8 lbs	1 gal 1 qts	
LETTUCE,LEAF,FRESH,HEAD	4 lbs		6-1/4 lbs

Method

1. Combine vegetables; toss lightly.
2. Combine cottage cheese with vegetables; mix well.
3. Place 1 lettuce leaf on each serving dish; add salad mixture. Cover; refrigerate until ready to serve. CCP: Hold for service at 41 F. or lower.

SALADS, SALAD DRESSINGS, AND RELISHES   No.M 019 00

GARDEN VEGETABLE SALAD

Yield 100   Portion   3/4 Cup

Calories	Carbohydrates	Protein	Fat	Cholesterol	Sodium	Calcium
12 cal	3 g	1 g	0 g	0 mg	18 mg	15 mg

Ingredient

Ingredient	Weight	Measure	Issue
CARROTS,FRESH,SLICED	2 lbs	1 qts 3-1/8 cup	2-1/2 lbs
CELERY,FRESH,CHOPPED	3 lbs	2 qts 3-3/8 cup	4-1/8 lbs
PEPPERS,GREEN,FRESH,CHOPPED	2 lbs	1 qts 2-1/8 cup	2-1/2 lbs
LETTUCE,ICEBERG,FRESH,CHOPPED	7 lbs	3 gal 2-1/2 qts	7-1/2 lbs

Method

1 Combine carrots, celery, and peppers. Tear prepared lettuce into small pieces; mix with other vegetables; toss lightly.
2 Cover; refrigerate until ready to serve. CCP: Hold for service at 41 F. or lower.

SALADS, SALAD DRESSINGS, AND RELISHES No.M 020 00
MARINATED CARROTS

Yield 100   Portion   1/2 Cup

Calories	Carbohydrates	Protein	Fat	Cholesterol	Sodium	Calcium
135 cal	24 g	1 g	5 g	0 mg	109 mg	28 mg

Ingredient	Weight	Measure	Issue
CARROTS,FRESH,STICKS	16 lbs	3 gal 2-1/8 qts	19-1/2 lbs
WATER,BOILING	20-7/8 lbs	2 gal 2 qts	
SOUP,CONDENSED,TOMATO	3-1/8 lbs	1 qts 1-3/4 cup	
SUGAR,GRANULATED	2-2/3 lbs	1 qts 2 cup	
VINEGAR,DISTILLED	3-1/8 lbs	1 qts 2 cup	
PEPPER,BLACK,GROUND	1/8 oz	1/3 tsp	
MUSTARD,DRY	3/4 oz	2 tbsp	
OIL,SALAD	1 lbs	2 cup	
ONIONS,FRESH,CHOPPED	2 lbs	1 qts 1-5/8 cup	2-1/4 lbs
PEPPERS,GREEN,FRESH,CHOPPED	2 lbs	1 qts 2-1/8 cup	2-1/2 lbs

Method

1. Add carrots to boiling water.
2. Bring back to a boil; reduce heat; simmer 5 minutes or until tender-crisp. Drain. Set aside for use in Step 6.
3. Combine soup, sugar, vinegar, pepper, and mustard in a mixer bowl. Blend at medium speed 3 minutes.
4. Add salad oil or olive oil slowly to mixture at low speed 2 minutes.
5. Add onions and peppers. Scrape down bowl. Blend 1 minute.
6. Pour mixture over warm carrots. Cover; refrigerate overnight or until flavors are well blended. Keep refrigerated until ready to serve. CCP: Hold for service at 41 F. or lower.

SALADS, SALAD DRESSINGS, AND RELISHES No.M 021 00

LOW CALORIE TANGY TARRAGON DRESSING

Yield 100   Portion   2 Tablespoons

Calories	Carbohydrates	Protein	Fat	Cholesterol	Sodium	Calcium
20 cal	6 g	0 g	0 g	0 mg	117 mg	5 mg

Ingredient	Weight	Measure	Issue
VINEGAR,DISTILLED	5-1/4 lbs	2 qts 2 cup	
WATER	2-5/8 lbs	1 qts 1 cup	
SUGAR,GRANULATED	14-1/8 oz	2 cup	
SALT	1 oz	1 tbsp	
PARSLEY,DEHYDRATED,FLAKED	3/8 oz	1/2 cup	
TARRAGON,GROUND	1/3 oz	2 tbsp	
GARLIC POWDER	3/8 oz	1 tbsp	
PEPPER,BLACK,GROUND	1/8 oz	1/3 tsp	

Method

1 Combine vinegar, water, sugar, salt, parsley, tarragon, garlic, and pepper; blend well.
2 Cover; refrigerate until ready to serve. CCP: Hold for service at 41 F. or lower.
3 Shake or beat well before using.

SALADS. SALAD DRESSINGS. AND RELISHES   No.M 021 01

LOW CALORIE BASIL DRESSING

Yield 100   Portion   2 Tablespoons

Calories	Carbohydrates	Protein	Fat	Cholesterol	Sodium	Calcium
20 cal	6 g	0 g	0 g	0 mg	117 mg	6 mg

Ingredient	Weight	Measure	Issue
VINEGAR,DISTILLED	5-1/4 lbs	2 qts 2 cup	
WATER	2-5/8 lbs	1 qts 1 cup	
SUGAR,GRANULATED	14-1/8 oz	2 cup	
SALT	1 oz	1 tbsp	
ONION POWDER	7/8 oz	1/4 cup	
BASIL,SWEET,WHOLE,CRUSHED	1/2 oz	3-1/3 tbsp	
PEPPER,BLACK,GROUND	1/8 oz	1/3 tsp	

Method

1 Combine vinegar, water, sugar, salt, pepper, onion powder, and sweet basil; blend well.
2 Cover; refrigerate until ready to serve. CCP: Hold for service at 41 F. or lower.
3 Shake or beat well before using.

SALADS, SALAD DRESSINGS, AND RELISHES No.M 022 00

TANGY YOGURT SALAD DRESSING

Yield 100   **Portion** 1 Tablespoon

Calories	Carbohydrates	Protein	Fat	Cholesterol	Sodium	Calcium
18 cal	3 g	1 g	0 g	1 mg	13 mg	28 mg

Ingredient   Weight   Measure   Issue

SUGAR,GRANULATED   7 oz   1 cup
VINEGAR,DISTILLED   8-1/3 oz   1 cup
MUSTARD,DRY   1/2 oz   1 tbsp
YOGURT,PLAIN,LOWFAT   3-1/4 lbs   1 qts 2 cup
HORSERADISH,PREPARED   2-7/8 oz   1/4 cup 1-2/3 tbsp

Method

1 Combine sugar, vinegar, and mustard; stir until sugar is dissolved.
2 Add yogurt and horseradish. Blend well.
3 Cover; refrigerate until ready to serve. CCP: Hold for service at 41 F. or lower.

SALADS. SALAD DRESSINGS. AND RELISHES No.M 023 00

JELLIED CRANBERRY AND ORANGE SALAD

Yield 100   Portion   4 Ounces

Calories	Carbohydrates	Protein	Fat	Cholesterol	Sodium	Calcium
106 cal	26 g	2 g	0 g	0 mg	50 mg	21 mg

Ingredient	Weight	Measure	Issue
WATER,BOILING	6-1/4 lbs	3 qts	
DESSERT POWDER,GELATIN,ORANGE	3-1/2 lbs	1 qts 3 cup	
WATER,COLD	10-1/2 lbs	1 gal 1 qts	
CRANBERRIES,FRESH	3-1/8 lbs	3 qts 2-7/8 cup	3-1/4 lbs
ORANGE,FRESH	1-3/4 lbs	6 each	
CELERY,FRESH,CHOPPED	1-5/8 lbs	1 qts 2-1/8 cup	2-1/4 lbs
SUGAR,GRANULATED	1-3/4 lbs	1 qts	
LETTUCE,LEAF,FRESH,HEAD	4 lbs		6-1/4 lbs

Method

1 Add gelatin to boiling water; stir until dissolved.
2 Add cold water; mix well.
3 Pour about 3 quarts gelatin mixture into each pan; chill until slightly thickened.
4 Grind cranberries. Set aside for use in Step 7.
5 Quarter oranges; remove seeds. DO NOT PEEL. Grind oranges.
6 Combine cranberries, oranges, celery, and sugar; mix well.
7 Add 2 quarts cranberry mixture to gelatin in each pan; stir to distribute evenly.
8 Chill until firm. Cut 5 by 7.
9 Place 1 lettuce leaf on each serving dish; add gelatin square. Cover; refrigerate at 41 F. or lower until ready to serve.

SALADS, SALAD DRESSINGS, AND RELISHES No.M 023 01

JELLIED CRANBERRY AND ORANGE SALAD (CANNED)

Yield 100   **Portion** 4 Ounces

Calories	Carbohydrates	Protein	Fat	Cholesterol	Sodium	Calcium
112 cal	27 g	2 g	0 g	0 mg	59 mg	19 mg

Ingredient	Weight	Measure	Issue
WATER,BOILING	6-1/4 lbs	3 qts	
DESSERT POWDER,GELATIN,ORANGE	3-1/2 lbs	1 qts 3 cup	
WATER,COLD	10-1/2 lbs	1 gal 1 qts	
ORANGE,FRESH		6 each	
CRANBERRY SAUCE,JELLIED	7 lbs	2 qts 3-1/2 cup	
CELERY,FRESH,CHOPPED	1-5/8 lbs	1 qts 2-1/8 cup	2-1/4 lbs
LETTUCE,LEAF,FRESH,HEAD	4 lbs		6-1/4 lbs

Method

1. Add gelatin to boiling water; stir until dissolved.
2. Add cold water; mix well.
3. Pour 3 quarts gelatin mixture into each pan; chill until slightly thickened.
4. Quarter oranges; remove seeds. DO NOT PEEL. Grind oranges.
5. Use canned Cranberry Sauce; beat with wire whip until smooth.
6. Combine oranges, cranberry sauce and celery; mix well.
7. Add 2 quarts cranberry mixture to gelatin in each pan; stir to distribute evenly.
8. Chill until firm. Cut 5 by 7.
9. Place 1 lettuce leaf on each serving dish; add gelatin square. Cover; refrigerate at 41 F. or lower until ready to serve.

SALADS, SALAD DRESSINGS, AND RELISHES   No.M 024 00

JELLIED CRANBERRY AND PINEAPPLE SALAD

Yield 100   Portion   5-1/2 Ounces

Calories	Carbohydrates	Protein	Fat	Cholesterol	Sodium	Calcium
149 cal	33 g	2 g	2 g	0 mg	54 mg	23 mg

Ingredient	Weight	Measure	Issue
PINEAPPLE,CANNED,CRUSHED,JUICE PACK,INCL LIQUIDS	6-5/8 lbs	3 qts	
CRANBERRY SAUCE,JELLIED	7-1/3 lbs	3 qts	
WATER,BOILING	8-1/3 lbs	1 gal	
DESSERT POWDER,GELATIN,CHERRY	3-1/2 lbs	1 qts 3 cup	
RESERVED LIQUID	6-1/4 lbs	3 qts	
JUICE,LEMON	8-5/8 oz	1 cup	
LEMON RIND,GRATED	3/8 oz	2 tbsp	
NUTS,UNSALTED,CHOPPED,COARSELY	10-1/3 oz	2 cup	
LETTUCE,LEAF,FRESH,HEAD	4 lbs		6-1/4 lbs

Method

1 Drain pineapple; reserve juice for use in Step 4 and pineapple for use in Step 6.
2 Using a wire whip, beat cranberry sauce at medium speed in mixer bowl until smooth. Set aside for use in Step 3.
3 Add gelatin to boiling water; stir until dissolved. Add cranberry sauce; mix well.
4 Add cold water and reserved juice, lemon juice, and rind; mix well.
5 Pour 1 gallon gelatin mixture into each pan; chill until slightly thickened.
6 Add 1 quart pineapple and 2/3 cup nuts to gelatin mixture in each pan; stir to distribute evenly.
7 Chill until firm. Cut 5 by 7.
8 Place 1 lettuce leaf on each serving dish; add gelatin square. Cover; refrigerate at 41 F. or lower until ready to serve.

SALADS, SALAD DRESSINGS, AND RELISHES No.M 025 00

JELLIED FRUIT SALAD

Yield 100					Portion 5 Ounces

Calories	Carbohydrates	Protein	Fat	Cholesterol	Sodium	Calcium
105 cal	26 g	2 g	0 g	0 mg	47 mg	19 mg

Ingredient	Weight	Measure	Issue
APRICOTS,CANNED,JUICE PACK,INCL LIQ,HALVES	6-1/2 lbs	3 qts	
DESSERT POWDER,GELATIN,ORANGE	3-1/2 lbs	1 qts 3 cup	
WATER,BOILING	8-1/3 lbs	1 gal	
RESERVED LIQUID	8-1/3 lbs	1 gal	
APPLES,FRESH,MEDIUM,UNPEELED,DICED	1-2/3 lbs	1 qts 2 cup	2 lbs
BANANA,FRESH,CHOPPED	2-1/4 lbs	1 qts 2-3/4 cup	3-1/2 lbs
MARSHMALLOWS,MINIATURE	1 lbs	2 qts 1 cup	
LETTUCE,LEAF,FRESH,HEAD	4 lbs		6-1/4 lbs

Method

1 Drain apricots; reserve juice for use in Step 3. Cut apricots in half; set aside for use in Step 5.
2 Add gelatin to boiling water; stir until dissolved.
3 Add cold water and reserved juice; mix well.
4 Pour 3 quarts gelatin mixture into each pan; chill until slightly thickened.
5 Add equal amounts of apricots, apples, and bananas to gelatin mixture in each pan; stir to distribute evenly.
6 Sprinkle 3 cups marshmallows over mixture in each pan.
7 Chill until firm. Cut 5 by 7.
8 Place 1 lettuce leaf on each serving dish; add gelatin square. Cover; refrigerate at 41 F. or lower until ready to serve.

SALADS. SALAD DRESSINGS. AND RELISHES No.M 025 01

JELLIED ORANGE SALAD

Yield 100   Portion   5 Ounces

Calories	Carbohydrates	Protein	Fat	Cholesterol	Sodium	Calcium
107 cal	26 g	2 g	0 g	0 mg	43 mg	31 mg

Ingredient	Weight	Measure	Issue
DESSERT POWDER,GELATIN,ORANGE	3-1/2 lbs	1 qts 3 cup	
WATER,BOILING	8-1/3 lbs	1 gal	
JUICE,ORANGE	6-5/8 lbs	3 qts	
PINEAPPLE,CANNED,CRUSHED	6-3/4 lbs	3 qts 3/8 cup	
ORANGE,FRESH,SECTIONS	6 lbs	3 qts 3-1/8 cup	8-1/4 lbs
LETTUCE,LEAF,FRESH,HEAD	4 lbs		6-1/4 lbs

Method

1. Add gelatin to boiling water; stir until dissolved.
2. Add orange juice and crushed pineapple; mix well.
3. Pour 3 quarts gelatin mixture into each pan; chill until slightly thickened.
4. Add peeled fresh oranges, sliced and cut into halves to gelatin mixture in each pan; stir to distribute evenly.
5. Chill until firm. Cut 5 by 7.
6. Place 1 lettuce leaf on each serving dish; add gelatin square. Cover; refrigerate at 41 F. or lower until ready to serve.

SALADS, SALAD DRESSINGS, AND RELISHES   No.M 025 02

JELLIED PEAR SALAD

Yield 100   **Portion** 5 Ounces

Calories	Carbohydrates	Protein	Fat	Cholesterol	Sodium	Calcium
93 cal	23 g	2 g	0 g	0 mg	46 mg	20 mg

Ingredient	Weight	Measure	Issue
PEARS,CANNED,HALVES,DRAINED,CHOPPED	13-1/4 lbs	1 gal 2 qts	
DESSERT POWDER,GELATIN,LIME	3-1/2 lbs	1 qts 3 cup	
WATER,BOILING	8-1/3 lbs	1 gal	
RESERVED LIQUID	8-1/3 lbs	1 gal	
LETTUCE,LEAF,FRESH,HEAD	4 lbs		6-1/4 lbs

Method

1 Drain pears; reserve juice for use in Step 3. Cut pear halves into 2 or 3 pieces; set aside for use in Step 5.
2 Add lime gelatin to boiling water; stir until dissolved.
3 Add cold water and reserved juice; mix well.
4 Pour 3 quarts gelatin mixture into each pan; chill until slightly thickened.
5 Add pears to gelatin mixture in each pan; stir to distribute evenly.
6 Chill until firm. Cut 5 by 7.
7 Place 1 lettuce leaf on each serving dish; add gelatin square. Cover; refrigerate at 41 F. or lower until ready to serve.

SALADS, SALAD DRESSINGS, AND RELISHES No. M 025 03

JELLIED PINEAPPLE, PEAR, AND BANANA SALAD

Yield 100　　　　　　　　　　　　　　　　　　**Portion** 5 Ounces

Calories	Carbohydrates	Protein	Fat	Cholesterol	Sodium	Calcium
120 cal	30 g	2 g	0 g	0 mg	47 mg	22 mg

Ingredient	Weight	Measure	Issue
PINEAPPLE,CANNED,CHUNKS,JUICE PACK,INCL LIQUIDS	6-3/4 lbs	3 qts 1/4 cup	
PEARS,CANNED,HALVES,JC PK,INCL LIQUIDS,CHOPPED	6-3/4 lbs	3 qts 3/8 cup	
DESSERT POWDER,GELATIN,STRAWBERRY	3-1/2 lbs	1 qts 3 cup	
WATER,BOILING	8-1/3 lbs	1 gal	
RESERVED LIQUID	8-1/3 lbs	1 gal	
BANANA,FRESH,CHOPPED	2-1/4 lbs	1 qts 2-3/4 cup	3-1/2 lbs
MARSHMALLOWS,MINIATURE	1 lbs	2 qts 1 cup	
LETTUCE,LEAF,FRESH,CHOPPED	4 lbs	2 gal 1/8 qts	6-1/4 lbs

Method

1. Drain pineapple chunks or tidbits, and canned pear halves; reserve juice for use in Step 3. Cut pear halves into 6 pieces; set aside for use in Step 5.
2. Add strawberry gelatin to boiling water; stir until dissolved.
3. Add cold water and reserved juice; mix well.
4. Pour 3 quarts gelatin mixture into each pan; chill until slightly thickened.
5. Add equal amounts of pineapple and bananas to gelatin mixture in each pan; stir to distribute evenly.
6. Sprinkle 3 cups marshmallows over mixture in each pan.
7. Chill until firm. Cut 5 by 7.
8. Place 1 lettuce leaf on each serving dish; add gelatin square. Cover. CCP; refrigerate at 41 F. or lower until ready to serve.

SALADS, SALAD DRESSINGS, AND RELISHES No.M 025 04
JELLIED STRAWBERRY SALAD

Yield 100   Portion   4-1/2 Ounces

Calories	Carbohydrates	Protein	Fat	Cholesterol	Sodium	Calcium
90 cal	22 g	2 g	0 g	0 mg	44 mg	21 mg

Ingredient / Weight / Measure / Issue

Ingredient	Weight	Measure	Issue
DESSERT POWDER,GELATIN,STRAWBERRY	3-1/2 lbs	1 qts 3 cup	
WATER,BOILING	8-1/3 lbs	1 gal	
WATER,COLD	4-1/8 lbs	2 qts	
STRAWBERRIES,FROZEN,THAWED	6 lbs	2 qts 2-5/8 cup	
PINEAPPLE,CANNED,CRUSHED,JUICE PACK,INCL LIQUIDS	3-3/8 lbs	1 qts 2-1/8 cup	
BANANA,FRESH,SLICED	2 lbs	1 qts 2 cup	3-1/8 lbs
LETTUCE,LEAF,FRESH,HEAD	4 lbs		6-1/4 lbs

Method

1. Add strawberry gelatin dessert powder to boiling water; stir until dissolved.
2. Add cold water, partially thawed strawberries, undrained canned crushed pineapple, and thinly sliced bananas. Stir to distribute evenly.
3. Pour 3 quarts gelatin mixture into each pan; chill until slightly thickened.
4. Chill until firm. Cut 5 by 7.
5. Place 1 lettuce leaf on each serving dish; add gelatin square. Cover; refrigerate at 41 F. or lower until ready to serve.

SALADS. SALAD DRESSINGS. AND RELISHES   No.M 025 05

JELLIED BANANA SALAD

Yield 100 **Portion** 5 Ounces

Calories	Carbohydrates	Protein	Fat	Cholesterol	Sodium	Calcium
93 cal	22 g	2 g	0 g	0 mg	44 mg	16 mg

Ingredient	Weight	Measure	Issue
DESSERT POWDER,GELATIN,ORANGE	3-1/2 lbs	1 qts 3 cup	
WATER,BOILING	8-1/3 lbs	1 gal	
WATER,COLD	8-1/3 lbs	1 gal	
BANANA,FRESH,CHOPPED	7-1/4 lbs	1 gal 1-1/2 qts	11-1/8 lbs
LETTUCE,LEAF,FRESH,HEAD	4 lbs		6-1/4 lbs

Method

1. Add gelatin to boiling water; stir until dissolved.
2. Add cold water; mix well.
3. Pour 3 quarts gelatin mixture into each pan; chill until slightly thickened.
4. Add 1/3 sliced bananas to gelatin in each pan; stir gently to distribute evenly.
5. Chill until firm. Cut 5 by 7.
6. Place 1 lettuce leaf on each serving dish; add gelatin square. Cover; refrigerate at 41 F. or lower until ready to serve.

SALADS, SALAD DRESSINGS, AND RELISHES   No.M 026 00

JELLIED FRUIT COCKTAIL SALAD

Yield 100   Portion   4-1/2 Ounces

Calories	Carbohydrates	Protein	Fat	Cholesterol	Sodium	Calcium
125 cal	23 g	3 g	3 g	0 mg	48 mg	25 mg

Ingredient	Weight	Measure	Issue
FRUIT COCKTAIL,CANNED,JUICE PACK,INCL LIQUIDS	12-1/2 lbs	1 gal 2 qts	
WATER,BOILING	8-1/3 lbs	1 gal	
DESSERT POWDER,GELATIN,CHERRY	3-1/2 lbs	1 qts 3 cup	
RESERVED LIQUID	8-1/3 lbs	1 gal	
JUICE,LEMON	1-5/8 lbs	3 cup	
NUTS,UNSALTED,CHOPPED,COARSELY	1-1/4 lbs	1 qts	
LETTUCE,LEAF,FRESH,HEAD	4 lbs		6-1/4 lbs

Method

1 Drain fruit cocktail; reserve juice for use in Step 3 and fruit for use in Step 5.
2 Add gelatin to boiling water; stir until dissolved.
3 Add cold water, reserved juice and lemon juice; mix well.
4 Pour 3 quarts gelatin mixture into each pan; chill until slightly thickened.
5 Add 2 quarts fruit cocktail and 1-1/3 cups nuts to gelatin mixture in each pan; stir to distribute evenly.
6 Chill until firm. Cut 5 by 7.
7 Place 1 lettuce leaf on each serving dish; add gelatin square. Cover; refrigerate at 41 F. or lower until ready to serve.

SALADS. SALAD DRESSINGS. AND RELISHES No.M 027 00

GERMAN COLE SLAW

Yield 100   Portion   1/2 Cup

Calories	Carbohydrates	Protein	Fat	Cholesterol	Sodium	Calcium
60 cal	5 g	1 g	4 g	0 mg	153 mg	29 mg

Ingredient	Weight	Measure	Issue
ONIONS,FRESH,CHOPPED	12 oz	2-1/8 cup	13-1/3 oz
PEPPERS,GREEN,FRESH,CHOPPED	12 oz	2-1/4 cup	14-5/8 oz
CELERY,FRESH,CHOPPED	12 oz	2-7/8 cup	1 lbs
CABBAGE,GREEN,FRESH,SHREDDED	12 lbs	4 gal 3-1/2 qts	15 lbs
VINEGAR,DISTILLED	1-5/8 lbs	3 cup	
OIL,SALAD	1 lbs	2 cup	
SUGAR,GRANULATED	3-1/2 oz	1/2 cup	
SALT	1-1/4 oz	2 tbsp	
PEPPER,BLACK,GROUND	1/8 oz	1/8 tsp	
PIMIENTO,CANNED,DRAINED,CHOPPED	4-1/4 oz	1/2 cup 2 tbsp	

Method

1 Combine onions, peppers, celery, and cabbage; toss lightly.
2 Combine vinegar, salad oil or olive oil, sugar, salt, and pepper; pour over cabbage mixture; mix well.
3 Garnish with pimientos. Cover; refrigerate until ready to serve. CCP: Hold for service at 41 F. or lower.

SALADS, SALAD DRESSINGS, AND RELISHES No. M 028 00
TACO SALAD

Yield 100   Portion 1-1/2 Cups

Calories	Carbohydrates	Protein	Fat	Cholesterol	Sodium	Calcium
307 cal	29 g	18 g	14 g	41 mg	908 mg	187 mg

Ingredient	**Weight**	**Measure**	**Issue**
CHILI CON CARNE		3 gal 1 qts	
LETTUCE,ROMAINE,FRESH,CHOPPED	8 lbs	4 gal 1/4 qts	8-1/2 lbs
ONIONS,FRESH,SLICED	12-1/8 oz	3 cup	13-1/2 oz
TOMATOES,FRESH,THIN WEDGES	5-1/8 lbs	3 qts 1 cup	5-1/4 lbs
CHEESE,CHEDDAR,SHREDDED	3-1/8 lbs	3 qts 1/2 cup	
CORN CHIPS	3-1/8 lbs		
TACO SAUCE		3 qts 2 cup	

Method
1. Prepare 1/2 recipe Chili Con Carne, Recipe No. L 028 00 or Chili Con Carne with Beans, Recipe No. L 059 00. Keep hot for use in Step 5. CCP: Internal temperature must reach 155 F. or higher for 15 seconds.
2. Tear lettuce into pieces.
3. Combine lettuce, onions, and tomatoes. Toss lightly.
4. Place 1 cup salad mixture in each soup bowl.
5. Ladle 1/2 cup hot chili over each salad.
6. Sprinkle 2 tablespoons cheese over each salad.
7. Sprinkle 6 to 9 corn chips and 2 tablespoons taco sauce over each salad.
8. Serve immediately. Taco Salad may be served with sour cream or guacamole.

Notes
1. In Steps 5 and 6, add chili and cheese just before serving to prevent wilted lettuce and unappetizing appearance.

SALADS. SALAD DRESSINGS. AND RELISHES   No.M 029 00

ITALIAN STYLE PASTA SALAD

Yield 100   **Portion**   1/2 Cup

Calories	Carbohydrates	Protein	Fat	Cholesterol	Sodium	Calcium
157 cal	15 g	6 g	8 g	13 mg	380 mg	93 mg

Ingredient	Weight	Measure	Issue
VINEGAR AND OIL DRESSING		1 qts	
BASIL,DRIED,CRUSHED	1 oz	1/4 cup 2-1/3 tbsp	
WATER	16-3/4 lbs	2 gal	
SALT	5/8 oz	1 tbsp	
OIL,SALAD	1/2 oz	1 tbsp	
MACARONI NOODLES,SHELLS,DRY	3-2/3 lbs	1 gal	
SALAMI,BEEF,CHOPPED	2 lbs		
CHEESE,PROVOLONE	2-1/8 lbs	1 qts 3-1/2 cup	
OLIVES,RIPE,PITTED,SLICED,INCL LIQUIDS	14-1/4 oz	3 cup	
CELERY,FRESH,CHOPPED	12 oz	2-7/8 cup	1 lbs
ONIONS,FRESH,CHOPPED	8 oz	1-3/8 cup	8-7/8 oz
PEPPERS,GREEN,FRESH,CHOPPED	8 oz	1-1/2 cup	9-3/4 oz
TOMATOES,FRESH,CHOPPED	2-1/2 lbs	1 qts 2-1/4 cup	2-1/2 lbs

Method

1 Prepare 1/2 recipe Vinegar and Oil Dressing, Recipe No. M 069 00. Add basil. Stir. Set aside for use in Step 4.
2 Add salt and salad oil to water; heat to rolling boil. Slowly add macaroni, rigatoni, rotini, or tortellini, stirring constantly or until water boils again. Cook 7 to 10 minutes or until just tender. Drain; rinse with cold water. Drain thoroughly.
3 Add salami, cheese, olives, celery, onions, peppers, and tomatoes to cooked pasta. Toss lightly.
4 Add Vinegar and Oil Dressing. Toss lightly. Cover; refrigerate at least 3 hours or until flavors are well blended. Keep refrigerated until ready to serve. CCP: Hold for service at 41 F. or lower.

Notes

1 In Step 1, 2 pounds prepared fat free Italian Salad Dressing may be used per 100 servings.

SALADS, SALAD DRESSINGS, AND RELISHES No.M 030 00
COBB SALAD

Yield 100   Portion 1 Cup

Calories	Carbohydrates	Protein	Fat	Cholesterol	Sodium	Calcium
213 cal	8 g	9 g	17 g	57 mg	351 mg	85 mg

Ingredient	Weight	Measure	Issue
BACON,RAW	5 lbs		
LETTUCE,LEAF,FRESH,HEAD	8 lbs		12-1/2 lbs
TOMATOES,FRESH,CHOPPED	2-1/2 lbs	1 qts 2-1/4 cup	2-1/2 lbs
AVOCADO,FRESH,DICED	3-1/8 lbs	2 qts 1-1/2 cup	4-1/2 lbs
ONIONS,FRESH,CHOPPED	1-1/2 lbs	1 qts 1/4 cup	1-2/3 lbs
CHEESE,BLUE-VEINED	1-7/8 lbs	1 qts 2-1/4 cup	
EGG,HARD COOKED,CHOPPED	1-7/8 lbs	1 qts 2-3/8 cup	
CHICKEN,COOKED,DICED	2-1/8 lbs		
JUICE,LEMON	2-1/8 oz	1/4 cup 1/3 tbsp	
GARLIC FRENCH DRESSING		3 qts 1/2 cup	

Method

1 Cook bacon until crisp; place cooked bacon on absorbent paper to eliminate excess fat. Chop bacon.
2 Trim, wash and prepare vegetables. Tear lettuce into pieces. Place 3/4 cup lettuce into each individual salad bowl.
3 Place 1 tablespoon blue cheese in the center on top of lettuce.
4 Arrange following ingredients around cheese in separate wedge-shaped sections: 1 tablespoon each of bacon, tomatoes, eggs, and chicken; 1-1/2 tablespoons avocado (toss avocado in lemon juice to prevent darkening); and 2 teaspoons onions.
5 CCP: Cover; refrigerate at 41 F. or lower.
6 Serve with Garlic French Dressing (Recipe No. M 060 00).

SALADS. SALAD DRESSINGS. AND RELISHES   No.M 031 00
KIDNEY BEAN SALAD

Yield 100   Portion   1/2 Cup

Calories	Carbohydrates	Protein	Fat	Cholesterol	Sodium	Calcium
152 cal	16 g	5 g	8 g	47 mg	404 mg	28 mg

Ingredient	Weight	Measure	Issue
SALAD DRESSING,MAYONNAISE TYPE	3 lbs	1 qts 2 cup	
PEPPER,BLACK,GROUND	1/4 oz	1 tbsp	
VINEGAR,DISTILLED	8-1/3 oz	1 cup	
BEANS,KIDNEY,DARK RED,CANNED,DRAINED	14 lbs	2 gal 1 qts	
CELERY,FRESH,SLICED	2-1/4 lbs	2 qts 1/2 cup	3-1/8 lbs
EGG,HARD COOKED,CHOPPED	2-1/4 lbs	20 Eggs	
ONIONS,FRESH,SLICED	12 oz	3 cup	13-1/3 oz
PEPPERS,GREEN,FRESH,MEDIUM,SLICED,THIN	1 lbs	3 cup	1-1/4 lbs
PICKLE RELISH,SWEET	2-1/8 lbs	1 qts	

Method

1. Combine Salad Dressing, pepper, and vinegar in mixer bowl; whip or mix well. Set aside for use in Step 3.
2. Drain beans; rinse; drain thoroughly.
3. Combine dressing, beans, celery, eggs, onions, peppers, and relish; mix carefully.
4. Cover; refrigerate until ready to serve. CCP: Hold for service at 41 F. or lower.

SALADS, SALAD DRESSINGS, AND RELISHES No.M 032 00

FRUIT MEDLEY SALAD

Yield 100   Portion   1/2 Cup

Calories	Carbohydrates	Protein	Fat	Cholesterol	Sodium	Calcium
101 cal	16 g	1 g	4 g	0 mg	23 mg	31 mg

Ingredient	Weight	Measure	Issue
COCONUT,PREPARED,SWEETENED FLAKES	5-3/4 oz	1-3/4 cup	
PINEAPPLE,CANNED,CHUNKS,JUICE PACK,DRAINED	2-5/8 lbs	1 qts 2 cup	
APPLES,FRESH,MEDIUM,UNPEELED,DICED	6-1/2 lbs	1 gal 1-7/8 qts	7-2/3 lbs
BANANA,FRESH,CHOPPED	2-3/4 lbs	2 qts 3/8 cup	4-1/4 lbs
CELERY,FRESH,CHOPPED	2 lbs	1 qts 3-1/2 cup	2-3/4 lbs
WALNUTS,SHELLED,CHOPPED	8-1/2 oz	2 cup	
RAISINS	7-2/3 oz	1-1/2 cup	
CHERRIES,MARASCHINO,CHOPPED	8-7/8 oz	1 cup	
WATER,COLD	2 lbs	3-3/4 cup	
WHIPPED TOPPING MIX,NONDAIRY,DRY	1 lbs	1 gal 1-5/8 qts	
MILK,NONFAT,DRY	1-5/8 oz	1/2 cup 2-2/3 tbsp	
EXTRACT,VANILLA	7/8 oz	2 tbsp	
LETTUCE,LEAF,FRESH,HEAD	4 lbs		6-1/4 lbs

Method

1 Toast coconut; cool; set aside for use in Step 3.
2 Combine pineapple, apples, and bananas in pineapple juice until ready to mix. Drain well.
3 Add celery, walnuts, raisins, chopped maraschino cherries, and coconut to drained pineapple, apples, and bananas mixture. Mix lightly; set aside for use in Step 6.
4 Place cold water in mixing bowl; add topping, milk, and vanilla. Whip at low speed 3 minutes or until blended. Scrape down
5 Whip at high speed 5 to 10 minutes or until stiff peaks are formed.
6 Add to fruit mixture tossing well to coat pieces. Cover, refrigerate until ready to serve. CCP: Hold for service at 41 F. or lower.
7 Place 1 lettuce leaf on each serving dish; add 1/2 cup salad mixture.

SALADS. SALAD DRESSINGS. AND RELISHES   No.M 033 00
LETTUCE AND TOMATO SALAD

Yield 100   Portion   3-1/2 Ounces

Calories	Carbohydrates	Protein	Fat	Cholesterol	Sodium	Calcium
24 cal	5 g	1 g	0 g	0 mg	11 mg	17 mg

Ingredient   Weight   Measure   Issue
LETTUCE,LEAF,FRESH,HEAD   4 lbs   6-1/4 lbs
TOMATOES,FRESH   21-7/8 lbs   22-1/3 lbs

Method
1 Separate leaves. Place 1 lettuce leaf on each serving dish.
2 Slice each tomato into 6 slices.
3 Arrange 4 slices tomatoes on each lettuce leaf. Cover; refrigerate until ready to serve. CCP: Hold for service at 41 F. or lower.

SALADS, SALAD DRESSINGS, AND RELISHES No.M 034 00

MACARONI SALAD

Yield 100　　　　　　　　　　　　　　　　　　**Portion** 1/2 Cup

Calories	Carbohydrates	Protein	Fat	Cholesterol	Sodium	Calcium
141 cal	20 g	3 g	6 g	26 mg	217 mg	14 mg

Ingredient	Weight	Measure	Issue
WATER,BOILING	29-1/4 lbs	3 gal 2 qts	
SALT	7/8 oz	1 tbsp	
OIL,SALAD	5/8 oz	1 tbsp	
MACARONI NOODLES,ELBOW,DRY	4-3/8 lbs	1 gal 3/4 qts	
EGG,HARD COOKED,CHOPPED	1-1/4 lbs	1 qts	
CELERY,FRESH,CHOPPED	1-7/8 lbs	1 qts 3-1/8 cup	2-5/8 lbs
ONIONS,FRESH,CHOPPED	1-3/8 lbs	3-7/8 cup	1-1/2 lbs
PICKLE RELISH,SWEET	1-1/3 lbs	2-1/2 cup	
SALAD DRESSING,MAYONNAISE TYPE	2 lbs	1 qts	
PIMIENTO,CANNED,DRAINED,CHOPPED	6-3/4 oz	1 cup	
PEPPER,BLACK,GROUND	1/8 oz	1/3 tsp	
VINEGAR,DISTILLED	6-1/4 oz	3/4 cup	
PAPRIKA,GROUND	1/8 oz	1/3 tsp	

Method

1. Add salt and salad oil to water; heat to a rolling boil.
2. Slowly add macaroni while stirring constantly, until water boils again. Cook about 15 minutes or until tender; stir occasionally. DO NOT OVERCOOK.
3. Drain. Rinse with cold water; drain thoroughly.
4. Combine macaroni, chopped eggs, celery, onions, pickle relish, Salad Dressing, pimientos, pepper, and vinegar. Toss lightly.
5. Garnish with paprika.
6. Cover; refrigerate until ready to serve. CCP: Hold for service at 41 F. or lower.

SALADS, SALAD DRESSINGS, AND RELISHES No. M 035 00

MIXED FRUIT SALAD

Yield 100   Portion 1/2 Cup

Calories	Carbohydrates	Protein	Fat	Cholesterol	Sodium	Calcium
79 cal	20 g	1 g	0 g	0 mg	3 mg	41 mg

Ingredient	Weight	Measure	Issue
PEACHES,CANNED,HALVES,JUICE PACK,DRAINED		1 gal 1/2 qts	
PEARS,CANNED,HALVES,DRAINED		1 gal 1/2 qts	
PINEAPPLE,CANNED,CHUNKS,JUICE PACK,DRAINED		1 gal 1/2 qts	
BANANA,FRESH,SLICED		1 gal 1-1/2 qts	
APPLES,FRESH,PEELED,DICED		2 gal 2 qts	
CANTELOUPE,FRESH,CUBED	8-3/4 lbs		
GRAPES,GREEN,FRESH,SEEDLESS		1 gal 1 qts	
APRICOTS,CANNED,JUICE PACK,DRAINED		1 gal 1/2 qts	
HONEYDEW MELON,DICED		3 gal	
GRAPEFRUIT,CANNED,LIGHT SYRUP,DRAINED,SECTIONED	13-1/2 lbs	1 gal 2 qts	
GRAPEFRUIT,FRESH,PARED,SECTIONS	19-3/4 lbs	35 each	38 lbs
ORANGE,FRESH,SECTIONS	5-1/2 lbs	3 qts 2 cup	7-5/8 lbs
LETTUCE,LEAF,FRESH,HEAD	4 lbs		6-1/4 lbs

Method

1. Combine any three fruits per 100 servings; cover; refrigerate.
2. Place 1 lettuce leaf on serving dish; arrange fruit on lettuce. Cover; refrigerate until ready to serve. CCP: Hold for service at 41 F. or lower.

SALADS, SALAD DRESSINGS, AND RELISHES   No.M 036 00
PERFECTION SALAD

Yield 100   Portion 3 Ounces

Calories	Carbohydrates	Protein	Fat	Cholesterol	Sodium	Calcium
72 cal	17 g	2 g	0 g	0 mg	56 mg	22 mg

Ingredient	Weight	Measure	Issue
DESSERT POWDER,GELATIN,LEMON	3-2/3 lbs	1 qts 3-1/2 cup	
WATER,BOILING	6-1/4 lbs	3 qts	
WATER,COLD	8-1/3 lbs	1 gal	
VINEGAR,DISTILLED	8-1/3 oz	1 cup	
CABBAGE,GREEN,FRESH,SHREDDED	1-1/2 lbs	2 qts 1-3/4 cup	1-7/8 lbs
CARROTS,FRESH,CHOPPED	6 oz	1-3/8 cup	7-1/3 oz
CELERY,FRESH,CHOPPED	2 lbs	1 qts 3-1/2 cup	2-3/4 lbs
PEPPERS,GREEN,FRESH,CHOPPED	8 oz	1-1/2 cup	9-3/4 oz
PIMIENTO,CANNED,DRAINED,CHOPPED	6-3/4 oz	1 cup	
LETTUCE,LEAF,FRESH,HEAD	4 lbs		6-1/4 lbs

Method
1. Dissolve gelatin in boiling water.
2. Add cold water and vinegar; mix well.
3. Pour 2-3/4 quarts into each pan.
4. Chill until slightly thickened.
5. Combine cabbage, carrots, celery, peppers, and pimientos.
6. Add 1-1/2 quarts vegetables to gelatin in each pan.
7. Chill until firm. Cut 5 by 7.
8. Place 1 lettuce leaf on each serving dish; add gelatin square. Cover; refrigerate until ready to serve. CCP: Hold for service at 41 F. or lower.

SALADS, SALAD DRESSINGS, AND RELISHES   No.M 036 01

GOLDEN GLOW SALAD

Yield 100   Portion   3-1/2 Ounces

Calories	Carbohydrates	Protein	Fat	Cholesterol	Sodium	Calcium
92 cal	22 g	2 g	0 g	0 mg	52 mg	23 mg

Ingredient	Weight	Measure	Issue
DESSERT POWDER,GELATIN,LEMON	3-2/3 lbs	1 qts 3-1/2 cup	
WATER,BOILING	6-1/4 lbs	3 qts	
WATER,COLD	6-1/4 lbs	3 qts	
VINEGAR,DISTILLED	8-1/3 oz	1 cup	
CARROTS,FRESH,SHREDDED	3-1/2 lbs	3 qts 2-1/2 cup	4-1/4 lbs
PINEAPPLE,CANNED,CRUSHED,JUICE PACK,INCL LIQUIDS	6-5/8 lbs	3 qts	
LETTUCE,LEAF,FRESH,HEAD	4 lbs		6-1/4 lbs

Method

1. Dissolve gelatin in boiling water.
2. Add cold water and vinegar; mix well.
3. Pour 2-1/2 quarts gelatin mixture into each pan.
4. Chill until slightly thickened.
5. Combine fresh carrots and undrained pineapple.
6. Add 1-3/4 quarts carrot-pineapple mixture to gelatin in each pan.
7. Chill until firm. Cut 5 by 7.
8. Place 1 lettuce leaf on each serving dish; add gelatin square. Cover; refrigerate until ready to serve. CCP: Hold for service at 41 F. or lower.

SALADS, SALAD DRESSINGS, AND RELISHES   No. M 036 02

JELLIED SPRING SALAD

Yield 100   Portion 3 Ounces

Calories	Carbohydrates	Protein	Fat	Cholesterol	Sodium	Calcium
70 cal	16 g	2 g	0 g	0 mg	47 mg	16 mg

Ingredient

Ingredient	Weight	Measure	Issue
DESSERT POWDER,GELATIN,LEMON	3-2/3 lbs	1 qts 3-1/2 cup	
WATER,BOILING	6-1/4 lbs	3 qts	
WATER,COLD	8-1/3 lbs	1 gal	
VINEGAR,DISTILLED	8-1/3 oz	1 cup	
CUCUMBER,FRESH,CHOPPED	2 lbs	1 qts 3-5/8 cup	2-3/8 lbs
ONIONS,GREEN,FRESH,CHOPPED	10-5/8 oz	3 cup	11-3/4 oz
PIMIENTO,CANNED,DRAINED,CHOPPED	1-2/3 oz	1/4 cup 1/3 tbsp	
LETTUCE,LEAF,FRESH,HEAD	4 lbs		6-1/4 lbs

Method

1. Dissolve gelatin in boiling water.
2. Add cold water and vinegar; mix well.
3. Pour 2-3/4 quarts into each pan.
4. Chill until slightly thickened.
5. Combine cucumbers, green onions, and pimientos.
6. Add 1-1/2 quarts vegetables to gelatin in each pan.
7. Chill until firm. Cut 5 by 7.
8. Place 1 lettuce leaf on each serving dish; add gelatin square. Cover; refrigerate until ready to serve. CCP: Hold for service at 41 F. or lower.

SALADS, SALAD DRESSINGS, AND RELISHES No. M 037 00

PICKLED BEET AND ONION SALAD

Yield 100 **Portion** 1/2 Cup

Calories	Carbohydrates	Protein	Fat	Cholesterol	Sodium	Calcium
71 cal	18 g	1 g	0 g	0 mg	417 mg	25 mg

Ingredient	Weight	Measure	Issue
BEETS,CANNED,SLICED,INCL LIQUIDS	26 lbs	3 gal	
RESERVED LIQUID	6-1/4 lbs	3 qts	
VINEGAR,DISTILLED	4-1/8 lbs	2 qts	
CINNAMON,GROUND	1/4 oz	1 tbsp	
CLOVES,GROUND	3/8 oz	1 tbsp	
SALT	1 oz	1 tbsp	
PEPPER,BLACK,GROUND	1/8 oz	1/3 tsp	
SUGAR,GRANULATED	12-1/3 oz	1-3/4 cup	
SUGAR,BROWN,PACKED	1 lbs	3-1/4 cup	
ONIONS,FRESH,SLICED	2 lbs	2 qts	2-1/4 lbs

Method

1. Drain beets; reserve juice for use in Step 2; beets for use in Step 4.
2. Combine reserved juice, vinegar, cinnamon, cloves, salt, pepper, and sugars.
3. Cover; bring to a boil; reduce heat; simmer 10 minutes. Cool.
4. Combine beets and onions.
5. Pour sauce over beets and onions. Cover; refrigerate at least 3 to 4 hours before serving. Keep refrigerated until ready to serve. CCP: Hold for service at 41 F. or lower.

SALADS, SALAD DRESSINGS, AND RELISHES No.M 038 00
PASTA SALAD

Yield 100   Portion   1/2 Cup

Calories	Carbohydrates	Protein	Fat	Cholesterol	Sodium	Calcium
172 cal	17 g	5 g	10 g	3 mg	300 mg	79 mg

Ingredient	Weight	Measure	Issue
WATER	16-3/4 lbs	2 gal	
SALT	5/8 oz	1 tbsp	
OIL,OLIVE	1/2 oz	1 tbsp	
SPAGHETTI NOODLES,DRY	3-1/2 lbs	3 qts 3-1/8 cup	
SALAD DRESSING,ITALIAN	3-1/8 lbs	1 qts 2 cup	
OIL,SALAD	1-7/8 oz	1/4 cup 1/3 tbsp	
BROCCOLI,FRESH,FLORETS	1-1/4 lbs	1 qts 2-1/2 cup	2 lbs
CARROTS,FRESH,SLICED	1 lbs	3-1/2 cup	1-1/4 lbs
COOKING SPRAY,NONSTICK	2 oz	1/4 cup 1/3 tbsp	
TOMATOES,FRESH,CHOPPED	3-1/2 lbs	2 qts 3/4 cup	3-5/8 lbs
SQUASH,ZUCCHINI,FRESH,SLICED	2-1/2 lbs	2 qts 2 cup	2-5/8 lbs
MUSHROOMS,FRESH,WHOLE,SLICED	1-7/8 lbs	3 qts 1/8 cup	2 lbs
ONIONS,FRESH,CHOPPED	1-1/2 lbs	1 qts 1/4 cup	1-2/3 lbs
OLIVES,RIPE,PITTED,SLICED,INCL LIQUIDS	14-1/4 oz	3 cup	
BASIL,SWEET,WHOLE,CRUSHED	1 oz	1/4 cup 2-1/3 tbsp	
PARSLEY,FRESH,BUNCH,CHOPPED	1-5/8 oz	3/4 cup 1/3 tbsp	1-3/4 oz
CHEESE,PARMESAN,GRATED	14-1/8 oz	1 qts	

Method

1 Add salt and oil to water; heat to a rolling boil. Slowly add vermicelli, stirring constantly until water boils again. Cook 7 to 10 minutes or until tender. Rinse with cold water; drain thoroughly.
2 Add dressing to vermicelli. Toss lightly. Set aside for use in Step 4.
3 Lightly spray griddle with non-stick cooking spray. Saute broccoli and carrots on lightly sprayed griddle for 8 to 10 minutes or until tender crisp.
4 Add tomatoes, squash, mushrooms, onions, olives, and sauteed vegetables to pasta mixture. Toss lightly.
5 Add basil, parsley, and parmesan cheese to pasta mixture. Toss lightly.

SALADS, SALAD DRESSINGS, AND RELISHES No.M 039 00
CORN RELISH

Yield 100 **Portion** 2-1/2 Tablespoons

Calories	Carbohydrates	Protein	Fat	Cholesterol	Sodium	Calcium
27 cal	6 g	1 g	0 g	0 mg	95 mg	4 mg

Ingredient	Weight	Measure	Issue
CORN,CANNED,WHOLE KERNEL,DRAINED	4-1/3 lbs	3 qts	
CELERY,FRESH,CHOPPED	1 lbs	3-3/4 cup	1-3/8 lbs
ONIONS,FRESH,CHOPPED	1 lbs	2-7/8 cup	1-1/8 lbs
PEPPERS,GREEN,FRESH,CHOPPED	8 oz	1-1/2 cup	9-3/4 oz
PIMIENTO,CANNED,DRAINED,CHOPPED	4-1/4 oz	1/2 cup 2 tbsp	
PEPPER,BLACK,GROUND	1/8 oz	1/8 tsp	
SALAD DRESSING,FRENCH,FAT FREE	1-1/8 lbs	2 cup	

Method

1. Combine corn, celery, onions, peppers, pimientos, and pepper.
2. Add Fat Free French Dressing or French Dressing Recipe No. M 058 00; mix well.
3. Cover; refrigerate 6 hours or until flavors are blended. Keep refrigerated until ready to serve. CCP: Hold for service at 41 F. or lower.

Notes

1. Serve as a relish with meat or fish. If served as a salad, double recipe. EACH PORTION: 1/3 Cup or 2 ounces.

SALADS, SALAD DRESSINGS, AND RELISHES   No. M 040 00
POTATO SALAD

Yield 100   Portion   2/3 Cup

Calories	Carbohydrates	Protein	Fat	Cholesterol	Sodium	Calcium
209 cal	22 g	3 g	13 g	45 mg	571 mg	19 mg

Ingredient	Weight	Measure	Issue
POTATOES,FRESH,PEELED,CUBED	18 lbs	3 gal 1-1/8 qts	22-1/4 lbs
WATER	16-3/4 lbs	2 gal	
SALT	2-1/2 oz	1/4 cup 1/3 tbsp	
ONIONS,FRESH,CHOPPED	1 lbs	2-7/8 cup	1-1/8 lbs
OIL,SALAD	9-5/8 oz	1-1/4 cup	
SALT	1 oz	1 tbsp	
PEPPER,BLACK,GROUND	1/8 oz	1/3 tsp	
VINEGAR,DISTILLED	5-5/8 oz	1/2 cup 2-2/3 tbsp	
CELERY,FRESH,CHOPPED	2-3/8 lbs	2 qts 1 cup	3-1/4 lbs
EGG,HARD COOKED,CHOPPED	2 lbs	18 Eggs	
PICKLE RELISH,SWEET	1-1/8 lbs	2 cup	
PIMIENTO,CANNED,DRAINED,CHOPPED	8-1/2 oz	1-1/4 cup	
SALAD DRESSING,MAYONNAISE TYPE	4 lbs	2 qts	
PARSLEY,FRESH,BUNCH,CHOPPED	1 oz	1/4 cup	1 oz
PAPRIKA,GROUND	1/2 oz	2 tbsp	

Method

1 Cover potatoes with water; bring to a boil; add salt; cover. Cook until tender.
2 Drain well. Cool slightly.
3 Combine onions, salad oil or olive oil, salt, pepper, and vinegar. Add to potatoes. Cover; refrigerate 1 hour.
4 Combine celery, eggs, relish, pimientos, and Salad Dressing; add to potato mixture.
5 Mix lightly but thoroughly to coat potatoes with Salad Dressing mixture.
6 Garnish with parsley and paprika.
7 Cover; refrigerate until ready to serve. CCP: Hold for service at 41 F. or lower.

SALADS. SALAD DRESSINGS. AND RELISHES   No.M 040 01

DEVILED POTATO SALAD

Yield 100   Portion   2/3 Cup

Calories	Carbohydrates	Protein	Fat	Cholesterol	Sodium	Calcium
234 cal	22 g	4 g	15 g	48 mg	650 mg	21 mg

Ingredient	Weight	Measure	Issue
POTATOES,FRESH,PEELED,CUBED	18 lbs	3 gal 1-1/8 qts	22-1/4 lbs
WATER	16-3/4 lbs	2 gal	
SALT	2-1/2 oz	1/4 cup 1/3 tbsp	
ONIONS,FRESH,CHOPPED	1 lbs	2-7/8 cup	1-1/8 lbs
OIL,SALAD	9-5/8 oz	1-1/4 cup	
SALT	1 oz	1 tbsp	
PEPPER,BLACK,GROUND	1/8 oz	1/3 tsp	
VINEGAR,DISTILLED	5-5/8 oz	1/2 cup 2-2/3 tbsp	
MUSTARD,PREPARED	6-5/8 oz	3/4 cup	
SUGAR,GRANULATED	7/8 oz	2 tbsp	
BACON,COOKED,CHOPPED	12 oz		
CELERY,FRESH,CHOPPED	2-3/8 lbs	2 qts 1 cup	3-1/4 lbs
EGG,HARD COOKED,CHOPPED	2 lbs	18 Eggs	
PICKLE RELISH,SWEET	1-1/8 lbs	2 cup	
PIMIENTO,CANNED,DRAINED,CHOPPED	8-1/2 oz	1-1/4 cup	
SALAD DRESSING,MAYONNAISE TYPE	4-1/8 lbs	2 qts 1/4 cup	
PARSLEY,FRESH,BUNCH,CHOPPED	1 oz	1/4 cup	1 oz
PAPRIKA,GROUND	1/2 oz	2 tbsp	

Method

1 Cover potatoes with cold water; bring to a boil; add salt; cover. Cook until tender.
2 Drain well. Cool slightly.
3 Combine onions, salad oil or olive oil, salt, pepper, and vinegar. Add to potatoes. Cover; refrigerate 1 hour.
4 Combine celery, eggs, relish, pimientos, mustard, sugar, bacon, and Salad Dressing.
5 Mix lightly but thoroughly to coat potatoes with Salad Dressing mixture.
6 Garnish with parsley and paprika.
7 Cover; refrigerate until ready to serve. CCP: Hold for service at 41 F. or lower.

SALADS, SALAD DRESSINGS, AND RELISHES   No.M 040 02

POTATO SALAD WITH VINEGAR DRESSING

Yield 100   Portion   2/3 Cup

Calories	Carbohydrates	Protein	Fat	Cholesterol	Sodium	Calcium
130 cal	31 g	2 g	0 g	0 mg	455 mg	18 mg

Ingredient	Weight	Measure	Issue
POTATOES,FRESH,PEELED,CUBED	22 lbs	4 gal	27-1/8 lbs
WATER	16-3/4 lbs	2 gal	
SALT	2-1/2 oz	1/4 cup 1/3 tbsp	
ONIONS,FRESH,CHOPPED	1 lbs	2-7/8 cup	1-1/8 lbs
CELERY,FRESH,CHOPPED	3 lbs	2 qts 3-3/8 cup	4-1/8 lbs
PICKLE RELISH,SWEET	1-1/8 lbs	2 cup	
PIMIENTO,CANNED,DRAINED,CHOPPED	8-1/2 oz	1-1/4 cup	
PEPPER,BLACK,GROUND	1/8 oz	1/3 tsp	
SALT	1 oz	1 tbsp	
SUGAR,GRANULATED	1-3/4 lbs	1 qts	
VINEGAR,DISTILLED	2-1/8 lbs	1 qts	
WATER	8-1/3 oz	1 cup	
PARSLEY,FRESH,BUNCH,CHOPPED	1 oz	1/4 cup	1 oz
PAPRIKA,GROUND	1/2 oz	2 tbsp	

Method

1. Cover potatoes with water; bring to a boil; add salt; cover. Cook until tender.
2. Drain well. Cool slightly.
3. Add onions to potatoes. Cover and refrigerate 1 hour.
4. Combine celery, relish, and pimientos. Add to potato mixture.
5. Combine pepper, salt, sugar, vinegar, and water to make vinegar dressing. Mix lightly but thoroughly to coat potatoes.
6. Garnish with parsley and paprika.
7. Cover; refrigerate until ready to serve. CCP: Hold for service at 41 F. or lower.

SALADS, SALAD DRESSINGS, AND RELISHES   No. M 041 00
POTATO SALAD (DEHYDRATED SLICED POTATOES)

Yield 100   Portion   2/3 Cup

Calories	Carbohydrates	Protein	Fat	Cholesterol	Sodium	Calcium
133 cal	10 g	2 g	10 g	41 mg	500 mg	17 mg

Ingredient	Weight	Measure	Issue
POTATO,WHITE,DEHYDRATED,SLICED	4-3/8 lbs		
WATER,BOILING	29-1/4 lbs	3 gal 2 qts	
SALT	1-7/8 oz	3 tbsp	
ONIONS,DEHYDRATED,CHOPPED	4 oz	2 cup	
PEPPERS,GREEN,FRESH,CHOPPED	7-7/8 oz	1-1/2 cup	9-5/8 oz
WATER,WARM	5-1/4 lbs	2 qts 2 cup	
PICKLE RELISH,SWEET	1-1/8 lbs	2 cup	
CELERY,FRESH,CHOPPED	2-3/8 lbs	2 qts 1 cup	3-1/4 lbs
PIMIENTO,CANNED,DRAINED,CHOPPED	8-1/2 oz	1-1/4 cup	
EGG,HARD COOKED,CHOPPED	1-3/4 lbs	1 qts 2 cup	
PEPPER,BLACK,GROUND	1/4 oz	1 tbsp	
SALT	1 oz	1 tbsp	
SALAD DRESSING,MAYONNAISE TYPE	4 lbs	2 qts	
VINEGAR,DISTILLED	5-5/8 oz	1/2 cup 2-2/3 tbsp	
PAPRIKA,GROUND	1/2 oz	2 tbsp	

Method

1. Add potatoes to boiling salted water; cover. Bring quickly to a boil. Reduce heat; simmer 20 to 25 minutes or until tender. Drain. Set aside for use in Step 3.
2. Rehydrate onions and peppers. Drain before using.
3. Carefully combine potatoes, onions, peppers, relish, celery, pimientos, eggs, salt, and pepper.
4. Combine Salad Dressing and vinegar; fold into potato mixture. Cover; refrigerate until ready to serve.
5. Garnish with paprika. CCP: Hold for service at 41 F. or lower.

SALADS, SALAD DRESSINGS, AND RELISHES   No.M 042 00
HOT POTATO SALAD

Yield 100   Portion   2/3 Cup

Calories	Carbohydrates	Protein	Fat	Cholesterol	Sodium	Calcium
168 cal	26 g	3 g	6 g	7 mg	355 mg	17 mg

Ingredient

Ingredient	Weight	Measure	Issue
POTATOES,FRESH,PEELED,CUBED	23 lbs	4 gal 3/4 qts	28-3/8 lbs
WATER	20-7/8 lbs	2 gal 2 qts	
SALT	1-1/4 oz	2 tbsp	
BACON,RAW	3 lbs		
ONIONS,FRESH,CHOPPED	1-3/8 lbs	3-7/8 cup	1-1/2 lbs
CELERY,FRESH,CHOPPED	2 lbs	1 qts 2-1/8 cup	2-3/4 lbs
SALT	1-1/4 oz	2 tbsp	
PEPPER,BLACK,GROUND	<1/16th oz	1/8 tsp	
BACON FAT,RENDERED	14-1/2 oz	2 cup	
WATER	1-5/8 lbs	3 cup	
VINEGAR,DISTILLED	3-1/8 lbs	1 qts 2 cup	
SUGAR,GRANULATED	12-1/3 oz	1-3/4 cup	
MUSTARD,DRY	3/8 oz	1 tbsp	

Method

1 Cover potatoes with water; bring to a boil; add salt; cover. Cook until tender. Drain well. Set aside for use in Step 2.
2 Cook bacon until crisp. Drain; combine bacon with potatoes. Set bacon fat aside for use in Step 4.
3 Carefully mix potato and bacon mixture with onions, celery, salt, and pepper.
4 Combine bacon fat, water, vinegar, sugar, and mustard; heat to boiling point.
5 Pour hot mixture over potato mixture; combine carefully.
6 Pour 2-1/8 gallon mixture into each pan.
7 Place in oven at 350 F. for 15 minutes or until thoroughly heated. Serve hot. CCP: Hold for service at 140 F. or higher.

SALADS. SALAD DRESSINGS. AND RELISHES   No.M 043 00
HOT POTATO SALAD (DEHYDRATED SLICED POTATOES)

Yield 100   Portion   2/3 Cup

Calories	Carbohydrates	Protein	Fat	Cholesterol	Sodium	Calcium
104 cal	14 g	2 g	5 g	6 mg	364 mg	9 mg

Ingredient	Weight	Measure	Issue
POTATO,WHITE,DEHYDRATED,SLICED	6-1/4 lbs		
WATER,BOILING	41-3/4 lbs	5 gal	
SALT	2-1/3 oz	1/4 cup	
ONIONS,DEHYDRATED,CHOPPED	3/4 oz	1/4 cup 2-1/3 tbsp	
WATER,WARM	14-5/8 oz	1-3/4 cup	
PICKLE RELISH,SWEET	1-1/8 lbs	2 cup	
PEPPER,BLACK,GROUND	1/2 oz	2 tbsp	
BACON,RAW	3 lbs		
VINEGAR,DISTILLED	2-1/8 lbs	1 qts	
WATER	3-1/8 lbs	1 qts 2 cup	
SUGAR,GRANULATED	1-1/4 lbs	2-3/4 cup	
BACON FAT,RENDERED	10-7/8 oz	1-1/2 cup	

Method

1. Add potatoes to boiling salted water. Cover. Bring quickly to a boil. Reduce heat; simmer 20 to 25 minutes or until potatoes are tender. Drain. Set aside for use in Step 3.
2. Rehydrate onions; drain well.
3. Combine onions, relish and pepper; mix well; add to potatoes. Set aside for use in Step 6.
4. Cook bacon until crisp. Remove bacon from fat; set bacon fat aside for use in Step 5. Set bacon aside for use in Step 7.
5. Combine vinegar, water, and sugar. Add gradually to bacon fat. Cook until sugar is dissolved stirring constantly.
6. Pour hot dressing over potato mixture; combine carefully.
7. Add bacon; reserve enough to sprinkle on top as a garnish. Serve hot. CCP: Hold for service at 140 F. or higher.

SALADS, SALAD DRESSINGS, AND RELISHES No.M 044 00
SPRING SALAD

Yield 100   **Portion** 3/4 Cup

Calories	Carbohydrates	Protein	Fat	Cholesterol	Sodium	Calcium
16 cal	4 g	1 g	0 g	0 mg	7 mg	27 mg

Ingredient

Ingredient	Weight	Measure	Issue
LETTUCE,LEAF,FRESH,HEAD	5 lbs		7-3/4 lbs
RADISH,FRESH,SLICES	1 lbs	3-7/8 cup	1-1/8 lbs
CUCUMBERS,FRESH,PEELED,SLICED	3 lbs	2 qts 3-1/2 cup	5-7/8 each
PEPPERS,GREEN,FRESH,CHOPPED	1-1/2 lbs	1 qts 1/2 cup	1-7/8 lbs
ONIONS,GREEN,FRESH,SLICED	2 lbs	2 qts 1-1/8 cup	2-1/4 lbs
TOMATOES,FRESH,THIN WEDGES	5 lbs	3 qts 5/8 cup	5-1/8 lbs

Method

1 Tear prepared lettuce into large pieces.
2 Combine lettuce with radishes, cucumbers, peppers, and onions; toss lightly.
3 Cover; CCP: Refrigerate at 41 F. or lower for use in Step 4.
4 Add tomatoes to other salad vegetables just before serving. Toss lightly. CCP: Hold for service at 41 F. or lower.

SALADS. SALAD DRESSINGS. AND RELISHES   No.M 045 00

THREE BEAN SALAD

Yield 100　　　　　　　　　　　　　　　　　　**Portion**　1/3 Cup

Calories	Carbohydrates	Protein	Fat	Cholesterol	Sodium	Calcium
120 cal	15 g	2 g	7 g	0 mg	366 mg	16 mg

Ingredient	**Weight**	**Measure**	**Issue**
BEANS,KIDNEY,DARK RED,CANNED,DRAINED	4-2/3 lbs	3 qts	
BEANS,WAX,CANNED,DRAINED	3-5/8 lbs	3 qts	
BEANS,GREEN,CANNED,DRAINED	3-5/8 lbs	3 qts	
ONIONS,FRESH,SLICED	1 lbs	3-3/4 cup	1 lbs
SUGAR,GRANULATED	2 lbs	1 qts 1/2 cup	
VINEGAR,DISTILLED	3-1/8 lbs	1 qts 2 cup	
OIL,SALAD	1-1/2 lbs	3 cup	
SALT	1-7/8 oz	3 tbsp	
PEPPER,BLACK,GROUND	1/8 oz	1/3 tsp	

Method

1. Drain all beans. Rinse kidney beans with cool water; drain.
2. Combine beans and onions. Set aside for use in Step 4.
3. Combine sugar, vinegar, salad oil or olive oil, salt, and pepper; whip or shake thoroughly.
4. Add dressing; blend well.
5. Cover; refrigerate at least 6 hours until flavors are well blended. Keep refrigerated until ready to serve. CCP: Hold for service at 41 F. or lower.

SALADS, SALAD DRESSINGS, AND RELISHES   No.M 045 01

PICKLED GREEN BEAN SALAD

Yield 100   Portion   1/3 Cup

Calories	Carbohydrates	Protein	Fat	Cholesterol	Sodium	Calcium
113 cal	14 g	1 g	7 g	0 mg	435 mg	24 mg

Ingredient	Weight	Measure	Issue
BEANS,GREEN,CANNED,DRAINED	19 lbs	3 gal 4 qts	
ONIONS,FRESH,SLICED	13-2/3 oz	3-3/4 cup	
SUGAR,GRANULATED	2 lbs	1 qts 1/2 cup	
VINEGAR,DISTILLED	3-1/8 lbs	1 qts 2 cup	
OIL,SALAD	1-1/2 lbs	3 cup	
SALT	1-7/8 oz	3 tbsp	
PEPPER,BLACK,GROUND	1/8 oz	1/3 tsp	

Method

1. Drain all beans.
2. Combine beans and onions. Set aside for use in Step 4.
3. Combine sugar, vinegar, salad oil, salt, and pepper; whip or shake thoroughly.
4. Add dressing; blend well.
5. Cover; refrigerate at least 6 hours until flavors are well blended. Keep refrigerated until ready to serve. CCP: Hold for service at 41 F. or lower.

SALADS. SALAD DRESSINGS. AND RELISHES No.M 046 00

TOSSED LETTUCE, CUCUMBER AND TOMATO SALAD

Yield 100   **Portion** 1 Cup

Calories	Carbohydrates	Protein	Fat	Cholesterol	Sodium	Calcium
13 cal	3 g	1 g	0 g	0 mg	6 mg	30 mg

Ingredient	**Weight**	**Measure**	**Issue**
LETTUCE,LEAF,FRESH,CHOPPED	8 lbs	4 gal 1/4 qts	12-1/2 lbs
ONIONS,GREEN,FRESH,SLICED	8 oz	2-1/4 cup	8-7/8 oz
CUCUMBERS,FRESH,PEELED,SLICED	4 lbs	3 qts 3-1/4 cup	7-7/8 each
TOMATOES,FRESH,THIN WEDGES	4 lbs	2 qts 2-1/8 cup	4-1/8 lbs

Method

1. Tear prepared lettuce into large pieces.
2. Combine lettuce with onions and cucumbers; toss lightly.
3. Cover;   Add tomatoes to other salad vegetables just before serving. Toss lightly. CCP: Hold for service at 41 F. or lower.

SALADS, SALAD DRESSINGS, AND RELISHES No.M 046 01
TOSSED GARDEN SALAD

Yield 100   Portion   1 Cup

Calories	Carbohydrates	Protein	Fat	Cholesterol	Sodium	Calcium
17 cal	4 g	1 g	0 g	0 mg	16 mg	32 mg

Ingredient

Ingredient	Weight	Measure	Issue
LETTUCE,LEAF,FRESH,HEAD	8 lbs		12-1/2 lbs
CARROTS,FRESH,CHOPPED	2 lbs	1 qts 3-1/8 cup	2-1/2 lbs
CELERY,FRESH,CHOPPED	2 lbs	1 qts 3-1/2 cup	2-3/4 lbs
ONIONS,FRESH,CHOPPED	8 oz	1-3/8 cup	8-7/8 oz
TOMATOES,FRESH,THIN WEDGES	4 lbs	2 qts 2-1/8 cup	4-1/8 lbs

Method

1. Tear prepared lettuce into large pieces.
2. Combine lettuce with carrots, celery, and onions; toss lightly.
3. Cover. Add tomatoes to other salad vegetables just before serving. Toss lightly. CCP: Hold for service at 41 F. or lower.

SALADS, SALAD DRESSINGS, AND RELISHES No.M 046 02

TOSSED CALICO GARDEN SALAD

Yield 100   Portion 1 Cup

Calories	Carbohydrates	Protein	Fat	Cholesterol	Sodium	Calcium
16 cal	3 g	1 g	0 g	0 mg	11 mg	31 mg

Ingredient	Weight	Measure	Issue
LETTUCE,LEAF,FRESH,HEAD	8 lbs		12-1/2 lbs
ONIONS,GREEN,FRESH,SLICED	8 oz	2-1/4 cup	8-7/8 oz
CUCUMBERS,FRESH,PEELED,SLICED	1 lbs	3-3/4 cup	2 each
CARROTS,FRESH,CHOPPED	1 lbs	3-1/2 cup	1-1/4 lbs
CELERY,FRESH,CHOPPED	1 lbs	3-3/4 cup	1-3/8 lbs
PEPPERS,GREEN,FRESH,CHOPPED	1 lbs	3 cup	1-1/4 lbs
TOMATOES,FRESH,THIN WEDGES	4 lbs	2 qts 2-1/8 cup	4-1/8 lbs

Method

1. Tear prepared lettuce into large pieces.
2. Combine lettuce with onions, cucumbers, carrots, celery and green peppers; toss lightly.
3. Cover; Add tomatoes to other salad vegetables just before serving. Toss lightly. CCP: Hold for service at 41 F. or lower.

SALADS, SALAD DRESSINGS, AND RELISHES No.M 046 03

TOSSED ROMAINE, CUCUMBER AND TOMATO SALAD

Yield 100   Portion   1 Cup

Calories	Carbohydrates	Protein	Fat	Cholesterol	Sodium	Calcium
12 cal	2 g	1 g	0 g	0 mg	5 mg	18 mg

Ingredient

Ingredient	Weight	Measure	Issue
LETTUCE,ROMAINE,FRESH,CHOPPED	8 lbs	4 gal 1/4 qts	8-1/2 lbs
ONIONS,GREEN,FRESH,SLICED	8 oz	2-1/4 cup	8-7/8 oz
CUCUMBER,FRESH,SLICED	4 lbs	3 qts 1-5/8 cup	4-3/4 lbs
TOMATOES,FRESH,THIN WEDGES	4 lbs	2 qts 2-1/8 cup	4-1/8 lbs

Method

1 Tear lettuce into large pieces. Combine lettuce with onions and cucumbers; toss lightly. Cover.
2 Add tomatoes to other salad vegetables just before serving. Toss lightly. CCP: Hold for service at 41 F. or lower.

SALADS. SALAD DRESSINGS. AND RELISHES   No.M 046 04
TOSSED RED LEAF LETTUCE, CUCUMBER AND TOMATO SALAD

Yield 100   Portion   1 Cup

Calories	Carbohydrates	Protein	Fat	Cholesterol	Sodium	Calcium
13 cal	3 g	1 g	0 g	0 mg	6 mg	30 mg

Ingredient	**Weight**	**Measure**	**Issue**
LETTUCE,FRESH,LEAF,RED	8 lbs	4 gal 1/4 qts	12-1/2 lbs
ONIONS,GREEN,FRESH,SLICED	8 oz	2-1/4 cup	8-7/8 oz
CUCUMBERS,FRESH,PEELED,SLICED	4 lbs	3 qts 3-1/4 cup	7-7/8 each
TOMATOES,FRESH,THIN WEDGES	4 lbs	2 qts 2-1/8 cup	4-1/8 lbs

Method

1 Combine lettuce with onions and cucumbers; toss lightly. Cover.
2 Add tomatoes to other salad vegetables just before serving. Toss lightly. CCP: Hold for service at 41 F. or lower.

SALADS, SALAD DRESSINGS, AND RELISHES No.M 046 05
GREEN LEAF LETTUCE, CUCUMBER AND TOMATO SALAD

Yield 100   Portion 1 Cup

Calories	Carbohydrates	Protein	Fat	Cholesterol	Sodium	Calcium
13 cal	3 g	1 g	0 g	0 mg	6 mg	30 mg

Ingredient

Ingredient	Weight	Measure	Issue
LETTUCE,LEAF,FRESH,HEAD	8 lbs		12-1/2 lbs
ONIONS,GREEN,FRESH,SLICED	8 oz	2-1/4 cup	8-7/8 oz
CUCUMBERS,FRESH,PEELED,SLICED	4 lbs	3 qts 3-1/4 cup	7-7/8 each
TOMATOES,FRESH,THIN WEDGES	4 lbs	2 qts 2-1/8 cup	4-1/8 lbs

Method

1 Tear lettuce into large pieces. Combine lettuce with onions and cucumbers; toss lightly. Cover.
2 Add tomatoes to other salad vegetables just before serving. Toss lightly. CCP: Hold for service at 41 F. or lower.

SALADS, SALAD DRESSINGS, AND RELISHES No.M 047 00

TOSSED GREEN SALAD

Yield 100   Portion 1 Cup

Calories	Carbohydrates	Protein	Fat	Cholesterol	Sodium	Calcium
8 cal	2 g	1 g	0 g	0 mg	6 mg	28 mg

Ingredient	Weight	Measure	Issue
ENDIVE,FRESH	2 lbs		2-1/4 lbs
LETTUCE,ROMAINE,FRESH	3 lbs	1 gal 2-1/8 qts	3-1/4 lbs
LETTUCE,LEAF,FRESH,HEAD	6 lbs		9-3/8 lbs

Method

1. Tear greens into large pieces. Combine greens; toss lightly.
2. Cover; refrigerate until ready to serve. CCP: Hold for service at 41 F. or lower.

Notes

1. In Step 1, per 100 servings: 2 pounds fresh escarole may be used for fresh endive and 3 pounds fresh spinach may be used for romaine.

SALADS, SALAD DRESSINGS, AND RELISHES   No. M 048 00
TOSSED VEGETABLE SALAD

Yield 100   Portion   1 Cup

Calories	Carbohydrates	Protein	Fat	Cholesterol	Sodium	Calcium
19 cal	4 g	1 g	0 g	0 mg	19 mg	34 mg

Ingredient	Weight	Measure	Issue
LETTUCE,LEAF,FRESH,HEAD	6 lbs		9-3/8 lbs
CABBAGE,GREEN,FRESH,SHREDDED	2 lbs	3 qts 1 cup	2-1/2 lbs
CELERY,FRESH,SLICED	3 lbs	2 qts 3-3/8 cup	4-1/8 lbs
CUCUMBERS,FRESH,PEELED,SLICED	3 lbs	2 qts 3-1/2 cup	5-7/8 each
ONIONS,FRESH,SLICED	2 lbs	1 qts 3-7/8 cup	2-1/4 lbs
RADISH,FRESH,SLICES	1 lbs	3-7/8 cup	1-1/8 lbs
TOMATOES,FRESH,THIN WEDGES	4 lbs	2 qts 2-1/8 cup	4-1/8 lbs

Method

1 Tear prepared lettuce into large pieces. Combine lettuce with cabbage, celery, cucumbers, onions and radishes; toss lightly.
2 Cover. Add tomatoes to other salad vegetables just before serving. Toss lightly. CCP: Hold for service at 41 F. or lower.

SALADS. SALAD DRESSINGS. AND RELISHES No.M 048 01
TOSSED CALICO VEGETABLE SALAD

Yield 100 **Portion** 1 Cup

Calories	Carbohydrates	Protein	Fat	Cholesterol	Sodium	Calcium
48 cal	11 g	2 g	0 g	0 mg	20 mg	56 mg

Ingredient | Weight | Measure | Issue

Ingredient	Weight	Measure	Issue
LETTUCE,LEAF,FRESH,HEAD	6 lbs		9-3/8 lbs
CABBAGE,RED,FRESH,SHREDDED	2 lbs	3 qts 1 cup	
CELERY,FRESH,SLICED	3 lbs	2 qts 3-3/8 cup	4-1/8 lbs
CUCUMBERS,FRESH,PEELED,SLICED	3 lbs	2 qts 3-1/2 cup	5-7/8 each
ONIONS,DEHYDRATED,CHOPPED	2 lbs	1 gal <1/16th qts	
RADISH,FRESH,SLICES	1 lbs	3-7/8 cup	1-1/8 lbs
TOMATOES,FRESH,THIN WEDGES	4 lbs	2 qts 2-1/8 cup	4-1/8 lbs

Method

1 Tear lettuce into large pieces. Combine lettuce with red cabbage, celery, cucumbers, onions, and radishes; toss lightly.
2 Cover. Add tomatoes to other salad vegetables just before serving. Toss lightly. CCP: Hold for service at 41 F. or lower.

SALADS, SALAD DRESSINGS, AND RELISHES No.M 049 00
VEGETABLE SALAD

Yield 100   Portion   1/2 Cup

Calories	Carbohydrates	Protein	Fat	Cholesterol	Sodium	Calcium
45 cal	9 g	2 g	0 g	0 mg	282 mg	26 mg

Ingredient	Weight	Measure	Issue
BEANS,GREEN,CANNED,DRAINED	3-5/8 lbs	3 qts	
CARROTS,CANNED,SLICED,DRAINED	6-1/2 lbs	3 qts	
PEAS,GREEN,CANNED,DRAINED	4-1/2 lbs	3 qts	
CELERY,FRESH,CHOPPED	5 lbs	1 gal 3/4 qts	6-7/8 lbs
ONIONS,FRESH,SLICED	1 lbs	4 cup	1-1/8 lbs
SALAD DRESSING,FRENCH,FAT FREE	2-1/4 lbs	1 qts	

Method

1 Drain beans, carrots, and peas thoroughly. Cut carrots into 1/2-inch pieces.
2 Combine beans, carrots, and peas with celery and onions; toss lightly.
3 Add French Dressing to vegetable mixture; toss lightly.
4 Cover; refrigerate until ready to serve. CCP: Hold for service at 41 F. or lower.

SALADS. SALAD DRESSINGS. AND RELISHES No.M 050 00

WALDORF SALAD

Yield 100   Portion   1/2 Cup

Calories	Carbohydrates	Protein	Fat	Cholesterol	Sodium	Calcium
119 cal	10 g	1 g	9 g	4 mg	83 mg	27 mg

Ingredient	Weight	Measure	Issue
MILK,NONFAT,DRY	1/2 oz	3 tbsp	
WATER,WARM	7-1/3 oz	3/4 cup 2 tbsp	
JUICE,LEMON	4-1/3 oz	1/2 cup	
SUGAR,GRANULATED	1-3/4 oz	1/4 cup 1/3 tbsp	
SALAD DRESSING,MAYONNAISE TYPE	2-1/2 lbs	1 qts 1 cup	
CELERY,FRESH,CHOPPED	3 lbs	2 qts 3-3/8 cup	4-1/8 lbs
WALNUTS,SHELLED,CHOPPED	1 lbs	1 qts	
APPLES,FRESH,MEDIUM,UNPEELED,DICED	9 lbs	2 gal 1/8 qts	10-5/8 lbs
LETTUCE,LEAF,FRESH,CHOPPED	4 lbs	2 gal 1/8 qts	6-1/4 lbs

Method

1 Reconstitute milk.
2 Combine lemon juice, sugar, and Regular Salad Dressing or Fat Free Dressing. Add to milk. Mix well.
3 Add celery, nuts, and apples to Salad Dressing mixture. Toss well to coat pieces.
4 Place 1 lettuce leaf on each serving dish; add salad. Cover; refrigerate until ready to serve. CCP: Hold for service at 41 F. or lower.

SALADS, SALAD DRESSINGS, AND RELISHES No.M 050 01

APPLE, CELERY, AND RAISIN SALAD

Yield 100   Portion 1/2 Cup

Calories	Carbohydrates	Protein	Fat	Cholesterol	Sodium	Calcium
115 cal	17 g	1 g	6 g	4 mg	87 mg	28 mg

Ingredient	Weight	Measure	Issue
MILK,NONFAT,DRY	1/2 oz	3 tbsp	
WATER,WARM	7-1/3 oz	3/4 cup 2 tbsp	
JUICE,LEMON	4-1/3 oz	1/2 cup	
SUGAR,GRANULATED	1-3/4 oz	1/4 cup 1/3 tbsp	
SALAD DRESSING,MAYONNAISE TYPE	2-1/2 lbs	1 qts 1 cup	
CELERY,FRESH,CHOPPED	4 lbs	3 qts 3-1/8 cup	5-1/2 lbs
RAISINS	1-7/8 lbs	1 qts 2 cup	
APPLES,FRESH,MEDIUM,UNPEELED,DICED	9 lbs	2 gal 1/8 qts	10-5/8 lbs
LETTUCE,LEAF,FRESH,HEAD	4 lbs		6-1/4 lbs

Method

1 Reconstitute milk.
2 Combine lemon juice, sugar, and Regular Salad Dressing or Fat Free Dressing. Add to milk. Mix well.
3 Add celery, raisins, and apples to Salad Dressing mixture. Toss well to coat pieces.
4 Place 1 lettuce leaf on each serving dish; add salad. Cover; refrigerate until ready to serve. CCP: Hold for service at 41 F. or lower.

SALADS. SALAD DRESSINGS. AND RELISHES   No.M 051 00

CRANBERRY ORANGE RELISH

Yield 100   **Portion** 5-1/4 Quarts

Calories	Carbohydrates	Protein	Fat	Cholesterol	Sodium	Calcium
8987 cal	2315 g	19 g	5 g	0 mg	37 mg	670 mg

Ingredient	**Weight**	**Measure**	**Issue**
CRANBERRIES,FRESH	4 lbs	1 gal 3/4 qts	4-1/4 lbs
ORANGE,FRESH	2-7/8 lbs	10 each	4 lbs
SUGAR,GRANULATED	4-1/4 lbs	2 qts 1-5/8 cup	

Method

1. Put cranberries through food grinder or chopper to grind fine. Set aside for use in Step 3.
2. Cut oranges into quarters; remove seeds. Coarse grind through food grinder or chopper.
3. Combine cranberries, oranges, and sugar; stir until sugar is dissolved.
4. Cover; refrigerate until ready to serve. CCP: Hold for service at 41 F. or lower.

SALADS. SALAD DRESSINGS. AND RELISHES No.M 052 00

GUACAMOLE

Yield 100   **Portion** 2 Tablespoons

Calories	Carbohydrates	Protein	Fat	Cholesterol	Sodium	Calcium
67 cal	3 g	1 g	6 g	2 mg	101 mg	4 mg

Ingredient	**Weight**	**Measure**	**Issue**
AVOCADO,FRESH,PUREED	6 lbs	2 qts 3-7/8 cup	8-2/3 lbs
ONIONS,FRESH,GRATED	6 oz	1 cup	6-2/3 oz
GARLIC POWDER	1/8 oz	1/8 tsp	
CHILI POWDER,DARK,GROUND	1/4 oz	1 tbsp	
SALT	5/8 oz	1 tbsp	
PEPPER,BLACK,GROUND	1/8 oz	1/8 tsp	
SALAD DRESSING,MAYONNAISE TYPE	1 lbs	2 cup	

Method

1. Combine avocados, onions, garlic, chili powder, salt, pepper, and Salad Dressing in mixer bowl. Whip at high speed until well blended.
2. Cover; refrigerate until ready to serve. CCP: Hold for service at 41 F. or lower.

Notes

1. For Salad: Serve 2 tablespoons guacamole on each lettuce leaf. Garnish with tomato wedge.
2. For Dip: Serve with potato chips, crackers, and corn chips.

SALADS, SALAD DRESSINGS, AND RELISHES No.M 053 00

GERMAN STYLE TOMATO SALAD

Yield 100 **Portion** 1/3 Cup

Calories	Carbohydrates	Protein	Fat	Cholesterol	Sodium	Calcium
68 cal	5 g	1 g	5 g	5 mg	171 mg	9 mg

Ingredient	Weight	Measure	Issue
TOMATOES,FRESH,CHOPPED	15-7/8 lbs	2 gal 2 qts	16-1/4 lbs
ONIONS,FRESH,CHOPPED	1 lbs	3 cup	1-1/8 lbs
PARSLEY,FRESH,BUNCH,CHOPPED	4-1/4 oz	2 cup	4-1/2 oz
SALT	1 oz	1 tbsp	
PEPPER,BLACK,GROUND	1/8 oz	1/3 tsp	
SALAD DRESSING,MAYONNAISE TYPE	2 lbs	1 qts	
CREAM,TABLE,HOMOGENIZED,HALF PINT	8-1/2 oz	1 cup	
BACON,RAW	3-1/4 oz		

Method

1. Cut tomatoes into 1/2-inch cubes.
2. Add chopped onions, parsley, salt, and pepper. Toss lightly. Cover; refrigerate until ready to serve.
3. Combine Salad Dressing and cream. Set aside for use in Step 5.
4. Cook bacon until crisp; drain. Set aside for use in Step 6.
5. Just before serving, add dressing to tomato mixture; toss gently.
6. Sprinkle bacon on top and serve. CCP: Hold for service at 41 F. or lower.

SALADS, SALAD DRESSINGS, AND RELISHES No. M 053 01

COUNTRY STYLE TOMATO SALAD

Yield 100   Portion 1/2 Cup

Calories	Carbohydrates	Protein	Fat	Cholesterol	Sodium	Calcium
53 cal	8 g	1 g	2 g	0 mg	119 mg	9 mg

Ingredient   **Weight**   **Measure**   **Issue**

TOMATOES,FRESH,CHOPPED   15 lbs   2 gal 1-1/2 qts   15-1/3 lbs
ONIONS,FRESH,CHOPPED   1 lbs   2-7/8 cup   1-1/8 lbs
PEPPERS,GREEN,FRESH,CHOPPED   1-1/2 lbs   1 qts 1/2 cup   1-7/8 lbs
CELERY,FRESH,CHOPPED   2 lbs   1 qts 3-1/2 cup   2-3/4 lbs
SALT   1 oz   1 tbsp
PEPPER,BLACK,GROUND   1/8 oz   1/3 tsp
VINEGAR,DISTILLED   1-5/8 lbs   3 cup
SUGAR,GRANULATED   12-1/3 oz   1-3/4 cup
OIL,SALAD   7-2/3 oz   1 cup

Method

1. Cut tomatoes into 1/2-inch cubes.
2. Add chopped onions, chopped fresh sweet peppers, chopped celery, salt, and pepper. Toss lightly. Cover; refrigerate until ready to serve.
3. Combine vinegar, granulated sugar, and salad oil. Add to salad; toss.
4. Cover; marinate in refrigerator at least 1 hour before serving. CCP: Hold for service at 41 F. or lower.

SALADS. SALAD DRESSINGS. AND RELISHES No.M 054 00
TOMATO FRENCH DRESSING

Yield 100   **Portion**   1 Tablespoon

Calories	Carbohydrates	Protein	Fat	Cholesterol	Sodium	Calcium
29 cal	2 g	0 g	2 g	0 mg	53 mg	2 mg

Ingredient	Weight	Measure	Issue
SOUP,CONDENSED,TOMATO	2-1/8 lbs	3-3/4 cup	
VINEGAR,DISTILLED	12-1/2 oz	1-1/2 cup	
SUGAR,GRANULATED	1-3/4 oz	1/4 cup 1/3 tbsp	
ONIONS,FRESH,GRATED	1-3/8 oz	1/4 cup 1/3 tbsp	1-5/8 oz
WORCESTERSHIRE SAUCE	3/4 oz	1 tbsp	
MUSTARD,DRY	3/8 oz	1 tbsp	
GARLIC POWDER	1/8 oz	1/8 tsp	
OIL,SALAD	7-2/3 oz	1 cup	

Method

1. Combine soup, vinegar, sugar, onions, Worcestershire sauce, mustard, and garlic powder in mixer bowl.
2. Using a wire whip, beat at medium speed about 2 minutes or until well blended.
3. Add salad oil or olive oil gradually while mixing at low speed for 3 minutes.
4. Beat at medium speed 2 minutes or until well blended.
5. Cover; refrigerate until ready to serve. CCP: Hold for service at 41 F. or lower.
6. Whip or stir well before using.

SALADS. SALAD DRESSINGS. AND RELISHES No.M 055 00
VINAIGRETTE DRESSING

Yield 100   Portion   1 Tablespoon

Calories	Carbohydrates	Protein	Fat	Cholesterol	Sodium	Calcium
42 cal	1 g	0 g	4 g	0 mg	117 mg	3 mg

Ingredient	Weight	Measure	Issue
SUGAR,GRANULATED	1 oz	2-1/3 tbsp	
SALT	1 oz	1 tbsp	
MUSTARD,DRY	1 oz	2-2/3 tbsp	
PEPPER,BLACK,GROUND	1/8 oz	1/4 tsp	
PEPPER,RED,GROUND	<1/16th oz	1/8 tsp	
VINEGAR,DISTILLED	1 lbs	2 cup	
WATER	1 lbs	2 cup	
OIL,SALAD	1 lbs	2 cup	
PEPPERS,GREEN,FRESH,CHOPPED	2-1/8 oz	1/4 cup 2-2/3 tbsp	2-5/8 oz
ONIONS,FRESH,GRATED	1-3/8 oz	1/4 cup 1/3 tbsp	1-5/8 oz
PARSLEY,FRESH,BUNCH,CHOPPED	3/4 oz	1/4 cup 2-1/3 tbsp	7/8 oz

Method

1. Combine sugar, salt, dry mustard, and black and red pepper in mixer bowl. Add vinegar and water.
2. Using a wire whip, beat at medium speed about 2 minutes or until well blended.
3. Add salad oil or olive oil gradually while mixing at low speed 3 minutes.
4. Add onions, peppers, and parsley; mix at medium speed 1 minute or until well blended.
5. Cover; refrigerate until ready to serve. CCP: Hold for service at 41 F. or lower.
6. Whip or stir well before using.

SALADS. SALAD DRESSINGS. AND RELISHES No.M 056 00

QUICK FRUIT DRESSING

Yield 100   Portion   1 Tablespoon

Calories	Carbohydrates	Protein	Fat	Cholesterol	Sodium	Calcium
16 cal	3 g	1 g	0 g	1 mg	8 mg	21 mg

Ingredient	Weight	Measure	Issue
YOGURT,PLAIN,LOWFAT	2-3/8 lbs	1 qts 1/2 cup	
SUGAR,GRANULATED	4-3/8 oz	1/2 cup 2 tbsp	
JUICE,PINEAPPLE,CANNED,UNSWEETENED	1-5/8 lbs	2-7/8 cup	

Method

1 Blend sugar into yogurt in mixer bowl at low speed.
2 Gradually add pineapple juice. Mix at low speed until just blended.
3 Cover and refrigerate until ready to serve. CCP: Hold for service at 41 F. or lower.
4 Stir well before using.

SALADS. SALAD DRESSINGS. AND RELISHES No.M 057 00
ZERO SALAD DRESSING

Yield 100   **Portion** 2 Tablespoons

Calories	Carbohydrates	Protein	Fat	Cholesterol	Sodium	Calcium
7 cal	2 g	0 g	0 g	0 mg	184 mg	4 mg

Ingredient	Weight	Measure	Issue
JUICE,TOMATO,CANNED	6-1/8 lbs	2 qts 3-1/2 cup	
VINEGAR,DISTILLED	1-5/8 lbs	3 cup	
SALT	3/4 oz	1 tbsp	
ONIONS,FRESH,GRATED	3/4 oz	2 tbsp	3/4 oz
PEPPERS,GREEN,FRESH,GROUND	2/3 oz	2 tbsp	3/4 oz
CARROTS,FRESH,CHOPPED	12 oz	2-5/8 cup	14-5/8 oz
PARSLEY,FRESH,BUNCH,CHOPPED	1/4 oz	2 tbsp	1/4 oz

Method
1 Combine tomato juice, vinegar, salt, onions, peppers, carrots and parsley; blend well.
2 Cover; refrigerate until ready to serve. CCP: Hold for service at 41 F. or lower.
3 Shake well before using.

Notes
1 In Step 2, 3 cups canned tomato juice concentrate mixed with 2-1/4 quarts water may be used for canned tomato juice per 1 gallon of dressing.

SALADS, SALAD DRESSINGS, AND RELISHES   No. M 058 00

FRENCH DRESSING

Yield 100   Portion   1 Tablespoon

Calories	Carbohydrates	Protein	Fat	Cholesterol	Sodium	Calcium
41 cal	2 g	0 g	4 g	0 mg	113 mg	2 mg

Ingredient	Weight	Measure	Issue
MUSTARD,DRY	5/8 oz	1 tbsp	
PAPRIKA,GROUND	1/4 oz	1 tbsp	
SALT	3/4 oz	1 tbsp	
SUGAR,GRANULATED	1-3/4 oz	1/4 cup 1/3 tbsp	
CATSUP	10-5/8 oz	1-1/4 cup	
ONIONS,FRESH,CHOPPED	3/4 oz	2 tbsp	3/4 oz
VINEGAR,DISTILLED	1 lbs	2 cup	
OIL,SALAD	13-1/2 oz	1-3/4 cup	

Method

1. Combine mustard flour, paprika, salt, sugar, catsup, and onions in mixer bowl.
2. Using a wire whip, beat at medium speed about 2 minutes or until well blended.
3. Continue beating; slowly add vinegar and salad oil alternately.
4. Cover; refrigerate until ready to serve. CCP: Hold for service at 41 F. or lower.
5. Shake or beat well before using.

SALADS, SALAD DRESSINGS, AND RELISHES No.M 058 01
LOW CALORIE FRENCH DRESSING

Yield 100   **Portion** 2 Tablespoons

Calories	Carbohydrates	Protein	Fat	Cholesterol	Sodium	Calcium
14 cal	3 g	0 g	0 g	0 mg	490 mg	4 mg

Ingredient	**Weight**	**Measure**	**Issue**
MUSTARD,DRY	1-1/4 oz	3 tbsp	
PAPRIKA,GROUND	1/2 oz	2 tbsp	
SALT	3-3/4 oz	1/4 cup 2-1/3 tbsp	
SUGAR,GRANULATED	3-1/2 oz	1/2 cup	
CATSUP	1-1/3 lbs	2-1/2 cup	
ONIONS,FRESH,GRATED	1-3/8 oz	1/4 cup 1/3 tbsp	1-5/8 oz
VINEGAR,DISTILLED	2-1/8 lbs	1 qts	
WATER	2-5/8 lbs	1 qts 1 cup	

Method

1 Combine mustard, paprika, salt, sugar, catsup, and onions in mixer bowl.
2 Using a wire whip, beat at medium speed about 2 minutes or until well blended.
3 Continue beating; slowly add vinegar and water alternately.
4 Cover; refrigerate until ready to serve. CCP: Hold for service at 41 F. or lower.
5 Shake or beat well before using.

SALADS. SALAD DRESSINGS. AND RELISHES   No.M 059 00
BLUE CHEESE DRESSING

Yield 100 **Portion** 1 Tablespoon

Calories	Carbohydrates	Protein	Fat	Cholesterol	Sodium	Calcium
19 cal	1 g	1 g	1 g	3 mg	47 mg	40 mg

Ingredient	Weight	Measure	Issue
YOGURT,PLAIN,LOWFAT	3 lbs	1 qts 1-1/2 cup	
VINEGAR,DISTILLED	1/2 oz	1 tbsp	
MUSTARD,DRY	3/8 oz	1 tbsp	
GARLIC POWDER	1/8 oz	1/4 tsp	
ONION POWDER	1/8 oz	1/4 tsp	
CHEESE,BLUE-VEINED	9-1/2 oz	2 cup	

Method

1. Place yogurt, vinegar, mustard flour, garlic powder, and onion powder in mixer bowl.
2. Using whip, mix at low speed 2 minutes until just blended.
3. Fold in cheese until just blended.
4. CCP: Cover; refrigerate product at 41 F. or lower until ready to serve.

SALADS, SALAD DRESSINGS, AND RELISHES No.M 060 00
GARLIC FRENCH DRESSING

Yield 100   **Portion** 1 Tablespoon

Calories	Carbohydrates	Protein	Fat	Cholesterol	Sodium	Calcium
40 cal	2 g	0 g	4 g	0 mg	53 mg	3 mg

Ingredient	**Weight**	**Measure**	**Issue**
SUGAR,GRANULATED	3-1/2 oz	1/2 cup	
GARLIC POWDER	1-1/4 oz	1/4 cup 1/3 tbsp	
MUSTARD,DRY	7/8 oz	2-1/3 tbsp	
ONION POWDER	1/2 oz	2 tbsp	
PAPRIKA,GROUND	1/4 oz	1 tbsp	
SALT	1/2 oz	3/8 tsp	
PEPPER,RED,GROUND	<1/16th oz	<1/16th tsp	
WATER	13-7/8 oz	1-5/8 cup	
JUICE,LEMON	10-3/4 oz	1-1/4 cup	
VINEGAR,DISTILLED	10-1/2 oz	1-1/4 cup	
OIL,SALAD	12-7/8 oz	1-5/8 cup	

Method

1. Combine sugar, garlic powder, dry mustard, onion powder, paprika, salt, and red pepper in mixer bowl. Add water, lemon juice, and vinegar.
2. Using a wire whip, beat at medium speed about 2 minutes or until well blended.
3. Add salad oil or olive oil gradually while mixing at low speed 3 minutes.
4. Beat at medium speed 2 minutes or until well blended.
5. Cover; refrigerate until ready to serve. CCP: Hold for service at 41 F. or lower.

SALADS. SALAD DRESSINGS. AND RELISHES No.M 061 00

TANGY SALAD DRESSING

Yield 100 **Portion** 1 Tablespoon

Calories	Carbohydrates	Protein	Fat	Cholesterol	Sodium	Calcium
41 cal	2 g	0 g	4 g	0 mg	167 mg	2 mg

Ingredient	Weight	Measure	Issue
CATSUP	14-7/8 oz	1-3/4 cup	
MUSTARD,DRY	3/8 oz	1 tbsp	
SALT	1 oz	1 tbsp	
PEPPER,BLACK,GROUND	1/8 oz	1/4 tsp	
WORCESTERSHIRE SAUCE	2-1/8 oz	1/4 cup 1/3 tbsp	
VINEGAR,DISTILLED	14-5/8 oz	1-3/4 cup	
SUGAR,GRANULATED	3/4 oz	1 tbsp	
OIL,SALAD	13-1/2 oz	1-3/4 cup	
PARSLEY,FRESH,BUNCH,CHOPPED	1/4 oz	2 tbsp	1/4 oz
PEPPERS,GREEN,FRESH,CHOPPED	6-5/8 oz	1-1/4 cup	8 oz

Method

1. Combine catsup, mustard, salt, pepper, vinegar, sugar, and Worcestershire sauce in mixer bowl. Beat at medium speed about 2 minutes or until blended.
2. Add salad oil or olive oil while mixing at medium speed for 3 minutes or until well blended.
3. Add parsley and peppers; mix at medium speed 1 minute or until blended.
4. Cover; refrigerate until ready to serve. CCP: Hold for service at 41 F. or lower.
5. Whip or stir well before using.

SALADS. SALAD DRESSINGS. AND RELISHES No.M 062 00
MEXICAN POTATO SALAD

Yield 100   Portion 3/4 Cup

Calories	Carbohydrates	Protein	Fat	Cholesterol	Sodium	Calcium
162 cal	20 g	3 g	8 g	0 mg	115 mg	28 mg

Ingredient	**Weight**	**Measure**	**Issue**
POTATOES,FRESH,PEELED,CUBED	17-7/8 lbs	3 gal 1 qts	22-1/8 lbs
WATER	25-1/8 lbs	3 gal	
SALT	5/8 oz	1 tbsp	
VINEGAR,RED WINE	1 lbs	2 cup	
WATER	1 lbs	2 cup	
PEPPERS,JALAPENOS,CANNED,CHOPPED	5/8 oz	2 tbsp	
SUGAR,GRANULATED	1-3/4 oz	1/4 cup 1/3 tbsp	
PAPRIKA,GROUND	3/8 oz	1 tbsp	
PEPPER,BLACK,GROUND	3/8 oz	1 tbsp	
CUMIN,GROUND	1/4 oz	1 tbsp	
MUSTARD,DRY	3/4 oz	2 tbsp	
OIL,SALAD	1-3/4 lbs	3-3/4 cup	
BROCCOLI,FRESH,CHOPPED	6-1/4 lbs	2 gal <1/16th qts	10-1/4 lbs
WATER	8-1/3 lbs	1 gal	
TOMATOES,FRESH,SLICED	1-7/8 lbs	1 qts 3/4 cup	1-7/8 lbs
BEANS,KIDNEY,DARK RED,CANNED,DRAINED	1-1/2 lbs	1 qts	

Method
1 Cover potatoes with water, bring to a boil; add salt; reduce heat to a simmer; cover. Cook 10 minutes or until just tender.
2 Drain well. Cool slightly. Set aside for use in Step 7.
3 Combine vinegar, water, and jalapeno peppers in mixer bowl; mix well.
4 Combine sugar, mustard, paprika, pepper, and cumin; blend well; add to vinegar mixture.
5 Mix at medium speed 2 minutes using a wire whip.
6 Add salad oil or olive oil gradually while mixing at low speed 3 minutes; scrape down bowl. Mix at medium speed 2 minutes or until well blended.
7 Pour dressing over potatoes. Mix lightly but thoroughly. Cover; refrigerate for use in Step 9.
8 Cover broccoli with water; bring to a boil; reduce heat. Simmer 4 minutes or until just tender. Drain thoroughly.
9 Add broccoli, tomatoes and beans to potato mixture. Toss lightly but thoroughly. Cover; refrigerate at least 3 hours or until flavors are well blended. CCP: Hold for service at 41 F. or lower.

Notes
1 In Step 1, For 100 portions: 7-1/4 pounds frozen broccoli may be used. Add to boiling water. Cook 1 minute or until thoroughly heated.

SALADS. SALAD DRESSINGS. AND RELISHES   No. M 063 00

THOUSAND ISLAND DRESSING

Yield 100   Portion   1 Tablespoon

Calories	Carbohydrates	Protein	Fat	Cholesterol	Sodium	Calcium
63 cal	3 g	0 g	6 g	10 mg	150 mg	1 mg

Ingredient	Weight	Measure	Issue
EGG,HARD COOKED,CHOPPED	4-3/4 oz	1 cup	
ONIONS,FRESH,GRATED	1/3 oz	1 tbsp	3/8 oz
PIMIENTO,CANNED,DRAINED,CHOPPED	3-3/8 oz	1/2 cup	
PICKLE RELISH,SWEET,DRAINED	4-1/3 oz	1/2 cup	
CATSUP	10-5/8 oz	1-1/4 cup	
SALAD DRESSING,MAYONNAISE TYPE	2-1/2 lbs	1 qts 1 cup	
SALT	1/3 oz	1/4 tsp	

Method

1 Combine eggs, onions, pimientos, relish, catsup, salad dressing, and salt; blend well.
2 Cover; refrigerate until ready to serve. CCP: Hold for service at 41 F. or lower.
3 Stir well before using.

SALADS, SALAD DRESSINGS, AND RELISHES No.M 064 00
CREAMY ITALIAN DRESSING

Yield 100　　　　　　　　　　　　　　　　　　　**Portion**　1 Tablespoon

Calories	Carbohydrates	Protein	Fat	Cholesterol	Sodium	Calcium
69 cal	2 g	0 g	7 g	5 mg	82 mg	7 mg

Ingredient　　　　　　　　　　　　　　　Weight　　　Measure　　　　　　Issue

SALAD DRESSING,MAYONNAISE TYPE	3 lbs	1 qts 2 cup
WATER	12-1/2 oz	1-1/2 cup
VINEGAR,DISTILLED	6-1/4 oz	3/4 cup
BASIL,SWEET,WHOLE,CRUSHED	3/8 oz	2-2/3 tbsp
OREGANO,CRUSHED	7/8 oz	1/4 cup 1-2/3 tbsp
GARLIC POWDER	1/8 oz	1/4 tsp
ONION POWDER	5/8 oz	2-2/3 tbsp
SUGAR,GRANULATED	1/8 oz	1/8 tsp

Method

1. Combine salad dressing, water, vinegar, basil, oregano, garlic, onion powder, and sugar in mixer bowl.
2. Beat at medium speed 3 to 5 minutes.
3. Cover; refrigerate at least 8 hours before serving. CCP: Hold for service at 41 F. or lower.

SALADS, SALAD DRESSINGS, AND RELISHES No.M 065 00
CREAMY HORSERADISH DRESSING

Yield 100 **Portion** 1 Tablespoon

Calories	Carbohydrates	Protein	Fat	Cholesterol	Sodium	Calcium
59 cal	4 g	0 g	5 g	3 mg	103 mg	2 mg

Ingredient — Weight — Measure — Issue

Ingredient	Weight	Measure	Issue
SALAD DRESSING,FRENCH,PREPARED,L/C	1-1/8 lbs	2 cup	
SALAD DRESSING,MAYONNAISE TYPE	2 lbs	1 qts	
HORSERADISH,PREPARED	8-1/2 oz	1 cup	
VINEGAR,DISTILLED	4-1/8 oz	1/2 cup	
SUGAR,GRANULATED	5-1/4 oz	3/4 cup	

Method

1. Combine French Dressing, Salad Dressing, horseradish, vinegar, and sugar; blend well.
2. Cover; refrigerate at least 1 to 2 hours for flavors to blend. Keep refrigerated until ready to serve. CCP: Hold for service at 41 F. or lower.

SALADS, SALAD DRESSINGS, AND RELISHES   No.M 066 00

LOW CALORIE TOMATO DRESSING

Yield 100　　　　　　　　　　　　　　　　　　　　　**Portion**   2 Tablespoons

Calories	Carbohydrates	Protein	Fat	Cholesterol	Sodium	Calcium
26 cal	6 g	1 g	0 g	0 mg	196 mg	4 mg

Ingredient　　　　　　　　　　　　　　　　Weight　　　Measure　　　　Issue

SOUP,CONDENSED,TOMATO	7-1/4 lbs	3 qts 1 cup	
GARLIC POWDER	7/8 oz	3 tbsp	
ONION POWDER	1/4 oz	1 tbsp	
PEPPER,BLACK,GROUND	<1/16th oz	1/8 tsp	
PICKLE RELISH,SWEET	6-1/2 oz	3/4 cup	
VINEGAR,WHITE WINE	1-1/4 lbs	2-1/4 cup	

Method

1 Combine soup, garlic, onion powder, pepper, relish, and wine vinegar; blend well.
2 Cover; refrigerate until ready to serve. Stir well before using. CCP: Hold for service at 41 F. or lower.

SALADS. SALAD DRESSINGS. AND RELISHES No.M 067 00

RUSSIAN DRESSING

Yield 100   **Portion** 1 Tablespoon

Calories	Carbohydrates	Protein	Fat	Cholesterol	Sodium	Calcium
72 cal	3 g	0 g	7 g	5 mg	106 mg	1 mg

Ingredient	Weight	Measure	Issue
SAUCE,CHILI	13-1/3 oz	1-1/2 cup	
ONIONS,FRESH,GRATED	1/3 oz	1 tbsp	3/8 oz
PAPRIKA,GROUND	1/4 oz	1 tbsp	
PEPPER,BLACK,GROUND	1/8 oz	1/4 tsp	
PIMIENTO,CANNED,DRAINED,CHOPPED	1-2/3 oz	1/4 cup 1/3 tbsp	
SALAD DRESSING,MAYONNAISE TYPE	3 lbs	1 qts 2 cup	

Method

1. Combine chili sauce, onions, paprika, pepper, pimientos, and Salad Dressing; blend well.
2. Cover; refrigerate until ready to serve. CCP: Hold for service at 41 F. or lower.
3. Stir well before using.

SALADS, SALAD DRESSINGS, AND RELISHES   No.M 068 00
SOUR CREAM DRESSING

Yield 100 **Portion** 1 Tablespoon

Calories	Carbohydrates	Protein	Fat	Cholesterol	Sodium	Calcium
39 cal	2 g	0 g	3 g	6 mg	89 mg	19 mg

Ingredient   Weight   Measure   Issue

Ingredient	Weight	Measure	Issue
SOUR CREAM, LOW FAT	3 lbs	1 qts 2 cup	
SALAD DRESSING, MAYONNAISE TYPE	1 lbs	2 cup	
VINEGAR, DISTILLED	2-1/8 oz	1/4 cup 1/3 tbsp	
SALT	1/2 oz	3/8 tsp	

Method
1. Carefully blend salad dressing into sour cream.
2. Combine vinegar and salt. Add to sour cream mixture stirring carefully.
3. Cover; refrigerate until ready to serve. CCP: Hold for service at 41 F. or lower.

SALADS. SALAD DRESSINGS. AND RELISHES   No.M 068 01

BLUE CHEESE AND SOUR CREAM DRESSING

Yield 100   **Portion**   1 Tablespoon

Calories	Carbohydrates	Protein	Fat	Cholesterol	Sodium	Calcium
54 cal	2 g	1 g	4 g	9 mg	146 mg	41 mg

Ingredient   Weight   Measure   Issue

SOUR CREAM,LOW FAT   3 lbs   1 qts 2 cup
SALAD DRESSING,MAYONNAISE TYPE   1 lbs   2 cup
VINEGAR,DISTILLED   2-1/8 oz   1/4 cup 1/3 tbsp
SALT   1/2 oz   3/8 tsp
CHEESE,BLUE-VEINED   14-1/4 oz   3 cup

Method

1. Carefully blend salad dressing into sour cream.
2. Combine vinegar, crumbled blue-veined cheese, and salt. Add to sour cream mixture stirring carefully. Stir with wire whip until blended.
3. Cover; refrigerate at least 2 hours before serving. Keep refrigerated until ready to serve. CCP: Hold for service at 41 F. or lower.

SALADS, SALAD DRESSINGS, AND RELISHES   No.M 069 00
VINEGAR AND OIL DRESSING

Yield 100   Portion   1 Tablespoon

Calories	Carbohydrates	Protein	Fat	Cholesterol	Sodium	Calcium
40 cal	0 g	0 g	4 g	0 mg	117 mg	1 mg

Ingredient	Weight	Measure	Issue
VINEGAR,DISTILLED	1-1/8 lbs	2-1/4 cup	
WATER	1-1/8 lbs	2-1/4 cup	
SALT	1 oz	1 tbsp	
GARLIC POWDER	1/4 oz	1/3 tsp	
PEPPER,BLACK,GROUND	1/8 oz	1/4 tsp	
OREGANO,CRUSHED	1/8 oz	1 tbsp	
OIL,SALAD	1 lbs	2 cup	

Method

1 Combine vinegar, water, salt, garlic powder, pepper, and oregano in mixer bowl.
2 Using a wire whip, beat at medium speed about 2 minutes or until well blended.
3 Add salad oil or olive oil gradually while mixing at low speed 3 minutes.
4 Mix at medium speed 1 minute or until well blended.
5 Cover; refrigerate until ready to serve. CCP: Hold for service at 41 F. or lower.
6 Whip or stir well before using.

SALADS. SALAD DRESSINGS. AND RELISHES No.M 070 00
ZESTY ROTINI PASTA SALAD

Yield 100   **Portion** 1/2 Cup

Calories	Carbohydrates	Protein	Fat	Cholesterol	Sodium	Calcium
106 cal	18 g	4 g	2 g	1 mg	382 mg	37 mg

Ingredient	**Weight**	**Measure**	**Issue**
WATER	20-7/8 lbs	2 gal 2 qts	
SALT	5/8 oz	1 tbsp	
OIL,OLIVE	1/2 oz	1 tbsp	
MACARONI NOODLES,ROTINI,DRY	4-3/8 lbs	1 gal 3/4 qts	
SALAD DRESSING,ITALIAN,DIET	3-3/8 lbs	1 qts 2 cup	
CHEESE,PARMESAN,GRATED	5-1/4 oz	1-1/2 cup	
SESAME SEEDS	2-1/2 oz	1/2 cup	
POPPY SEEDS	7/8 oz	3 tbsp	
PAPRIKA,GROUND	1 oz	1/4 cup 1/3 tbsp	
TOMATOES,FRESH,CHOPPED	3-1/2 lbs	2 qts 3/4 cup	3-5/8 lbs
CUCUMBER,FRESH,CHOPPED	3-1/2 lbs	3 qts 1-3/8 cup	4-1/8 lbs
PEPPERS,GREEN,FRESH,CHOPPED	2 lbs	1 qts 2-1/8 cup	2-1/2 lbs
ONIONS,FRESH,CHOPPED	1-1/2 lbs	1 qts 1/4 cup	1-2/3 lbs

Method

1. Add salt and salad oil or olive oil to water; heat to a rolling boil.
2. Add rotini slowly while stirring constantly, until water boils again. Cook about 10 to 12 minutes or until tender; stir occasionally. DO NOT OVER COOK.
3. Drain. Rinse with cold water.
4. Combine dressing with cheese, sesame seeds, poppy seeds, and paprika. Add to rotini. Toss lightly.
5. Add tomatoes, cucumbers, peppers, and onions. Toss lightly. Cover and refrigerate at least 3 hours or until flavors are blended. Keep refrigerated until ready to serve.

SALADS, SALAD DRESSINGS, AND RELISHES No.M 071 00

SALSA PASTA SALAD

Yield 100   Portion 1/2 Cup

Calories	Carbohydrates	Protein	Fat	Cholesterol	Sodium	Calcium
98 cal	19 g	4 g	1 g	0 mg	221 mg	18 mg

Ingredient	Weight	Measure	Issue
WATER	20-7/8 lbs	2 gal 2 qts	
SALT	5/8 oz	1 tbsp	
OIL,SALAD	1/2 oz	1 tbsp	
MACARONI NOODLES,ROTINI,DRY	3-1/8 lbs	3 qts 1-1/2 cup	
TOMATOES,FRESH,CHOPPED	4 lbs	2 qts 2 cup	4 lbs
ONIONS,FRESH,CHOPPED	4-1/4 oz	3/4 cup	4-2/3 oz
PEPPERS,JALAPENOS,CANNED,CHOPPED	2-3/8 oz	1/2 cup	
SALT	5/8 oz	1 tbsp	
SUGAR,GRANULATED	1/2 oz	1 tbsp	
CARROTS,FROZEN,SLICED	1-2/3 lbs	1 qts 2 cup	
WATER,BOILING	3-1/8 lbs	1 qts 2 cup	
PEAS,GREEN,FROZEN	2-1/4 lbs	1 qts 3 cup	
CORN,FROZEN,WHOLE KERNEL	2-1/2 lbs	1 qts 3 cup	
WATER,BOILING	1 lbs	2 cup	
CHICKPEAS	2-2/3 lbs	1 qts 1 cup	
OLIVES,RIPE,PITTED,SLICED,INCL LIQUIDS	9-1/2 oz	2 cup	

Method

1. Add salt and oil to water; heat to rolling boil. Slowly add rotini while stirring constantly until water boils again. Cook 10 to 12 minutes. DO NOT OVERCOOK. Drain, rinse with cold water; drain thoroughly.
2. Combine tomatoes, onions, peppers, salt, and sugar; mix well.
3. Add tomato mixture to rotini. Toss lightly but thoroughly.
4. Refrigerate for use in Step 8.
5. Add carrots to boiling water. Return to a boil; reduce heat; cover; simmer 8 to 10 minutes or until just tender.
6. Drain; set aside for use in Step 8.
7. Add peas and corn to water. Bring to a boil; reduce heat; cover; simmer 4 minutes. Drain.
8. Add chick peas, olives, carrots, peas, and corn to rotini mixture. Toss lightly. Cover; refrigerate at least 3 hours or until flavors are well blended. Keep refrigerated until ready to serve. CCP: Hold for service at 41 F. or lower.

Notes

1. In Step 2, 2-3/4 quarts prepared salsa may be used.

SALADS, SALAD DRESSINGS, AND RELISHES No.M 072 00

CONFETTI RICE SALAD

Yield 100　　　　　　　　　　　　　　　　　**Portion**　1/2 Cup

Calories	Carbohydrates	Protein	Fat	Cholesterol	Sodium	Calcium
97 cal	18 g	2 g	2 g	2 mg	250 mg	44 mg

Ingredient	Weight	Measure	Issue
RICE,LONG GRAIN	3-3/4 lbs	2 qts 1-3/8 cup	
WATER,COLD	10-1/2 lbs	1 gal 1 qts	
SALT	7/8 oz	1 tbsp	
YOGURT,PLAIN,LOWFAT	3-1/4 lbs	1 qts 2 cup	
SALAD DRESSING,MAYONNAISE TYPE	10-1/2 oz	1-3/8 cup	
VINEGAR,DISTILLED	2-1/8 oz	1/4 cup 1/3 tbsp	
SALT	1 oz	1 tbsp	
PARSLEY,FRESH,BUNCH,CHOPPED	1/2 oz	1/4 cup	1/2 oz
GARLIC POWDER	1/4 oz	1/3 tsp	
PEPPER,BLACK,GROUND	1/8 oz	1/3 tsp	
TOMATOES,FRESH,CHOPPED	3 lbs	1 qts 3-1/2 cup	3 lbs
CELERY,FRESH,CHOPPED	2 lbs	1 qts 3-1/2 cup	2-3/4 lbs
PIMIENTO,CANNED,DRAINED,CHOPPED	1 lbs	2-1/4 cup	
PEPPERS,GREEN,FRESH,CHOPPED	1-1/2 lbs	1 qts 1/2 cup	1-7/8 lbs
ONIONS,FRESH,CHOPPED	8 oz	1-3/8 cup	8-7/8 oz

Method

1. Combine rice, water, and salt; bring to a boil. Stir occasionally. Cover tightly; simmer 20 to 25 minutes. DO NOT STIR.
2. Remove from heat; transfer to shallow serving pans. Cover, refrigerate for 1 hour.
3. Combine yogurt, salad dressing, vinegar, salt, parsley, garlic powder, and pepper; mix thoroughly. Combine with chilled rice.
4. Add tomatoes, celery, pimientos, peppers, and onions. Mix lightly.
5. Cover; refrigerate until ready to serve. CCP: Hold for service at 41 F. or lower.

Notes

1. In Step 5, rice salad may be served on lettuce. Use 4 pounds fresh lettuce, trimmed and cored for 100 portions.

SALADS. SALAD DRESSINGS. AND RELISHES No.M 072 01
CREAMY CUCUMBER RICE SALAD

Yield 100   Portion   1/2 Cup

Calories	Carbohydrates	Protein	Fat	Cholesterol	Sodium	Calcium
105 cal	18 g	3 g	2 g	3 mg	260 mg	75 mg

Ingredient | Weight | Measure | Issue

Ingredient	Weight	Measure	Issue
RICE,LONG GRAIN	3-3/4 lbs	2 qts 1-3/8 cup	
WATER,COLD	10-1/2 lbs	1 gal 1 qts	
SALT	7/8 oz	1 tbsp	
YOGURT,PLAIN,LOWFAT	6-1/2 lbs	3 qts	
CUCUMBER,FRESH,CHOPPED	12-5/8 oz	3 cup	15 oz
SALAD DRESSING,MAYONNAISE TYPE	10-1/2 oz	1-3/8 cup	
SALT	1 oz	1 tbsp	
DILL WEED,DRIED	1/2 oz	1/4 cup 2/3 tbsp	
GARLIC POWDER	1/4 oz	1/3 tsp	
PEPPER,BLACK,GROUND	1/8 oz	1/3 tsp	
CUCUMBER,FRESH,SLICED	2-1/2 lbs	2 qts 1/2 cup	3 lbs
CELERY,FRESH,CHOPPED	2 lbs	1 qts 3-1/2 cup	2-3/4 lbs
PIMIENTO,CANNED,DRAINED,CHOPPED	1 lbs	2-1/4 cup	
PEPPERS,GREEN,FRESH,CHOPPED	1-1/2 lbs	1 qts 1/2 cup	1-7/8 lbs
ONIONS,FRESH,CHOPPED	8 oz	1-3/8 cup	8-7/8 oz

Method
1 Combine rice, water, and salt; bring to a boil. Stir occasionally. Cover tightly; simmer 20 to 25 minutes. DO NOT STIR.
2 Remove from heat; transfer to shallow serving pans. Cover, refrigerate for 1 hour.
3 Combine yogurt, cucumbers, dressing, dill weed, garlic powder, and black pepper; mix thoroughly. Combine with chilled rice.
4 Add celery, pimientos, peppers, and onions. Mix lightly.
5 Cover; refrigerate until ready to serve. CCP: Hold for service at 41 F. or lower.

Notes
1 In Step 5, rice salad may be served on lettuce. Use 4 pounds fresh lettuce, trimmed and cored for 100 portions.

SALADS, SALAD DRESSINGS, AND RELISHES No.M 073 00
KIWI FRUIT SALAD

Yield 100   **Portion** 1/2 Cup

Calories	Carbohydrates	Protein	Fat	Cholesterol	Sodium	Calcium
106 cal	25 g	2 g	1 g	1 mg	22 mg	79 mg

Ingredient	**Weight**	**Measure**	**Issue**
YOGURT,PLAIN,LOWFAT	5-3/8 lbs	2 qts 2 cup	
SUGAR,GRANULATED	1-1/2 lbs	3-1/2 cup	
JUICE,LIME	1-1/4 lbs	2-1/2 cup	
NUTMEG,GROUND	1/4 oz	3/8 tsp	
GINGER,GROUND	1/8 oz	3/8 tsp	
APPLES,FRESH,MEDIUM,UNPEELED,DICED	7-1/4 lbs	1 gal 2-5/8 qts	8-1/2 lbs
KIWIFRUIT,FRESH,CHOPPED	7-1/4 lbs	1 gal 5/8 qts	8-3/8 lbs
ORANGE,FRESH,SLICED	3-7/8 lbs	2 qts 1-3/4 cup	5-1/3 lbs
PINEAPPLE,CANNED,CHUNKS,JUICE PACK,DRAINED	4 lbs	2 qts 1 cup	
LETTUCE,LEAF,FRESH,HEAD	4 lbs		6-1/4 lbs

Method

1 Combine yogurt, sugar, lime juice, nutmeg, and ginger in mixer bowl. Blend at medium speed until smooth, about 2 minutes.
2 Combine apples, kiwi fruit, oranges, and pineapple. Mix lightly.
3 Place 1 lettuce leaf on each serving dish. Place 1/2 cup fruit mixture on lettuce. CCP: Cover; refrigerate product at 41 F. or lower.
4 Serve each portion with 1 ounce of dressing.

SALADS, SALAD DRESSINGS, AND RELISHES No.M 074 00

MARINATED BLACK BEAN SALAD

Yield 100   Portion   3/4 Cup

Calories	Carbohydrates	Protein	Fat	Cholesterol	Sodium	Calcium
179 cal	36 g	10 g	1 g	0 mg	149 mg	53 mg

Ingredient	Weight	Measure	Issue
BEANS,BLACK,CANNED,DRAINED	18-5/8 lbs	2 gal 1/4 qts	
CORN,FROZEN,WHOLE KERNEL	11 lbs	1 gal 3-5/8 qts	
TOMATOES,CANNED,DICED,DRAINED	6-5/8 lbs	3 qts	
PEPPERS,GREEN,FRESH,CHOPPED	2-1/4 lbs	1 qts 2-7/8 cup	2-3/4 lbs
PEPPERS,RED FRESH,DICED	2-1/4 lbs	1 qts 2-7/8 cup	2-3/4 lbs
ONIONS,RED,FRESH,CHOPPED	1-1/4 lbs	3-1/2 cup	1-3/8 lbs
JUICE,LEMON	12-7/8 oz	1-1/2 cup	
PEPPERS,JALAPENOS,CANNED,CHOPPED	7-1/4 oz	1-1/2 cup	
CILANTRO,FRESH,DICED	2 oz	1-1/4 cup	2-1/8 oz
CUMIN,GROUND	1/2 oz	2-2/3 tbsp	
SUGAR,GRANULATED	3/4 oz	1 tbsp	
SALT	5/8 oz	1 tbsp	
GARLIC POWDER	5/8 oz	2 tbsp	
LETTUCE,LEAF,FRESH,HEAD	4 lbs		6-1/4 lbs

Method

1. Combine beans, corn, tomatoes, green peppers, red peppers, onions, lemon juice, jalapeno peppers, cilantro, cumin, sugar, salt, and garlic powder. Mix well to thoroughly combine all ingredients.
2. CCP: Cover; refrigerate product at 41 F. or lower.
3. Place lettuce leaf in serving dish. Top with 3/4 cup salad. CCP: Refrigerate product at 41 F. or lower.

Notes

1. In Step 1, 7 pounds dry black beans and 3 gallons water may be used. Follow Steps 1 through 5 of Recipe No. Q 030 00, Boston Baked Beans.
2. In Step 1, 1/4 ounce or 6-2/3 tablespoons dry cilantro may be used.

SALADS, SALAD DRESSINGS, AND RELISHES No.M 504 00

BROCCOLI SALAD

Yield 100   **Portion** 1/2 Cup

Calories	Carbohydrates	Protein	Fat	Cholesterol	Sodium	Calcium
266 cal	33 g	4 g	15 g	6 mg	129 mg	51 mg

Ingredient	Weight	Measure	Issue
BROCCOLI,FRESH,BUNCH	13-1/2 lbs		22-1/8 lbs
SALAD DRESSING,MAYONNAISE TYPE	4 lbs	2 qts	
SUGAR,GRANULATED	1-3/4 lbs	1 qts	
VINEGAR,DISTILLED	4-1/8 oz	1/2 cup	
MILK,NONFAT,DRY	1/4 oz	1 tbsp	
WATER	4-1/8 oz	1/2 cup	
RAISINS	4-3/4 lbs	3 qts 3 cup	
WALNUTS,SHELLED,HALVES AND PIECES	2-1/8 lbs	2 qts	
ONIONS,FRESH,CHOPPED	11-1/4 oz	2 cup	12-1/2 oz

Method

1. Cut broccoli heads into florets. Dice stems.
2. Reconstitute milk. Combine fat free salad dressing, sugar, vinegar, and milk. Mix well. Add to broccoli.
3. Add raisins, walnuts (optional), and onions to broccoli mixture. Stir to coat all pieces with dressing.
4. CCP: Refrigerate product at 41F. or lower. Hold for service at 41 F. or lower.

N. SANDWICHES No. 0(1)

INDEX

Card No.		Card No.	
N 001 01	Bacon, Lettuce, and Tomato Sandwich	N 008 02	Turkey Salad Sandwich
N 002 00	Gyros	N 009 00	Corned Beef Sandwich
N 002 01	Gyros (RTU)	N 009 01	Corned Beef and Cheese Sandwich
N 003 00	Steak and Cheese Submarine	N 010 00	Egg Salad Sandwich
N 003 01	Steak, Cheese and Onion Submarine	N 010 01	New York Egg Salad Sandwich (Egg and Tomato)
N 003 02	Steak and Onion Submarine		
N 004 00	Roast Beef Sandwich	N 011 00	Ham Sandwich
N 004 01	Roast Pork Sandwich	N 011 01	Fried Ham Sandwich
N 004 02	Roast Turkey Sandwich	N 011 02	Ham and Cheese Sandwich
N 004 05	Turkey Croissant	N 011 03	Ham and Tomato Sandwich
N 005 00	Sausage and Biscuit	N 012 00	Grilled Hamburger (Beef Patties)
N 005 01	Ham and Biscuit	N 012 01	Cheeseburger (Beef Patties)
N 006 00	Grilled Cheese Sandwich	N 012 02	Cheesy Baconburger (Beef Patties)
N 006 01	German Style Hamwich	N 012 03	Double Decker Cheeseburger (Beef Patties)
N 006 03	Grilled Ham and Cheese Sandwich	N 012 04	Chiliburger (Beef Patties)
N 007 00	English Muffin with Bacon, Egg and Cheese	N 012 05	Deluxe Hamburger (Beef Patties)
N 007 01	English Muffin with Ham, Egg, and Cheese	N 012 06	Pizzaburger (Beef Patties)
N 007 02	English Muffin with Canadian Bacon, Egg, & Cheese	N 012 07	Deluxe Cheeseburger (Beef Patties)
		N 013 00	Ham Salad Sandwich
N 007 03	English Muffin with Sausage, Egg, & Cheese	N 014 00	Peanut Butter and Jelly Sandwich
N 008 00	Chicken Salad Sandwich	N 015 00	Tuna Salad Sandwich
N 008 01	Chicken Salad Sandwich (Canned Chicken)	N 015 01	Grilled Tuna and Cheese Sandwich

N. SANDWICHES No. 0(1)

Card No.		Card No.	
N 015 02	Salmon Salad Sandwich (Canned Salmon)	N 025 00	Monte Cristo Sandwich
N 015 03	Tuna and Tomato Sandwich	N 026 00	Italian Veal Cutlet Submarine
N 016 00	Cream Cheese Bagel	N 027 00	Barbecued Beef Sandwich (Sloppy Joe)
N 016 01	Cream Cheese and Tomato Bagel	N 027 01	Barbecued Pork Sandwich (Pork Butt)
N 016 02	Cream Cheese and Olive Bagel	N 027 02	Barbecued Pork Sandwich (Frozen Barbecued Pork)
N 017 00	Cold Cut Sandwich	N 027 03	Turkey Barbecue Sandwich
N 017 01	Cold Cut Sandwich with Cheese	N 028 00	Italian Pepper Beef Sandwich
N 018 00	Western Sandwich (Denver)	N 030 00	Simmered Frankfurter on Roll
N 019 00	Submarine Sandwich	N 030 01	Grilled Polish Sausage Sandwich
N 019 01	Italian Style Submarine	N 030 02	Simmered Knockwurst on Roll
N 020 00	Grilled Reuben Sandwich	N 030 03	Simmered Quarter Pound Frankfurter
N 020 02	Grilled Pastrami Reuben Sandwich	N 030 05	Grilled Frankfurter with Fried Peppers and Onions
N 021 00	Baked Chicken Fillet Sandwich (Breaded)	N 031 00	Monte Carlo Sandwich (Open-Faced Turkey and Ham)
N 021 01	Deep Fat Fried Chicken Fillet Sandwich (Breaded)	N 032 00	Fishwich
N 021 02	Chicken Fillet and Cheese Sandwich (Breaded)	N 032 01	Cheese Fishwich
N 021 03	Grilled Chicken Fillet Sandwich (Unbreaded)	N 032 03	Fishwich (Batter Dipped)
N 022 00	Cannonball Sandwich (Meatball)	N 033 00	Hot Roast Turkey Sandwich
N 022 01	Hot Italian Sausage Sandwich	N 034 00	Corn Dog
N 022 02	Cannonball Sandwich (Canned Meatballs)	N 034 01	Corn Dog (Corn Bread Mix)
N 023 00	Hot Pastrami Sandwich	N 035 00	Hot Roast Beef Sandwich (Oven Roast)
N 024 00	Barbecued Beef Sandwich (Canned)		
N 024 01	Barbecued Beef Sandwich (Diced Beef)		

INDEX

Card No.		Card No.	
N 035 01	Hot Roast Beef Sandwich (Precooked Roast Beef)	N 048 00	Jamaican Jerk Chicken Sandwich
N 036 00	Hot Roast Pork Sandwich	N 049 00	Mexican Beef Wrap
N 036 01	Hot Roast Pork Sandwich (Fresh Ham)	N 049 01	Mexican Turkey Wrap
N 037 00	Grilled Ham, Egg and Cheese Sandwich	N 050 00	Crunchy Vegetable Burrito
N 037 01	Grilled Bacon, Egg, and Cheese Sandwich	N 051 00	Vegetarian Hearty Burger
N 037 02	Grilled Ham and Egg Sandwich	N 052 00	Cajun Chicken Sandwich
N 037 03	Grilled Sausage, Egg, and Cheese Sandwich	N 502 00	Baked Turkey Melt
N 038 00	Moroccan Pockets	N 800 00	Meatball Hoagie using Precooked Meatballs
N 039 00	Cheese Deli Sandwich	N 805 00	Double Decker Beef & Turkey Sandwich
N 040 00	Taco Burger		
N 041 00	Chili Dog with Cheese and Onions		
N 041 01	Chili Dog (Canned Chili Con Carne)		
N 041 02	Chili Dog		
N 042 00	Beef Fajita Pita		
N 042 01	Chicken Fajita Pita		
N 043 00	Chicken Pita Pocket Sandwich		
N 044 00	Grilled Chicken Breast Sandwich		
N 045 00	Chicken Caesar Rollup Sandwich		
N 046 00	Garden Vegetable Wrap		
N 047 00	Roast Beef & Cheese Rollup Sandwich		
N 047 01	Hot Roast Beef & Cheese Rollup Sandwich		

SANDWICHES No.N 001 01

BACON, LETTUCE, AND TOMATO SANDWICH

Yield 100   **Portion**   1 Sandwich

Calories	Carbohydrates	Protein	Fat	Cholesterol	Sodium	Calcium
281 cal	29 g	10 g	14 g	17 mg	573 mg	74 mg

Ingredient	Weight	Measure	Issue
BACON,SLICED,RAW	12 lbs		
BREAD,WHITE,SLICE	11 lbs	200 sl	
TOMATOES,FRESH,SLICED	11-1/8 lbs	1 gal 3 qts	11-1/3 lbs
LETTUCE,LEAF,FRESH,HEAD	5 lbs		7-3/4 lbs
SALAD DRESSING,MAYONNAISE TYPE	2 lbs	1 qts	

Method

1. Prepare bacon according to package directions.
2. Place 2 slices bacon, 2 slices tomato, and lettuce leaf on 1 slice of bread; spread second slice of bread with about 2 teaspoons Salad Dressing. Top with second slice of bread.
3. Cut each sandwich in half. Prepare in 25 portion batches. Notes: Toast bread slices if desired.

SANDWICHES No.N 002 00

GYROS

Yield 100   Portion   1 Sandwich

Calories	Carbohydrates	Protein	Fat	Cholesterol	Sodium	Calcium
447 cal	49 g	37 g	11 g	79 mg	725 mg	244 mg

Ingredient	Weight	Measure	Issue
BEEF,OVEN ROAST,PRE COOKED	20 lbs		
ONIONS,FRESH,SLICED	2 lbs	1 qts 3-7/8 cup	2-1/4 lbs
TOMATOES,FRESH,CHOPPED	8-1/2 lbs	1 gal 1-3/8 qts	8-2/3 lbs
LETTUCE,ICEBERG,FRESH,SHREDDED	3-1/8 lbs	1 gal 2-1/2 qts	3-1/3 lbs
YOGURT,PLAIN,NONFAT	6-1/2 lbs	3 qts	
CUCUMBER,FRESH,CHOPPED	2-5/8 lbs	2 qts 2 cup	3-1/8 lbs
DILL WEED,DRIED	1/2 oz	1/4 cup 1-1/3 tbsp	
GARLIC POWDER	5/8 oz	2 tbsp	
BREAD,PITA,GYROS,8-INCH	21 lbs	100 each	

Method

1. Slice beef thin, about 20 slices per pound. Place 4 pounds, about 80 slices, on each pan for use in Step 5. CCP: Internal temperature must reach 165 F. or higher for 15 seconds. Hold for service at 140 F. or higher.
2. Separate onion slices into rings. Chop tomatoes and shred lettuce. Set aside for use in Step 6.
3. Peel, seed and chop cucumbers. Combine yogurt, cucumbers, dill weed and garlic powder. Mix well. CCP: Refrigerate at 41 F. or lower for use in Step 6.
4. Cut pita bread in half, forming 2 pockets. Place on pans. Using a convection oven, bake at 325 F. for 5 minutes or until warm and pliable on high fan, closed vent.
5. Place 2 slices beef in each pita pocket.
6. Top each pita with 1 tablespoon yogurt-cucumber sauce, 2 onion rings, 2 tablespoons diced tomatoes and 2 tablespoons lettuce.

SANDWICHES No.N 002 01

GYROS (RTU)

Yield 100   **Portion** 1 Sandwich

Calories	Carbohydrates	Protein	Fat	Cholesterol	Sodium	Calcium
520 cal	54 g	37 g	17 g	79 mg	916 mg	198 mg

Ingredient   Weight   Measure   Issue

Ingredient	Weight	Measure	Issue
BEEF,OVEN ROAST,PRE COOKED	20 lbs		
ONIONS,FRESH,SLICED	1-1/2 lbs	1 qts 2 cup	1-2/3 lbs
TOMATOES,FRESH,CHOPPED	9-1/8 lbs	1 gal 1-3/4 qts	9-1/3 lbs
LETTUCE,ICEBERG,FRESH,CHOPPED	3-1/8 lbs	1 gal 2-1/2 qts	3-3/8 lbs
BREAD,PITA,GYROS,8-INCH	21 lbs	100 each	
SAUCE,TZATZIKI	10 lbs	1 gal 3/4 qts	

Method

1. Slice beef thin, about 20 slices per pound. Place 4 pounds, about 80 slices, on each pan for use in Step 5. CCP: Internal temperature must reach 165 F. or higher for 15 seconds. Hold for service at 140 F. or higher.
2. Separate onion slices into rings. Chop tomatoes and shred lettuce. Set aside for use in Step 6.
3. Place bread on 5 pans. Bake 10 minutes or until warm and pliable.
4. Using a convection oven, bake meat in 350 F. on high fan, closed vent.
5. Place about 3 tablespoons of prepared tzatziki sauce on each gyro bread. Top with 6 slices of meat (3 oz). Top with tomatoes, lettuce, and onion rings. Fold in half; secure with toothpick or roll up bread around filling and wrap with 3-inch wide strip of aluminum foil.

SANDWICHES No.N 003 00

STEAK AND CHEESE SUBMARINE

Yield 100 **Portion** 1 Sandwich

Calories	Carbohydrates	Protein	Fat	Cholesterol	Sodium	Calcium
475 cal	20 g	32 g	29 g	102 mg	468 mg	218 mg

Ingredient | Weight | Measure | Issue

Ingredient	Weight	Measure	Issue
BEEF,STEAK,SANDWICH,THIN SLICES,RAW	25 lbs		
COOKING SPRAY,NONSTICK	2 oz	1/4 cup 1/3 tbsp	
CHEESE,AMERICAN,SLICED	6-1/4 lbs	100 sl	
ROLL,FRENCH	8-3/8 lbs	100 each	

Method

1. Grill steaks on one side 30 seconds on lightly sprayed grill at 350 F.
2. Turn steaks; cover half of steaks with cheese slices. Grill steaks 30 seconds. CCP: Internal temperature must reach 145 F. or higher for 15 seconds.
3. Slice rolls in half lengthwise. Place 1 steak on bottom half of each roll. Add steak with cheese on top.
4. Cover with top half of roll. CCP: Hold at 140 F. or higher for service.

SANDWICHES No.N 003 01

STEAK, CHEESE AND ONION SUBMARINE

Yield 100   Portion 1 Sandwich

Calories	Carbohydrates	Protein	Fat	Cholesterol	Sodium	Calcium
510 cal	27 g	33 g	29 g	102 mg	471 mg	236 mg

Ingredient	Weight	Measure	Issue
ONIONS,FRESH,SLICED	20 lbs	4 gal 3-3/4 qts	22-1/4 lbs
COOKING SPRAY,NONSTICK	2 oz	1/4 cup 1/3 tbsp	
BEEF,STEAK,SANDWICH,THIN SLICES,RAW	25 lbs		
CHEESE,AMERICAN,SLICED	6-1/4 lbs	100 sl	
ROLL,FRENCH	8-3/8 lbs	100 each	

Method

1. Lightly spray grill with non-stick cooking spray. Grill thinly sliced onions 5 to 6 minutes.
2. Grill steaks on one side 30 seconds on lightly sprayed grill at 350 F. Turn steaks; cover half of steaks with cheese slices. Grill steaks 30 seconds. CCP: Internal temperature must reach 145 F. or higher for 15 seconds.
3. Slice rolls in half lengthwise. Place 1 steak on bottom half of each roll. Add steak with cheese on top. Add 1/3 cup grilled onions on each sandwich.
4. Cover with top half of roll. CCP: Hold at 140 F. or higher for service.

SANDWICHES No.N 003 02

STEAK AND ONION SUBMARINE

Yield 100 **Portion** 1 Sandwich

Calories	Carbohydrates	Protein	Fat	Cholesterol	Sodium	Calcium
403 cal	27 g	26 g	20 g	75 mg	287 mg	61 mg

Ingredient | Weight | Measure | Issue

Ingredient	Weight	Measure	Issue
ONIONS,FRESH,SLICED	20 lbs	4 gal 3-3/4 qts	22-1/4 lbs
COOKING SPRAY,NONSTICK	2 oz	1/4 cup 1/3 tbsp	
BEEF,STEAK,SANDWICH,THIN SLICES,RAW	25 lbs		
ROLL,FRENCH	8-3/8 lbs	100 each	

Method

1. Lightly spray grill with non-stick cooking spray. Grill thinly sliced onions 5 to 6 minutes.
2. Grill steaks on one side 1/2 minute on lightly sprayed grill at 350 F. CCP: Internal temperature must reach 145 F. or higher for 15 seconds.
3. Slice rolls in half lengthwise. Place 1 steak on bottom half of each roll. Top with 1/3 cup onions on each sandwich.
4. Cover with top half of roll. CCP: Hold at 140 F. or higher for service.

SANDWICHES No. N 004 00

ROAST BEEF SANDWICH

Yield 100   Portion 1 Sandwich

Calories	Carbohydrates	Protein	Fat	Cholesterol	Sodium	Calcium
368 cal	29 g	31 g	14 g	76 mg	430 mg	81 mg

Ingredient	Weight	Measure	Issue
BEEF,OVEN ROAST,PRE COOKED	18-3/4 lbs		
MUSTARD,PREPARED	8-7/8 oz	1 cup	
SALAD DRESSING,MAYONNAISE TYPE	1-5/8 lbs	3-1/4 cup	
BREAD,WHEAT	12-1/2 lbs	200 sl	
LETTUCE,LEAF,FRESH,HEAD	4 lbs		6-1/4 lbs

Method

1. Slice beef into thin slices, about 16 to 22 slices per pound.
2. Combine mustard and salad dressing; blend well.
3. Spread 1 slice of bread with 2 teaspoons dressing mixture. Place 3 ounces or 3 to 4 slices, beef on bread. Top with lettuce if desired, and second slice of bread.
4. Cut each sandwich in half. CCP: Hold at 41 F. or lower until ready to serve.

Notes

1. 100 Crossiants may be substituted for wheat bread.

SANDWICHES No.N 004 01

ROAST PORK SANDWICH

Yield 100   **Portion** 1 Sandwich

Calories	Carbohydrates	Protein	Fat	Cholesterol	Sodium	Calcium
411 cal	31 g	29 g	19 g	72 mg	451 mg	86 mg

Ingredient	Weight	Measure	Issue
PORK,LOIN,BONELESS,COOKED	18-3/4 lbs		
MUSTARD,PREPARED	8-7/8 oz	1 cup	
SALAD DRESSING,MAYONNAISE TYPE	1-5/8 lbs	3-1/4 cup	
BREAD,WHEAT,SLICED	12-1/2 lbs	200 sl	
LETTUCE,ICEBERG,FRESH	4 lbs		4-1/3 lbs

Method

1. Slice meat into thin slices.
2. Combine mustard and salad dressing; blend well.
3. Spread 1 slice bread with 2 teaspoons salad dressing mixture. Place 3 ounces or 3 to 4 slices, meat on bread; top with second slice of bread. Add lettuce if desired.
4. Cut each sandwich in half. CCP: Hold at 41 F. or lower until ready to serve.

SANDWICHES No.N 004 02

ROAST TURKEY SANDWICH

Yield 100   **Portion** 1 Sandwich

Calories	Carbohydrates	Protein	Fat	Cholesterol	Sodium	Calcium
343 cal	33 g	23 g	13 g	55 mg	959 mg	109 mg

Ingredient

Ingredient	Weight	Measure	Issue
TURKEY,BNLS,WHITE AND DARK MEAT	21 lbs		
MUSTARD,PREPARED	8-7/8 oz	1 cup	
SALAD DRESSING,MAYONNAISE TYPE	1-5/8 lbs	3-1/4 cup	
BREAD,WHEAT,SLICED	12-1/2 lbs	200 sl	
LETTUCE,LEAF,FRESH,HEAD	4 lbs		6-1/4 lbs

Method

1. Slice turkey into thin slices, 16 to 22 slices per pound.
2. Combine mustard and salad dressing; blend well.
3. Spread 1 slice of bread with 2 teaspoons salad dressing mixture. Place 3 ounces or 3 to 4 slices, meat on bread; top with second slice of bread. Add lettuce if desired.
4. Cut each sandwich in half. CCP: Hold at 41 F. or lower until ready to serve.

SANDWICHES No.N 004 05

TURKEY CROISSANT

Yield 100 **Portion** 1 Sandwich

Calories	Carbohydrates	Protein	Fat	Cholesterol	Sodium	Calcium
414 cal	30 g	22 g	22 g	93 mg	1056 mg	57 mg

Ingredient	Weight	Measure	Issue
TURKEY,BNLS,WHITE AND DARK MEAT	21 lbs		
MUSTARD,PREPARED	8-7/8 oz	1 cup	
SALAD DRESSING,MAYONNAISE TYPE	1-5/8 lbs	3-1/4 cup	
CROISSANT,HALVED	12-5/8 lbs	100 each	
LETTUCE,ICEBERG,FRESH	4 lbs		4-1/3 lbs

Method

1. Slice turkey into thin slices, 16 to 22 slices per pound.
2. Combine mustard and salad dressing; blend well.
3. Slice croissants in half. Overlap croissants on sheet pans. Bake until crisp in 300 F. convection oven, about 3 minutes on high fan, open vent. Remove from oven.
4. Spread bottom half of each hot croissant with 2 teaspoons salad dressing mixture. Place 3 ounces, 3 to 4 slices meat on each croissant. Top with other half of croissant. Add lettuce if desired. CCP: Hold at 41 F. or lower for service.

SANDWICHES No.N 005 00

SAUSAGE AND BISCUIT

Yield 100   **Portion** 1 Sandwich

Calories	Carbohydrates	Protein	Fat	Cholesterol	Sodium	Calcium
242 cal	24 g	9 g	12 g	22 mg	690 mg	123 mg

Ingredient	Weight	Measure	Issue
BAKING POWDER BISCUITS		100 each	
SAUSAGE PATTY,PORK,RAW,2 OZ	5-7/8 lbs	100 each	

Method

1. Prepare Baking Powder Biscuits, Recipe No. D 001 00 or D 001 01. Split biscuits in half. Keep hot for use in Step 3.
2. Place 25 sausage patties on each sheet pan. Using a convection oven, bake uncovered at 325 F. for 7 minutes or until done on low fan, open vent. Drain. CCP: Internal temperature must reach 145 F. or higher for 15 seconds.
3. Place 1 patty on bottom of each split biscuit. Add top biscuit. CCP: Hold for service at 140 F. or higher.

SANDWICHES No.N 005 01

HAM AND BISCUIT

Yield 100   **Portion** 1 Sandwich

Calories	Carbohydrates	Protein	Fat	Cholesterol	Sodium	Calcium
196 cal	24 g	9 g	7 g	14 mg	707 mg	117 mg

Ingredient   Weight   Measure   Issue

Ingredient	Weight	Measure	Issue
BAKING POWDER BISCUITS		100 each	
HAM, COOKED, BONELESS	6-1/4 lbs		
COOKING SPRAY, NONSTICK	2 oz	1/4 cup 1/3 tbsp	

Method

1. Prepare Baking Powder Biscuits, Recipe No. D 001 00 or D 001 01. Split biscuits in half. Keep hot for use in Step 3.
2. Cut ham into 1-ounce slices. Grill on lightly sprayed 350 F. griddle until lightly browned. CCP: Internal temperature must reach 145 F. or higher for 15 seconds.
3. Place 1 slice of ham on bottom of each split biscuit. Add top biscuit. CCP: Hold for service at 140 F. or higher.

SANDWICHES No.N 006 00

GRILLED CHEESE SANDWICH

Yield 100   Portion   1 Sandwich

Calories	Carbohydrates	Protein	Fat	Cholesterol	Sodium	Calcium
411 cal	26 g	17 g	27 g	74 mg	713 mg	405 mg

Ingredient	Weight	Measure	Issue
CHEESE,AMERICAN,SLICED	12-1/2 lbs	200 sl	
BREAD,WHITE	11 lbs	200 sl	
BUTTER,MELTED	2 lbs	1 qts	

Method

1 Place 2 slices cheese between 2 slices bread.
2 Brush lightly top and bottom of sandwiches with butter or margarine.
3 Grill on 400 F. griddle until sandwiches are lightly browned on each side and cheese is melted.
4 Cut each sandwich in half. Serve hot. CCP: Hold at 140 F. or higher for service.

Notes

1 In Step 3, sandwiches may be browned in a convection oven at 425 F. for 5 minutes.

SANDWICHES No.N 006 01

GERMAN STYLE HAMWICH

Yield 100   Portion   1 Sandwich

Calories	Carbohydrates	Protein	Fat	Cholesterol	Sodium	Calcium
418 cal	26 g	16 g	28 g	61 mg	1032 mg	244 mg

Ingredient	Weight	Measure	Issue
MARGARINE,SOFTENED	2 lbs	1 qts	
MUSTARD,PREPARED	1-1/8 lbs	2 cup	
HORSERADISH,PREPARED	1 oz	2 tbsp	
POPPY SEEDS	7/8 oz	3 tbsp	
ONIONS,FRESH,CHOPPED	3-3/4 oz	1/2 cup 2-2/3 tbsp	4-1/8 oz
CHEESE,AMERICAN,SLICED	6-1/4 lbs	100 sl	
HAM,COOKED,BONELESS	6-1/4 lbs		
BREAD,WHITE	11 lbs	200 sl	
BUTTER,MELTED	2 lbs	1 qts	

Method

1. Combine softened margarine, mustard, horseradish, poppy seed, and onions; mix well.
2. Spread 1 tablespoon filling on 1 slice bread. Slice ham into 100 slices. Place 1 slice ham and 1 slice of cheese over filling in each sandwich. Top with second slice bread.
3. Brush lightly top and bottom of sandwiches with butter or margarine.
4. Grill on 400 F. griddle until sandwiches are lightly browned on each side and cheese is melted. CCP: Internal temperature must reach 145 F. or higher for 15 seconds.
5. Cut each sandwich in half. Serve hot. CCP: Hold at 140 F. or higher for service.

Notes

1. In Step 4, hamwich may be browned in a 425 F. convection oven for 5 minutes.

SANDWICHES No.N 006 03

GRILLED HAM AND CHEESE SANDWICH

Yield 100   **Portion** 1 Sandwich

Calories	Carbohydrates	Protein	Fat	Cholesterol	Sodium	Calcium
391 cal	25 g	21 g	22 g	76 mg	1251 mg	235 mg

Ingredient

Ingredient	Weight	Measure	Issue
HAM,COOKED,BONELESS	12-1/2 lbs		
CHEESE,AMERICAN,SLICED	6-1/4 lbs	100 sl	
BREAD,WHITE	11 lbs	200 sl	
BUTTER,MELTED	2 lbs	1 qts	

Method

1. Slice ham into 100 slices. Place 1 slice cheese and 2 slices ham on each sandwich.
2. Brush lightly top and bottom of sandwiches with butter or margarine.
3. Grill until sandwiches are lightly browned on each side and cheese is melted on 400 F. griddle. CCP: Internal temperature must reach 145 F. or higher for 15 seconds.
4. Cut each sandwich in half. Serve hot. CCP: Hold at 140 F. or higher for service.

Notes

1. In Step 3, sandwiches may be oven toasted or browned in a 425 F. convection oven for 5 minutes.

SANDWICHES No.N 007 00

ENGLISH MUFFIN WITH BACON, EGG AND CHEESE

Yield 100 **Portion** 1 Sandwich

Calories	Carbohydrates	Protein	Fat	Cholesterol	Sodium	Calcium
345 cal	27 g	18 g	18 g	228 mg	601 mg	301 mg

Ingredient   **Weight**   **Measure**   **Issue**

BACON,SLICED,RAW   1-1/4 lbs   100 sl
ENGLISH MUFFINS,SPLIT OR CUT   12-5/8 lbs   100 each
EGGS,WHOLE,FROZEN   10 lbs   1 gal 2/3 qts
COOKING SPRAY,NONSTICK   2 oz   1/4 cup 1/3 tbsp
CHEESE,AMERICAN,SLICED   6-1/4 lbs   100 sl

Method

1. Cook bacon according to Recipe No. L 002 00 or L 002 02. Drain. Set aside for use in Step 4.
2. Place muffin halves on sheet pans in rows, 5 by 7; Using a convection oven, toast 2 to 3 minutes at 325 F. Set aside for use in Step 6.
3. Place thawed eggs into a bowl; ladle individual 1/4 cup portions of beaten eggs on 325 F. greased griddle. Cook until firm. CCP: Internal temperature must reach 145 F. or higher for 15 seconds.
4. Place 1 slice bacon on egg.
5. Place 1 slice cheese on top of bacon. Continue to cook until cheese melts.
6. Place 1 cheese and bacon-topped fried egg on bottom half of each split muffin. Top with second half of muffin. CCP: Hold for service at 140 F. or higher.

SANDWICHES No.N 007 01

ENGLISH MUFFIN WITH HAM, EGG, AND CHEESE

Yield 100   Portion   1 Sandwich

Calories	Carbohydrates	Protein	Fat	Cholesterol	Sodium	Calcium
355 cal	27 g	22 g	17 g	237 mg	871 mg	302 mg

Ingredient

Ingredient	Weight	Measure	Issue
HAM,COOKED,BONELESS	6-1/4 lbs		
ENGLISH MUFFINS,SPLIT OR CUT	12-5/8 lbs	100 each	
EGGS,WHOLE,FROZEN	10 lbs	1 gal 2/3 qts	
COOKING SPRAY,NONSTICK	2 oz	1/4 cup 1/3 tbsp	
CHEESE,AMERICAN,SLICED	6-1/4 lbs	100 sl	

Method

1. Grill 1-ounce slices of ham until lightly browned on 325 F. griddle.
2. Place muffin halves on sheet pans in rows, 5 by 7; Using a convection oven, toast 2 to 3 minutes at 325 F. Set aside for use in Step 6.
3. Place thawed eggs into a bowl; ladle individual 1/4 cup portions of beaten eggs on 325 F. greased griddle. Cook until firm. CCP: Internal temperature must reach 145 F. or higher for 15 seconds.
4. Place 1 slice ham on egg.
5. Place 1 slice cheese on top of ham. Continue to cook until cheese melts.
6. Place 1 cheese and ham-topped fried egg on bottom half of each split muffin. Top with second half of muffin. CCP: Hold for service at 140 F. or higher.

SANDWICHES No.N 007 02

ENGLISH MUFFIN WITH CANADIAN BACON, EGG, & CHEESE

Yield 100   Portion   1 Sandwich

Calories	Carbohydrates	Protein	Fat	Cholesterol	Sodium	Calcium
356 cal	28 g	22 g	17 g	237 mg	909 mg	303 mg

Ingredient

Ingredient	Weight	Measure	Issue
BACON,CANADIAN,SLICED,1 OZ	6-1/4 lbs	100 sl	
ENGLISH MUFFINS,SPLIT OR CUT	12-5/8 lbs	100 each	
EGGS,WHOLE,FROZEN	10 lbs	1 gal 2/3 qts	
COOKING SPRAY,NONSTICK	2 oz	1/4 cup 1/3 tbsp	
CHEESE,AMERICAN,SLICED	6-1/4 lbs	100 sl	

Method

1. Cook thawed Canadian bacon according to Recipe No. L 002 01 or L 002 03. Drain. Set aside for use in Step 4.
2. Place muffin halves on sheet pans in rows, 5 by 7; Using a convection oven, toast 2 to 3 minutes in 325 F. oven. Set aside for use in Step 6.
3. Place thawed eggs into a bowl; ladle individual 1/4 cup portions of beaten eggs on 325 F. greased griddle. Cook until firm. CCP: Internal temperature must reach 145 F. or higher for 15 seconds.
4. Place 1 slice Canadian bacon on egg.
5. Place 1 slice cheese on top of bacon. Continue to cook until cheese melts.
6. Place 1 cheese and bacon-topped fried egg on bottom half of each split muffin. Top with second half of muffin. CCP: Hold for service at 140 F. or higher.

SANDWICHES No.N 007 03

ENGLISH MUFFIN WITH SAUSAGE, EGG, AND CHEESE

Yield 100   Portion   1 Sandwich

Calories	Carbohydrates	Protein	Fat	Cholesterol	Sodium	Calcium
459 cal	28 g	24 g	27 g	256 mg	1026 mg	313 mg

Ingredient	Weight	Measure	Issue
SAUSAGE PATTY,PORK,RAW,2 OZ	18-3/4 lbs		
ENGLISH MUFFINS,SPLIT OR CUT	12-5/8 lbs	100 each	
EGGS,WHOLE,FROZEN	10 lbs	1 gal 2/3 qts	
COOKING SPRAY,NONSTICK	2 oz	1/4 cup 1/3 tbsp	
CHEESE,AMERICAN,SLICED	6-1/4 lbs	100 sl	

Method

1. Cook sausage patties according to instructions on package. CCP: Internal temperature must reach 145 F. or higher for 15 seconds.
2. Place muffin halves on sheet pans in rows, 5 by 7; Using a convection oven, toast 2 to 3 minutes at 325 F. Set aside for use in Step 6.
3. Place thawed eggs into a small bowl; pour on 325 F. greased griddle. Fry 2 minutes; turn. CCP: Internal temperature must reach 145 F. or higher for 15 seconds.
4. Place 1 sausage patty on egg.
5. Place 1 slice cheese on top of sausage patty. Continue to cook until cheese melts.
6. Place 1 cheese and sausage-topped fried egg on bottom half of each split muffin. Top with second half of muffin. CCP: Hold for service at 140 F. or higher.

SANDWICHES No.N 008 00

CHICKEN SALAD SANDWICH

Yield 100   Portion   1 Sandwich

Calories	Carbohydrates	Protein	Fat	Cholesterol	Sodium	Calcium
364 cal	30 g	28 g	14 g	78 mg	650 mg	102 mg

Ingredient	Weight	Measure	Issue
CHICKEN,COOKED,DICED	18 lbs		
CELERY,FRESH,CHOPPED	12 lbs	2 gal 3-1/3 qts	16-1/2 lbs
SALAD DRESSING,MAYONNAISE TYPE	2-3/4 lbs	1 qts 1-1/2 cup	
ONIONS,FRESH,CHOPPED	14 oz	2-1/2 cup	1 lbs
JUICE,LEMON	8-5/8 oz	1 cup	
SALT	1-2/3 oz	2-2/3 tbsp	
PEPPER,BLACK,GROUND	1/4 oz	1 tbsp	
BREAD,WHITE	11 lbs	200 sl	
LETTUCE,LEAF,FRESH,HEAD	4 lbs		6-1/4 lbs

Method

1 Combine chicken, celery, salad dressing, onions, lemon juice, salt and pepper. Mix lightly but thoroughly.
2 Spread 1 slice bread with 3/4 cup filling; top with lettuce and second slice of bread.
3 Cut each sandwich in half. CCP: Cover and refrigerate sandwiches at 41 F. or lower.

SANDWICHES No.N 008 01

CHICKEN SALAD SANDWICH (CANNED CHICKEN)

Yield 100   Portion   1 Sandwich

Calories	Carbohydrates	Protein	Fat	Cholesterol	Sodium	Calcium
385 cal	31 g	26 g	17 g	65 mg	901 mg	107 mg

Ingredient	Weight	Measure	Issue
CHICKEN,BONED,CANNED,PIECES	23-1/4 lbs	2 gal 1-1/8 qts	
CELERY,FRESH,CHOPPED	14-1/2 lbs	3 gal 1-3/4 qts	19-7/8 lbs
SALAD DRESSING,MAYONNAISE TYPE	3-1/2 lbs	1 qts 3 cup	
ONIONS,FRESH,CHOPPED	1 lbs	2-7/8 cup	1-1/8 lbs
JUICE,LEMON	8-5/8 oz	1 cup	
PEPPER,BLACK,GROUND	1/3 oz	1 tbsp	
BREAD,WHITE	11 lbs	200 sl	
LETTUCE,LEAF,FRESH,CHOPPED	4 lbs	2 gal 1/8 qts	6-1/4 lbs

Method

1 Drain chicken. Cut chicken into 1 inch pieces.
2 Combine chicken, celery, salad dressing, onions, lemon juice and pepper. Mix lightly but thoroughly.
3 Spread 1 slice bread with 3/4 cup filling; top with lettuce and second slice of bread.
4 Cut each sandwich in half. CCP: Cover and refrigerate sandwiches at 41 F. or lower.

SANDWICHES No.N 008 02

TURKEY SALAD SANDWICH

Yield 100   **Portion** 1 Sandwich

Calories	Carbohydrates	Protein	Fat	Cholesterol	Sodium	Calcium
331 cal	31 g	20 g	14 g	50 mg	1058 mg	116 mg

Ingredient	Weight	Measure	Issue
TURKEY,BNLS,WHITE AND DARK MEAT,DICED	18 lbs		
CELERY,FRESH,CHOPPED	12 lbs	2 gal 3-1/3 qts	16-1/2 lbs
SALAD DRESSING,MAYONNAISE TYPE	2-3/4 lbs	1 qts 1-1/2 cup	
ONIONS,FRESH,CHOPPED	14 oz	2-1/2 cup	1 lbs
JUICE,LEMON	8-5/8 oz	1 cup	
SALT	1-2/3 oz	2-2/3 tbsp	
PEPPER,BLACK,GROUND	1/4 oz	1 tbsp	
BREAD,WHITE	11 lbs	200 sl	
LETTUCE,LEAF,FRESH,HEAD	4 lbs		6-1/4 lbs

Method

1. Combine turkey, celery, salad dressing, onions, lemon juice, salt and pepper. Mix lightly but thoroughly.
2. Spread 1 slice bread with 3/4 cup filling; top with lettuce and second slice of bread.
3. Cut each sandwich in half. CCP: Cover and refrigerate sandwiches at 41 F. or lower.

SANDWICHES No. N 009 00

CORNED BEEF SANDWICH

Yield 100   **Portion** 1 Sandwich

Calories	Carbohydrates	Protein	Fat	Cholesterol	Sodium	Calcium
358 cal	33 g	19 g	16 g	71 mg	1359 mg	73 mg

Ingredient	**Weight**	**Measure**	**Issue**
BEEF,CORNED,COOKED,SLICED	16 lbs		
BREAD,RYE,SLICE	14-1/8 lbs	200 sl	
MUSTARD,PREPARED	2-1/4 lbs	1 qts	
LETTUCE,LEAF,FRESH,HEAD	4 lbs		6-1/4 lbs

Method

1. Slice corned beef across the grain into 1/16-inch slices.
2. Spread 1 slice bread with 2 teaspoons mustard. Place 3 to 4 slices corned beef on bread; top with lettuce and second slice bread.
3. Cut each sandwich in half. CCP: Hold for service at 41 F. or lower.

SANDWICHES No.N 009 01

CORNED BEEF AND CHEESE SANDWICH

Yield 100   Portion 1 Sandwich

Calories	Carbohydrates	Protein	Fat	Cholesterol	Sodium	Calcium
464 cal	34 g	27 g	24 g	97 mg	1433 mg	345 mg

Ingredient	Weight	Measure	Issue
BEEF,CORNED,COOKED,SLICED	16 lbs		
CHEESE,SWISS,SLICED	6-1/4 lbs	100 sl	
BREAD,RYE,SLICE	14-1/8 lbs	200 sl	
MUSTARD,PREPARED	2-1/4 lbs	1 qts	
LETTUCE,LEAF,FRESH,HEAD	4 lbs		6-1/4 lbs

Method

1. Slice corned beef across the grain 1/16-inch slices.
2. Spread 1 slice bread with 2 teaspoons mustard. Place 2 to 3 slices corned beef and 1 slice cheese on bread; top with lettuce and second slice bread.
3. Cut each sandwich in half. CCP: Hold for service at 41 F. or lower.

SANDWICHES No.N 010 00

EGG SALAD SANDWICH

Yield 100   **Portion** 1 Sandwich

Calories	Carbohydrates	Protein	Fat	Cholesterol	Sodium	Calcium
335 cal	32 g	14 g	17 g	323 mg	578 mg	108 mg

Ingredient	Weight	Measure	Issue
EGG,HARD COOKED	16-1/2 lbs	150 Eggs	
ONIONS,FRESH,CHOPPED	12-2/3 oz	2-1/4 cup	14-1/8 oz
PICKLES,CUCUMBER,SWEET,CHOPPED	2-1/8 lbs	1 qts 1-5/8 cup	
MUSTARD,PREPARED	13-1/4 oz	1-1/2 cup	
SALAD DRESSING,MAYONNAISE TYPE	3 lbs	1 qts 2 cup	
BREAD,WHITE	11 lbs	200 sl	
LETTUCE,LEAF,FRESH,HEAD	4 lbs		6-1/4 lbs

Method

1 Cook eggs according to Recipe No. F 004 00. Cool. Shell; finely chop eggs.
2 Combine eggs, onions, pickles, mustard, and salad dressing; mix together lightly.
3 Spread 1 slice bread with 1/2 cup egg salad; top with lettuce if desired and second slice bread.
4 Cut each sandwich in half. CCP: Hold for service at 41 F. or lower.

SANDWICHES No.N 010 01

NEW YORK EGG SALAD SANDWICH (EGG AND TOMATO)

Yield 100   Portion   1 Sandwich

Calories	Carbohydrates	Protein	Fat	Cholesterol	Sodium	Calcium
345 cal	34 g	14 g	17 g	323 mg	582 mg	110 mg

Ingredient	Weight	Measure	Issue
EGG,HARD COOKED,CHOPPED	16-1/2 lbs	150 Eggs	
PICKLES,CUCUMBER,SWEET,CHOPPED	2-1/8 lbs	1 qts 1-5/8 cup	
MUSTARD,PREPARED	13-1/4 oz	1-1/2 cup	
SALAD DRESSING,MAYONNAISE TYPE	3 lbs	1 qts 2 cup	
TOMATOES,FRESH,SLICED	11-1/8 lbs	1 gal 3 qts	11-1/3 lbs
BREAD,WHITE	11 lbs	200 sl	
LETTUCE,LEAF,FRESH,HEAD	4 lbs		6-1/4 lbs

Method

1 Cook eggs according to Recipe No. F 004 00. Cool. Shell; finely chop eggs.
2 Combine eggs, pickles, mustard, and salad dressing; mix together lightly.
3 Spread 1 slice bread with 1/2 cup egg salad; top each with 2 slices tomato, lettuce if desired, and second slice bread.
4 Cut each sandwich in half. CCP: Hold for service at 41 F. or lower.

SANDWICHES No.N 011 00

HAM SANDWICH

Yield 100   Portion   1 Sandwich

Calories	Carbohydrates	Protein	Fat	Cholesterol	Sodium	Calcium
304 cal	32 g	22 g	9 g	43 mg	1620 mg	73 mg

Ingredient	Weight	Measure	Issue
HAM,COOKED,BONELESS	18-3/4 lbs		
BREAD,RYE,SLICE	14-1/8 lbs	200 sl	
MUSTARD,PREPARED	2-1/4 lbs	1 qts	
LETTUCE,LEAF,FRESH,HEAD	4 lbs		6-1/4 lbs

Method

1 Slice ham into thin slices, 20 to 24 slices per pound.
2 Spread 1 slice bread with mustard. Place 3 ounces or 3 slices of ham on bread; top with lettuce and second slice of bread.
3 Cut each sandwich in half. CCP: Hold for service at 41 F. or lower.

SANDWICHES No.N 011 01

FRIED HAM SANDWICH

Yield 100   Portion   1 Sandwich

Calories	Carbohydrates	Protein	Fat	Cholesterol	Sodium	Calcium
344 cal	32 g	22 g	14 g	43 mg	1620 mg	73 mg

Ingredient

Ingredient	Weight	Measure	Issue
HAM,COOKED,BONELESS	18-3/4 lbs		
BREAD,RYE,SLICE	14-1/8 lbs	200 sl	
MUSTARD,PREPARED	2-1/4 lbs	1 qts	
LETTUCE,LEAF,FRESH,HEAD	4 lbs		6-1/4 lbs

Method

1 Slice ham into 1/4-inch slices, about 3 ounces each. Grill on lightly greased 350 F. griddle about 1 minute on each side until lightly browned.
2 Spread 1 slice bread with mustard. Place 3 ounces or 3 slices of ham on bread; top with lettuce and second slice of bread.
3 Cut each sandwich in half. Serve hot. CCP: Hold at 140 F. or higher for service.

SANDWICHES No.N 011 02

HAM AND CHEESE SANDWICH

Yield 100   **Portion** 1 Sandwich

Calories	Carbohydrates	Protein	Fat	Cholesterol	Sodium	Calcium
368 cal	33 g	25 g	15 g	54 mg	1333 mg	343 mg

Ingredient

Ingredient	Weight	Measure	Issue
HAM,COOKED,BONELESS	12-1/2 lbs		
CHEESE,SWISS,SLICED	6-1/4 lbs	100 sl	
BREAD,RYE,SLICE	14-1/8 lbs	200 sl	
MUSTARD,PREPARED	2-1/4 lbs	1 qts	
LETTUCE,LEAF,FRESH,HEAD	4 lbs		6-1/4 lbs

Method

1. Slice ham into thin slices, 20 to 24 slices per pound.
2. Spread 1 slice bread with mustard. Place 2 slices ham on bread. Place 1 slice cheese on top of ham. Top with lettuce and second slice bread.
3. Cut each sandwich in half. CCP: Hold for service at 41 F. or lower.

SANDWICHES No.N 011 03

HAM AND TOMATO SANDWICH

Yield 100   Portion   1 Sandwich

Calories	Carbohydrates	Protein	Fat	Cholesterol	Sodium	Calcium
315 cal	35 g	23 g	9 g	43 mg	1625 mg	75 mg

Ingredient

Ingredient	Weight	Measure	Issue
HAM,COOKED,BONELESS	18-3/4 lbs		
BREAD,RYE,SLICE	14-1/8 lbs	200 sl	
TOMATOES,FRESH,SLICED	11-1/8 lbs	1 gal 3 qts	11-1/3 lbs
MUSTARD,PREPARED	2-1/4 lbs	1 qts	
LETTUCE,LEAF,FRESH,HEAD	4 lbs		6-1/4 lbs

Method

1 Slice ham into thin slices, 20 to 24 slice per pound.
2 Spread 1 slice bread with mustard. Place 3 ounces ham on bread; top with lettuce, 2 slices tomato, and second slice of bread.
3 Cut each sandwich in half. CCP: Hold for service at 41 F. or lower until ready to serve.

SANDWICHES No.N 012 00

GRILLED HAMBURGER (BEEF PATTIES)

Yield 100   Portion 1 Burger

Calories	Carbohydrates	Protein	Fat	Cholesterol	Sodium	Calcium
294 cal	22 g	19 g	14 g	50 mg	277 mg	66 mg

Ingredient   Weight   Measure   Issue

BEEF PATTY,10% FAT,RAW,3 OZ   14 lbs   100 each
BUN,HAMBURGER   9-1/2 lbs   100 each

Method

1 Grill patties 4 minutes or until browned on 350 F. griddle. Turn; grill on other side for 4 minutes. CCP: Internal temperature must reach 155 F. or higher for 15 seconds.
2 Serve hot on buns. CCP: Hold for service at 140 F. or higher.

SANDWICHES No.N 012 01

CHEESEBURGER (BEEF PATTIES)

Yield 100   Portion   1 Burger

Calories	Carbohydrates	Protein	Fat	Cholesterol	Sodium	Calcium
400 cal	22 g	25 g	23 g	77 mg	461 mg	240 mg

Ingredient   Weight   Measure   Issue

BEEF PATTY,10% FAT,RAW,3 OZ   14 lbs   100 each
CHEESE,AMERICAN,SLICED   6-1/4 lbs   100 sl
BUN,HAMBURGER   9-1/2 lbs   100 each

Method

1 Grill patties 4 minutes on each side or until browned on 350 F. griddle. CCP: Internal temperature must reach 155 F. or higher for 15 seconds. Place 1 slice cheese on each patty. Continue to grill until cheese melts.
2 Serve hot on buns. CCP: Hold for service at 140 F. or higher.

SANDWICHES No.N 012 02

CHEESY BACONBURGER (BEEF PATTIES)

Yield 100   Portion   1 Burger

Calories	Carbohydrates	Protein	Fat	Cholesterol	Sodium	Calcium
444 cal	22 g	28 g	26 g	83 mg	582 mg	241 mg

Ingredient	Weight	Measure	Issue
BACON,SLICED,RAW	6 lbs		
BEEF PATTY,10% FAT,RAW,3 OZ	14 lbs	100 each	
CHEESE,AMERICAN,SLICED	6-1/4 lbs	100 sl	
BUN,HAMBURGER	9-1/2 lbs	100 each	

Method

1. Cook bacon according to directions on Recipe No. L 002 00 or L 002 02. Cut bacon in half.
2. Grill patties on 350 F. griddle for 4 minutes or until browned; turn; CCP: Internal temperature must reach 155 F. or higher for 15 seconds. Place 1 slice cheese on each patty, continue to grill until cheese melts. Top melted cheese with 2 half slices bacon.
3. Serve hot on buns. CCP: Hold for service at 140 F. or higher.

SANDWICHES No.N 012 03

DOUBLE DECKER CHEESEBURGER (BEEF PATTIES)

Yield 100 **Portion** 1 Burger

Calories	Carbohydrates	Protein	Fat	Cholesterol	Sodium	Calcium
683 cal	24 g	47 g	43 g	153 mg	863 mg	424 mg

Ingredient	**Weight**	**Measure**	**Issue**
CHEESE,AMERICAN,SLICED	12-1/2 lbs	200 sl	
BEEF PATTY,10% FAT,RAW,3 OZ	28-1/8 lbs	200 each	
LETTUCE,ICEBERG,FRESH,SHREDDED	2 lbs	1 gal	2-1/8 lbs
ONIONS,FRESH,CHOPPED	12-2/3 oz	2-1/4 cup	14-1/8 oz
PICKLES,DILL,SLICES	3-1/8 lbs	2 qts 1-1/8 cup	
BUN,HAMBURGER	9-1/2 lbs	100 each	

Method

1 Grill patties 4 minutes or until browned on 350 F. griddle. Turn; grill on other side for 4 minutes. CCP: Internal temperature must reach 155 F. or higher for 15 seconds. Place 1 slice of cheese on each patty. Continue to grill until cheese melts.

2 Place 1 tablespoon shredded lettuce and 1 slice pickle on bottom and middle bun slices. Place a cheeseburger on top of bottom and middle slices of bun; place 1/2 teaspoon finely chopped onion on top of cheeseburger. Assemble layers; cover with top of bun. CCP: Hold for service at 140 F. or higher.

SANDWICHES No.N 012 04

CHILIBURGER (BEEF PATTIES)

Yield 100   Portion 1 Burger

Calories	Carbohydrates	Protein	Fat	Cholesterol	Sodium	Calcium
329 cal	25 g	22 g	15 g	53 mg	419 mg	75 mg

Ingredient	Weight	Measure	Issue
CHILI CON CARNE,CANNED,NO BEANS	6-3/4 lbs	3 qts	
BEEF PATTY,10% FAT,RAW,3 OZ	14 lbs	100 each	
BUN,HAMBURGER	9-1/2 lbs	100 each	

Method

1 Prepare canned chili con carne without beans. Grill patties 4 minutes or until browned on 350 F. griddle. Turn; grill on other side 4 minutes. CCP: Internal temperature must reach 155 F. or higher for 15 seconds.
2 Place hamburger patty on 1/2 of bun. Spread 2 tablespoons chili on each grilled hamburger. Cover with top bun.
3 Serve hot on buns. CCP: Hold for service at 140 F. or higher.

SANDWICHES No.N 012 05

DELUXE HAMBURGER (BEEF PATTIES)

Yield 100   **Portion**   1 Burger

Calories	Carbohydrates	Protein	Fat	Cholesterol	Sodium	Calcium
329 cal	30 g	20 g	14 g	50 mg	700 mg	81 mg

Ingredient	Weight	Measure	Issue
BEEF PATTY,10% FAT,RAW,3 OZ	14 lbs	100 each	
LETTUCE,ICEBERG,FRESH,SHREDDED	3 lbs	1 gal 2-1/4 qts	3-1/4 lbs
TOMATOES,FRESH,SLICED	6-1/2 lbs	1 gal 1/8 qts	6-5/8 lbs
ONIONS,FRESH,SLICED	3 lbs	2 qts 3-7/8 cup	3-1/3 lbs
PICKLES,DILL CHIPS	3-1/8 lbs	1 qts 2-1/4 cup	
CATSUP	3-1/3 lbs	1 qts 2-1/4 cup	
MUSTARD,PREPARED	1-1/8 lbs	2-1/8 cup	
BUN,HAMBURGER	9-1/2 lbs	100 each	

Method

1 Grill patties on 350 F. griddle 4 minutes or until browned. Turn; grill on other side 4 minutes. CCP: Internal temperature must reach 155 F. or higher for 15 seconds.

2 On each burger, spread 1 teaspoon mustard on bottom bun and 1 tablespoon catsup on top bun. Add 2 slices onion, 1 lettuce leaf, 1 slice tomato, and 2 slices pickle on bottom bun. Place grilled hamburger on bottom bun. Cover with top bun. CCP: Hold for service at 140 F. or higher.

SANDWICHES No.N 012 06

PIZZABURGER (BEEF PATTIES)

Yield 100   **Portion** 1 Burger

Calories	Carbohydrates	Protein	Fat	Cholesterol	Sodium	Calcium
389 cal	27 g	25 g	20 g	70 mg	464 mg	216 mg

Ingredient	Weight	Measure	Issue
CHEESE,MOZZARELLA,SLICED	5-1/4 lbs	1 gal 1-1/4 qts	
SAUCE,PIZZA,CANNED	11-7/8 lbs	1-1/2 #10cn	
BEEF PATTY,10% FAT,RAW,3 OZ	14 lbs	100 each	
BUN,HAMBURGER	9-1/2 lbs	100 each	

Method

1. Grill patties on 350 F. griddle 4 minutes or until browned. Turn; grill on other side 4 minutes. CCP: Internal temperature must reach 155 F. or higher for 15 seconds. Place 1 slice cheese on each patty. Continue to grill until cheese melts. Heat sauce to boiling.
2. Spread 3 tbsp pizza sauce on each hamburger. Cover with top bun. CCP: Hold for service at 140 F. or higher.

SANDWICHES No.N 012 07

DELUXE CHEESEBURGER (BEEF PATTIES)

Yield 100   **Portion** 1 Burger

Calories	Carbohydrates	Protein	Fat	Cholesterol	Sodium	Calcium
435 cal	30 g	26 g	23 g	77 mg	884 mg	255 mg

Ingredient	**Weight**	**Measure**	**Issue**
BEEF PATTY,10% FAT,RAW,3 OZ	14 lbs	100 each	
CHEESE,AMERICAN,SLICED	6-1/4 lbs	100 sl	
LETTUCE,ICEBERG,FRESH,SHREDDED	3 lbs	1 gal 2-1/4 qts	3-1/4 lbs
TOMATOES,FRESH,SLICED	6-1/2 lbs	1 gal 1/8 qts	6-5/8 lbs
ONIONS,FRESH,SLICED	3 lbs	2 qts 3-7/8 cup	3-1/3 lbs
PICKLES,DILL,SLICES	3-1/8 lbs	2 qts 1-1/8 cup	
CATSUP	3-1/3 lbs	1 qts 2-1/4 cup	
MUSTARD,PREPARED	1-1/8 lbs	2-1/8 cup	
BUN,HAMBURGER	9-1/2 lbs	100 each	

Method

1. Grill patties on 350 F. griddle 4 minutes or until browned. Turn and grill on the other side for 4 minutes. CCP: Internal temperature must reach 155 F. or higher for 15 seconds. Place 1 slice cheese on each patty, continue to grill until cheese melts.
2. Spread 1 tsp mustard on bottom bun and 1 tbsp catsup on top bun. Add 2 slices onion, 1 lettuce leaf, 1 slice tomato and 2 slices pickle on bottom bun. Place grilled cheeseburger on bottom bun; cover with top bun. Serve hot on buns. CCP: Hold for service at 140 F. or higher.

SANDWICHES No.N 013 00

HAM SALAD SANDWICH

Yield 100   Portion   1 Sandwich

Calories	Carbohydrates	Protein	Fat	Cholesterol	Sodium	Calcium
347 cal	31 g	20 g	16 g	116 mg	1338 mg	80 mg

Ingredient	Weight	Measure	Issue
PORK,HAM,CURED,GROUND	15 lbs	2 gal 2-1/3 qts	
EGG,HARD COOKED,CHOPPED	4 lbs	36 Eggs	
PICKLES,CUCUMBER,SWEET,CHOPPED	2-1/4 lbs	1 qts 2 cup	
SALAD DRESSING,MAYONNAISE TYPE	3 lbs	1 qts 2 cup	
BREAD,WHITE	11 lbs	200 sl	
LETTUCE,LEAF,FRESH,HEAD	4 lbs		6-1/4 lbs

Method

1 Combine cooked ham, eggs, pickles and Salad Dressing; mix together lightly.
2 Spread one slice of bread with 3/4 cup of ham salad. Top with lettuce if desired, and second slice of bread.
3 Cut each sandwich in half. CCP: Hold for service at 41 F. or lower.

SANDWICHES No.N 014 00

PEANUT BUTTER AND JELLY SANDWICH

Yield 100 **Portion** 1 Sandwich

Calories	Carbohydrates	Protein	Fat	Cholesterol	Sodium	Calcium
376 cal	44 g	12 g	18 g	1 mg	426 mg	68 mg

Ingredient	**Weight**	**Measure**	**Issue**
BREAD,WHITE	11 lbs	200 sl	
PEANUT BUTTER	7-1/8 lbs	3 qts 1/2 cup	
JELLY,GRAPE	4-1/8 lbs	1 qts 2-1/4 cup	

Method

1 Spread each slice of bread with 1 tablespoon peanut butter. Spread 1 slice bread with 1 tablespoon jelly. Top with second slice.
2 Cut each sandwich in half.

Notes

1 In Step 1, jam may be used.

SANDWICHES No.N 015 00

TUNA SALAD SANDWICH

Yield 100 **Portion** 1 Sandwich

Calories	Carbohydrates	Protein	Fat	Cholesterol	Sodium	Calcium
339 cal	35 g	20 g	13 g	102 mg	702 mg	98 mg

Ingredient	Weight	Measure	Issue
FISH,TUNA,CANNED,WATER PACK,INCL LIQUIDS	10-7/8 lbs	2 gal	
CELERY,FRESH,CHOPPED	8 lbs	1 gal 3-5/8 qts	11 lbs
ONIONS,FRESH,CHOPPED	1-3/8 lbs	1 qts	1-5/8 lbs
PICKLE RELISH,SWEET,DRAINED	2-2/3 lbs	1 qts 1 cup	
SALAD DRESSING,MAYONNAISE TYPE	4 lbs	2 qts	
PEPPER,BLACK,GROUND	1/3 oz	1 tbsp	
JUICE,LEMON	1-1/4 lbs	2-3/8 cup	
EGG,HARD COOKED,CHOPPED	4-1/4 lbs	38 Eggs	
BREAD,WHITE	11 lbs	200 each	
LETTUCE,LEAF,FRESH,HEAD	4 lbs		6-1/4 lbs

Method

1. Drain and flake tuna.
2. Combine tuna, celery and onions. Mix lightly and thoroughly.
3. Combine salad dressing, pickle relish, lemon juice and pepper. Stir to blend thoroughly.
4. Add chopped eggs and salad dressing mixture to tuna mixture. Mix lightly.
5. Spread 1-slice bread with 3/4 cup tuna salad. Top with lettuce if desired and second slice of bread. Cut each sandwich in half. CCP: Refrigerate product at 41 F. or lower until ready to serve.

SANDWICHES No.N 015 01

GRILLED TUNA AND CHEESE SANDWICH

Yield 100 **Portion** 1 Sandwich

Calories	Carbohydrates	Protein	Fat	Cholesterol	Sodium	Calcium
470 cal	33 g	23 g	27 g	48 mg	902 mg	243 mg

Ingredient	Weight	Measure	Issue
FISH,TUNA,CANNED,WATER PACK,INCL LIQUIDS	10-7/8 lbs	2 gal	
ONIONS,FRESH,CHOPPED	2 lbs	1 qts 1-5/8 cup	2-1/4 lbs
CELERY,FRESH,CHOPPED	2 lbs	1 qts 3-1/2 cup	2-3/4 lbs
PICKLE RELISH,SWEET,DRAINED	2-1/8 lbs	1 qts	
JUICE,LEMON	1-1/8 lbs	2 cup	
SALAD DRESSING,MAYONNAISE TYPE	4 lbs	2 qts	
BREAD,WHITE	11 lbs	200 sl	
CHEESE,AMERICAN,SLICED	6-1/4 lbs	100 sl	
MARGARINE	2 lbs	1 qts	

Method

1 Drain and flake tuna.
2 Combine tuna, onions, celery, relish, lemon juice, and salad dressing. Mix together lightly.
3 Spread 1-slice bread with 3/4 cup tuna filling. Top each with 1 slice cheese and second slice of bread.
4 Brush top and bottom of sandwiches lightly with melted margarine. Grill on 400 F. griddle until bread is golden brown on each side and cheese is melted. Serve hot. CCP: Hold at 140 F. or higher for service.

SANDWICHES No.N 015 02

SALMON SALAD SANDWICH (CANNED SALMON)

Yield 100   **Portion** 1 Sandwich

Calories	Carbohydrates	Protein	Fat	Cholesterol	Sodium	Calcium
397 cal	35 g	24 g	18 g	118 mg	601 mg	288 mg

Ingredient

Ingredient	Weight	Measure	Issue
SALMON,CANNED,PINK	17-1/4 lbs	2 gal 2-5/8 qts	
ONIONS,FRESH,CHOPPED	1-1/3 lbs	3-3/4 cup	1-1/2 lbs
CELERY,FRESH,CHOPPED	8 lbs	1 gal 3-5/8 qts	11 lbs
PICKLE RELISH,SWEET,DRAINED	2-2/3 lbs	1 qts 1 cup	
JUICE,LEMON	1-1/4 lbs	2-3/8 cup	
PEPPER,BLACK,GROUND	1/3 oz	1 tbsp	
SALAD DRESSING,MAYONNAISE TYPE	4-1/4 lbs	2 qts 1/2 cup	
EGG,HARD COOKED,CHOPPED	4-1/4 lbs	38 Eggs	
BREAD,WHITE	11 lbs	200 each	
LETTUCE,LEAF,FRESH,HEAD	4 lbs		6-1/4 lbs

Method

1. Remove and discard skin and bones from salmon. Flake salmon.
2. Combine salmon, onions and celery. Mix lightly but thoroughly.
3. Combine salad dressing, pickle relish, lemon juice and pepper. Stir to blend thoroughly.
4. Add chopped eggs and salad dressing mixture to salmon mixture. Mix lightly.
5. Spread 1 slice bread with 3/4 cups salmon salad. Top with lettuce if desired and second slice of bread; cover. Cut each sandwich in half. CCP: Refrigerate product at 41 F. or lower until ready to serve.

SANDWICHES No.N 015 03

TUNA AND TOMATO SANDWICH

Yield 100   Portion   1 Sandwich

Calories	Carbohydrates	Protein	Fat	Cholesterol	Sodium	Calcium
342 cal	35 g	20 g	14 g	102 mg	663 mg	90 mg

Ingredient	Weight	Measure	Issue
FISH,TUNA,CANNED,WATER PACK,INCL LIQUIDS	10-7/8 lbs	2 gal	
ONIONS,FRESH,CHOPPED	2 lbs	1 qts 1-5/8 cup	2-1/4 lbs
CELERY,FRESH,CHOPPED	2 lbs	1 qts 3-1/2 cup	2-3/4 lbs
PICKLE RELISH,SWEET,DRAINED	2-1/8 lbs	1 qts	
JUICE,LEMON	1-1/8 lbs	2 cup	
SALAD DRESSING,MAYONNAISE TYPE	4 lbs	2 qts	
EGG,HARD COOKED,CHOPPED	4-1/4 lbs	38 Eggs	
BREAD,WHITE	11 lbs	200 sl	
LETTUCE,LEAF,FRESH,HEAD	4 lbs		6-1/4 lbs
TOMATOES,FRESH,SLICED	11-1/8 lbs	1 gal 3 qts	11-1/3 lbs

Method

1 Drain and flake tuna.
2 Combine tuna, onions, celery, relish, lemon juice, and salad dressing. Mix together lightly.
3 Add chopped eggs and salad dressing mixture to tuna mixture. Mix lightly.
4 Spread 1 slice bread with 3/4 cup tuna filling; top each with 2 slices tomato, lettuce if desired, and second slice of bread.
5 Cut each sandwich in half. CCP: Hold for service at 41 F. or lower.

SANDWICHES No.N 016 00

CREAM CHEESE BAGEL

Yield 100   **Portion** 1 Bagel

Calories	Carbohydrates	Protein	Fat	Cholesterol	Sodium	Calcium
346 cal	48 g	12 g	12 g	32 mg	561 mg	89 mg

Ingredient	Weight	Measure	Issue
CHEESE, CREAM	6-3/8 lbs	3 qts 1/2 cup	
BAGEL	19-5/8 lbs	100 each	

Method

1. Place cream cheese in mixer bowl; beat at medium speed 3 to 4 minutes.
2. Cut bagels in half. Place in rows, 5 by 7, on ungreased sheet pans. Using a convection oven, toast 3 to 4 minutes at 325 F. until warm but still soft.
3. Spread each bagel half with 1 tablespoon cream cheese.

SANDWICHES No.N 016 01

CREAM CHEESE AND TOMATO BAGEL

Yield 100   **Portion** 1 Bagel

Calories	Carbohydrates	Protein	Fat	Cholesterol	Sodium	Calcium
357 cal	51 g	12 g	12 g	32 mg	565 mg	92 mg

Ingredient | Weight | Measure | Issue

Ingredient	Weight	Measure	Issue
CHEESE,CREAM	6-3/8 lbs	3 qts 1/2 cup	
BAGEL	19-5/8 lbs	100 each	
TOMATOES,FRESH,SLICED	11-1/8 lbs	1 gal 3 qts	11-1/3 lbs

Method

1. Place cream cheese in mixer bowl; beat at medium speed 3 to 4 minutes.
2. Cut bagels in half. Place in rows, 5 by 7, on ungreased sheet pans. Using a convection oven, toast 3 to 4 minutes at 325 F. until warm but still soft.
3. Spread each bagel half with 1 tablespoon cream cheese. Place 1 slice tomato on each half.

SANDWICHES No.N 016 02

CREAM CHEESE AND OLIVE BAGEL

Yield 100   Portion 1 Bagel

Calories	Carbohydrates	Protein	Fat	Cholesterol	Sodium	Calcium
350 cal	49 g	12 g	12 g	32 mg	602 mg	93 mg

Ingredient	Weight	Measure	Issue
CHEESE,CREAM	6-3/8 lbs	3 qts 1/2 cup	
OLIVES,GREEN,STUFFED,CHOPPED	1 lbs	3-3/8 cup	
BAGEL	19-5/8 lbs	100 each	

Method

1. Place cream cheese in mixer bowl; beat at medium speed 3 to 4 minutes. Finely chop olives. Add to cream cheese; beat an additional 2 minutes.
2. Cut bagels in half. Place in rows, 5 by 7, on ungreased sheet pans. Using a convection oven, toast 3 to 4 minutes at 325 F. until warm but still soft.
3. Spread each bagel half with 1 tablespoon cream cheese.

SANDWICHES No.N 017 00

COLD CUT SANDWICH

Yield 100 **Portion** 1 Sandwich

Calories	Carbohydrates	Protein	Fat	Cholesterol	Sodium	Calcium
332 cal	31 g	17 g	16 g	44 mg	1104 mg	79 mg

Ingredient	**Weight**	**Measure**	**Issue**
BREAD,WHITE	11 lbs	200 sl	
SALAD DRESSING,MAYONNAISE TYPE	2 lbs	1 qts	
TURKEY,BNLS,WHITE AND DARK MEAT	3 lbs		
HAM,COOKED,1 OZ SLICE	5-1/3 lbs	100 sl	
SALAMI,SLICED	6-1/4 lbs	100 sl	
LETTUCE,LEAF,FRESH,HEAD	4 lbs		6-1/4 lbs
TOMATOES,FRESH,SLICED	11-1/8 lbs	1 gal 3 qts	11-1/3 lbs
ONIONS,FRESH,SLICED	1-1/2 lbs	1 qts 2 cup	1-2/3 lbs

Method

1 Spread 1 slice of bread with salad dressing.
2 Add 3 slices of meat. Top with lettuce, 2 slices of tomato, sliced onions, and second slice bread.
3 Cut each sandwich in half. CCP: Hold for service at 41 F. or lower.

SANDWICHES No.N 017 01

COLD CUT SANDWICH WITH CHEESE

Yield 100   **Portion** 1 Sandwich

Calories	Carbohydrates	Protein	Fat	Cholesterol	Sodium	Calcium
365 cal	30 g	20 g	18 g	55 mg	1004 mg	217 mg

Ingredient

Ingredient	Weight	Measure	Issue
BREAD,WHITE	11 lbs	200 sl	
SALAD DRESSING,MAYONNAISE TYPE	2 lbs	1 qts	
TURKEY,BNLS,WHITE AND DARK MEAT	6 lbs		
HAM,COOKED,1 OZ SLICE	5-1/3 lbs	100 sl	
CHEESE,AMERICAN	5-1/4 lbs	1 gal 1-1/4 qts	
LETTUCE,ICEBERG,FRESH	4 lbs		4-1/3 lbs
TOMATOES,FRESH,SLICED	11-1/8 lbs	1 gal 3 qts	11-1/3 lbs
ONIONS,FRESH,SLICED	1-1/2 lbs	1 qts 2 cup	1-2/3 lbs

Method

1 Spread 1 slice of bread with salad dressing.
2 Add 2 slices meat. Top with 1 slice cheese, 2 slices tomato and sliced onions if desired. Top with second slice bread.
3 Cut each sandwich in half. CCP: Hold for service at 41 F. or lower.

SANDWICHES No.N 018 00

WESTERN SANDWICH (DENVER)

Yield 100   Portion 1 Sandwich

Calories	Carbohydrates	Protein	Fat	Cholesterol	Sodium	Calcium
245 cal	26 g	15 g	8 g	165 mg	750 mg	90 mg

Ingredient	Weight	Measure	Issue
PORK,HAM,CURED,CHOPPED	7-1/2 lbs		
EGGS,WHOLE,FROZEN	7-1/2 lbs	3 qts 2 cup	
ONIONS,FRESH,CHOPPED	1 lbs	3 cup	1-1/8 lbs
PEPPERS,GREEN,FRESH,CHOPPED	7-7/8 oz	1-1/2 cup	9-5/8 oz
COOKING SPRAY,NONSTICK	2 oz	1/4 cup 1/3 tbsp	
BREAD,WHITE,SLICE	11 lbs	200 sl	
LETTUCE,LEAF,FRESH,HEAD	4 lbs		6-1/4 lbs

Method

1 Combine ham, eggs, onions, and peppers; stir to mix well.
2 Pour 1/3 cup mixture on lightly sprayed griddle. Cook until both sides are lightly browned. CCP: Internal temperature must reach 145 F. or higher for 15 seconds.
3 Place omelet on 1 slice of bread; top with lettuce and second slice of bread.
4 Cut each sandwich in half. Serve hot. CCP: Hold for service at 140 F. or higher.

SANDWICHES No. N 019 00

SUBMARINE SANDWICH

Yield 100 **Portion** 1 Sandwich

Calories	Carbohydrates	Protein	Fat	Cholesterol	Sodium	Calcium
519 cal	26 g	33 g	31 g	90 mg	1642 mg	481 mg

Ingredient	Weight	Measure	Issue
ROLL,FRENCH	8-3/8 lbs	100 each	
SALAD DRESSING,MAYONNAISE TYPE	2 lbs	1 qts	
SALAMI,SLICED	6-1/4 lbs	100 sl	
HAM,COOKED,1 OZ SLICE	5-1/3 lbs	100 sl	
TURKEY,BNLS,WHITE AND DARK MEAT	6 lbs		
CHEESE,PROVOLONE	12-1/2 lbs	200 sl	
TOMATOES,FRESH,SLICED	11-1/8 lbs	1 gal 3 qts	11-1/3 lbs
LETTUCE,ICEBERG,FRESH	3 lbs		3-1/4 lbs

Method

1. Cut rolls in half lengthwise; spread each half with Salad Dressing.
2. Slice Provolone cheese. On bottom half of each roll, arrange 3 slices meat, 2 slices cheese and 2 slices tomato.
3. Sprinkle shredded lettuce on top.
4. Cover with top half of roll. CCP: Hold for service at 41 F. or lower.

SANDWICHES No. N 019 01

ITALIAN STYLE SUBMARINE

Yield 100   **Portion** 1 Sandwich

Calories	Carbohydrates	Protein	Fat	Cholesterol	Sodium	Calcium
537 cal	26 g	33 g	33 g	87 mg	1588 mg	491 mg

Ingredient	Weight	Measure	Issue
ROLL,FRENCH	8-3/8 lbs	100 each	
HAM,COOKED,1 OZ SLICE	5-1/3 lbs	100 sl	
TURKEY,BNLS,WHITE AND DARK MEAT	6 lbs		
SALAMI,SLICED	6-1/4 lbs	100 sl	
CHEESE,PROVOLONE	12-1/2 lbs	200 sl	
TOMATOES,FRESH,SLICED	11-1/8 lbs	1 gal 3 qts	11-1/3 lbs
LETTUCE,ICEBERG,FRESH,SHREDDED	3 lbs	1 gal 2-1/4 qts	3-1/4 lbs
OIL,OLIVE	1-3/8 lbs	3 cup	
VINEGAR,DISTILLED	8-1/3 oz	1 cup	
OREGANO,CRUSHED	1-5/8 oz	1/2 cup 2 tbsp	
ONIONS,FRESH,SLICED	2 lbs	1 qts 3-7/8 cup	2-1/4 lbs

Method

1. Cut rolls in half lengthwise; spread each half with salad dressing.
2. On bottom half of each roll, arrange 3 slices meat, 2 slices cheese and 2 slices tomato.
3. Sprinkle shredded lettuce on top. Mix oil and vinegar. Sprinkle over lettuce. If desired, add crushed oregano and thinly sliced onions.
4. Cover with top half of roll. CCP: Hold for service at 41 F. or lower.

SANDWICHES No.N 020 00

GRILLED REUBEN SANDWICH

Yield 100   **Portion** 1 Sandwich

Calories	Carbohydrates	Protein	Fat	Cholesterol	Sodium	Calcium
608 cal	37 g	27 g	39 g	130 mg	1699 mg	334 mg

Ingredient	Weight	Measure	Issue
BEEF,CORNED,COOKED	16 lbs		
THOUSAND ISLAND DRESSING		2 qts	
BREAD,RYE,SLICE	14-1/8 lbs	200 sl	
SAUERKRAUT,SHREDDED,CANNED,DRAINED	3-3/4 lbs	3 qts	
CHEESE,SWISS,SLICED	6-1/4 lbs	100 sl	
BUTTER,MELTED	2 lbs	1 qts	
COOKING SPRAY,NONSTICK	2 oz	1/4 cup 1/3 tbsp	

Method

1. Slice corned beef across the grain into 1/16-inch slices.
2. Spread each slice of bread with about 2 teaspoons Thousand Island dressing.
3. Place 3 to 6 slices corned beef on 1 slice bread, 2 tablespoons sauerkraut, and 1 slice cheese; top with second slice of bread.
4. Brush lightly outside of sandwich with melted butter.
5. Place sandwiches with cheese side up on lightly sprayed griddle. Grill 4 minutes or until lightly browned at 375 F.; turn. Grill 6 minutes or until lightly browned and cheese is melted.
6. Cut each sandwich in half. Serve hot. CCP: Hold for service at 140 F. or higher.

Notes

1. In Step 6, sandwiches may be baked in a 400 F. convection oven, for 10 minutes on high fan, closed vent.

SANDWICHES No.N 020 02

GRILLED PASTRAMI REUBEN SANDWICH

Yield 100   **Portion** 1 Sandwich

Calories	Carbohydrates	Protein	Fat	Cholesterol	Sodium	Calcium
679 cal	39 g	26 g	46 g	126 mg	1767 mg	335 mg

Ingredient	Weight	Measure	Issue
PASTRAMI,PRECOOKED	16 lbs		
THOUSAND ISLAND DRESSING		2 qts	
BREAD,RYE,SLICE	14-1/8 lbs	200 sl	
SAUERKRAUT,SHREDDED,CANNED,DRAINED	3-3/4 lbs	3 qts	
CHEESE,SWISS,SLICED	6-1/4 lbs	100 sl	
BUTTER,MELTED	2 lbs	1 qts	
COOKING SPRAY,NONSTICK	2 oz	1/4 cup 1/3 tbsp	

Method

1. Slice pastrami across the grain into thin slices.
2. Spread each slice bread with about 2 teaspoons dressing.
3. Place about 3 to 6 slices of meat on 1 slice bread, 2 tablespoons sauerkraut, and 1 slice cheese; top with second slice of bread.
4. Brush lightly outside of sandwich with melted butter.
5. Place sandwiches with cheese side up on lightly greased griddle at 375 F. Grill 4 minutes or until lightly browned; turn. Grill 6 minutes or until lightly browned and cheese is melted.
6. Cut each sandwich in half. Serve hot. CCP: Hold for service at 140 F. or higher.

SANDWICHES No.N 021 00

BAKED CHICKEN FILLET SANDWICH (BREADED)

Yield 100 **Portion** 1 Sandwich

Calories	Carbohydrates	Protein	Fat	Cholesterol	Sodium	Calcium
460 cal	36 g	18 g	27 g	40 mg	891 mg	98 mg

Ingredient

Ingredient	Weight	Measure	Issue
CHICKEN FILLET,BREADED,PRECOOKED,3 OZ	18-3/4 lbs		
SALAD DRESSING,MAYONNAISE TYPE	2-1/3 lbs	1 qts 5/8 cup	
BUN,HAMBURGER	9-1/2 lbs	100 each	
LETTUCE,LEAF,FRESH,HEAD	4 lbs		6-1/4 lbs
TOMATOES,FRESH,SLICED	6-1/2 lbs	1 gal 1/8 qts	6-5/8 lbs

Method

1. Place chicken fillets on pans.
2. Using a convection oven, bake at 375 F. 12 to 14 minutes or until thoroughly heated on high fan, closed vent. CCP: Internal temperature must reach 165 F. or higher for 15 seconds.
3. Place 1 fillet on bottom half of bun. Spread 2 teaspoons salad dressing on top half of bun (optional).
4. Place lettuce leaf and 1 tomato slice over fillet (optional). Cover with top half of bun. Serve hot. CCP: Hold for service at 140 F. or higher.

SANDWICHES No.N 021 01

DEEP FAT FRIED CHICKEN FILLET SANDWICH (BREADED)

Yield 100   Portion 5-1/2 Ounces

Calories	Carbohydrates	Protein	Fat	Cholesterol	Sodium	Calcium
460 cal	36 g	18 g	27 g	40 mg	891 mg	98 mg

Ingredient

Ingredient	Weight	Measure	Issue
CHICKEN FILLET,BREADED,PRECOOKED,3 OZ	18-3/4 lbs		
SALAD DRESSING,MAYONNAISE TYPE	2-1/3 lbs	1 qts 5/8 cup	
BUN,HAMBURGER	9-1/2 lbs	100 each	
LETTUCE,LEAF,FRESH,HEAD	4 lbs		6-1/4 lbs
TOMATOES,FRESH,SLICED	6-1/2 lbs	1 gal 1/8 qts	6-5/8 lbs

Method

1 Fry chicken fillets in 350 F. deep fat for 3 to 4 minutes or until heated. Drain in basket or on absorbent paper. CCP: Internal temperature must reach 165 F. or higher for 15 seconds.
2 Place 1 fillet on bottom half of bun. Spread 2 teaspoons salad dressing (optional) on top half of bun.
3 Place lettuce leaf and 1 tomato slice over fillet (optional). Cover with top half of bun. Serve hot. CCP: Hold for service at 140 F. or higher.

SANDWICHES No.N 021 02

CHICKEN FILLET AND CHEESE SANDWICH (BREADED)

Yield 100   Portion 1 Sandwich

Calories	Carbohydrates	Protein	Fat	Cholesterol	Sodium	Calcium
505 cal	36 g	21 g	31 g	52 mg	969 mg	172 mg

Ingredient	Weight	Measure	Issue
CHICKEN FILLET,BREADED,PRECOOKED,3 OZ	18-3/4 lbs		
CHEESE,AMERICAN,SLICED	2-5/8 lbs		
SALAD DRESSING,MAYONNAISE TYPE	2-1/3 lbs	1 qts 5/8 cup	
BUN,HAMBURGER	9-1/2 lbs	100 each	
LETTUCE,LEAF,FRESH,HEAD	4 lbs		6-1/4 lbs
TOMATOES,FRESH,SLICED	6-1/2 lbs	1 gal 1/8 qts	6-5/8 lbs

Method

1. Place chicken fillets on pans.
2. Using a convection oven, bake 12 to 14 minutes at 375 F. or until thoroughly heated on high fan, closed vent. CCP: Internal temperature must reach 165 F. or higher for 15 seconds.
3. Cut cheese slices in half. Place 1/2 slice cheese on top of each fillet. Return to oven; heat 1 minute or until cheese begins to melt.
4. Place 1 fillet on bottom half of bun. Spread 2 teaspoons salad dressing on top half of bun (optional).
5. Place lettuce leaf and 1 tomato slice over fillet (optional). Cover with top half of bun. Serve hot. CCP: Hold for service at 140 F. or higher.

SANDWICHES No.N 021 03

GRILLED CHICKEN FILLET SANDWICH (UNBREADED)

Yield 100 **Portion** 1 Sandwich

Calories	Carbohydrates	Protein	Fat	Cholesterol	Sodium	Calcium
350 cal	25 g	29 g	14 g	79 mg	381 mg	86 mg

Ingredient	**Weight**	**Measure**	**Issue**
CHICKEN FILLET,UNBREADED,PRECOOKED,3 OZ	18-3/4 lbs		
COOKING SPRAY,NONSTICK	2 oz	1/4 cup 1/3 tbsp	
SALAD DRESSING,MAYONNAISE TYPE	2-1/3 lbs	1 qts 5/8 cup	
BUN,HAMBURGER	9-1/2 lbs	100 each	
LETTUCE,LEAF,FRESH,HEAD	4 lbs		6-1/4 lbs
TOMATOES,FRESH,SLICED	5-3/4 lbs	3 qts 2-1/2 cup	5-7/8 lbs

Method

1. Grill chicken fillets 6 minutes on each side or until thoroughly heated on a 350 F. lightly sprayed griddle. CCP: Internal temperature must reach 165 F. or higher for 15 seconds.
2. Place 1 fillet on bottom half of bun. Spread 2 teaspoons salad dressing on top half of bun (optional).
3. Place lettuce leaf and 1 tomato slice over fillet (optional). Cover with top half of bun. Serve hot. CCP: Hold for service at 140 F. or higher.

SANDWICHES No.N 022 00

CANNONBALL SANDWICH (MEATBALL)

Yield 100   **Portion** 1 Sandwich

Calories	Carbohydrates	Protein	Fat	Cholesterol	Sodium	Calcium
417 cal	40 g	28 g	15 g	92 mg	980 mg	150 mg

Ingredient	Weight	Measure	Issue
BEEF,GROUND,BULK,RAW,90% LEAN	20 lbs		
ONIONS,FRESH,CHOPPED	2-1/3 lbs	1 qts 2-5/8 cup	2-5/8 lbs
BREADCRUMBS,DRY,GROUND,FINE	2-1/8 lbs	2 qts 1 cup	
EGGS,WHOLE,FROZEN	12-7/8 oz	1-1/2 cup	
SALT	3 oz	1/4 cup 1 tbsp	
PEPPER,BLACK,GROUND	1/4 oz	1 tbsp	
ROLL,FRENCH	8-3/8 lbs	100 each	
SAUCE,PIZZA,CANNED	38-1/2 lbs	4 gal	

Method

1. Combine beef, onions, bread crumbs, eggs, salt and pepper; mix lightly but thoroughly.
2. Shape into 300 1-1/3 ounce meatballs. Place 100 meatballs on each pan.
3. Using a convection oven, bake 12-14 minutes at 350 F. on high fan, closed vent or until browned. CCP: Internal temperature must reach 155 F. or higher for 15 seconds. Discard fat.
4. Slice rolls in half lengthwise with bottom half thicker than top. Place 3 meatballs on bottom half of each roll. Pour 4 ounces of pizza sauce over meatballs. Cover with top half of roll.
5. Serve hot. CCP: Hold for service at 140 F. or higher.

SANDWICHES No.N 022 01

HOT ITALIAN SAUSAGE SANDWICH

Yield 100 **Portion** 1 Sandwich

Calories	Carbohydrates	Protein	Fat	Cholesterol	Sodium	Calcium
400 cal	35 g	20 g	20 g	54 mg	1127 mg	144 mg

Ingredient

Ingredient	Weight	Measure	Issue
SAUSAGE,ITALIAN,HOT	18-3/4 lbs		
ROLL,FRENCH	8-3/8 lbs	100 each	
SAUCE,PIZZA,CANNED	38-1/2 lbs	4 gal	

Method

1. Place Italian sausage links in single layers on sheet pans. Pierce each sausage. Pour 1 cup hot water over sausages in each pan. Cover; bake in 400 F. oven 20 minutes. Remove cover; bake 15 minutes or until browned. CCP: Internal temperature must reach 145 F. or higher for 15 seconds.
2. Slice rolls in half lengthwise with bottom half thicker than top.
3. Split sausages lengthwise. Place one sausage on bottom half of each roll.
4. Pour 4 ounces of pizza sauce over sausage. Cover with top half of roll.
5. Serve hot. CCP: Hold for service at 140 F. or higher.

SANDWICHES No.N 022 02

CANNONBALL SANDWICH (CANNED MEATBALLS)

Yield 100   **Portion** 1 Sandwich

Calories	Carbohydrates	Protein	Fat	Cholesterol	Sodium	Calcium
481 cal	38 g	23 g	26 g	66 mg	915 mg	142 mg

Ingredient	Weight	Measure	Issue
ROLL,FRENCH	8-3/8 lbs	100 each	
BEEF,MEATBALLS,CANNED	18-2/3 lbs	2 gal 1-1/3 qts	
SAUCE,PIZZA,CANNED	38-1/2 lbs	4 gal	

Method

1. Slice rolls in half lengthwise with bottom half thicker than top.
2. Heat meatballs and sauce thoroughly. CCP: Internal temperature must reach 165 F. or higher for 15 seconds. Place 3 meatballs on bottom half of each roll.
3. Serve hot. CCP: Hold for service at 140 F. or higher.

SANDWICHES No.N 023 00

HOT PASTRAMI SANDWICH

Yield 100   **Portion** 1 Sandwich

Calories	Carbohydrates	Protein	Fat	Cholesterol	Sodium	Calcium
426 cal	34 g	18 g	24 g	67 mg	1425 mg	61 mg

Ingredient   Weight   Measure   Issue

PASTRAMI,PRECOOKED   16 lbs
BREAD,RYE,SLICE   14-1/8 lbs   200 sl
MUSTARD,PREPARED   2-1/4 lbs   1 qts

Method

1 Slice pastrami across grain into thin slices, 19 to 25 slices per pound.
2 Steam until thoroughly heated. CCP: Internal temperature must reach 145 F. or higher for 15 seconds.
3 Spread 1 slice of bread with mustard; add 3 to 4 slices pastrami; top with second slice bread. Cut in half; serve hot. CCP: Hold for service at 140 F. or higher.

SANDWICHES No.N 024 00

BARBECUED BEEF SANDWICH (CANNED)

Yield 100   Portion   1 Sandwich

Calories	Carbohydrates	Protein	Fat	Cholesterol	Sodium	Calcium
457 cal	30 g	40 g	19 g	104 mg	486 mg	79 mg

Ingredient	Weight	Measure	Issue
BEEF,CANNED,CHUNKS,W/NATURAL JUICE,DRAINED	29 lbs	6 gal 2-1/2 qts	
RESERVED STOCK	8-1/3 lbs	1 gal	
ONIONS,FRESH,CHOPPED	1-3/4 lbs	1 qts 1 cup	2 lbs
PEPPERS,GREEN,FRESH,CHOPPED	1 lbs	3 cup	1-1/4 lbs
TOMATO PASTE,CANNED	4-1/3 lbs	1 qts 3-1/2 cup	
CHILI POWDER,DARK,GROUND	1/2 oz	2 tbsp	
PEPPER,BLACK,GROUND	1/8 oz	1/8 tsp	
SUGAR,BROWN,PACKED	10-7/8 oz	2-1/8 cup	
VINEGAR,DISTILLED	8-1/3 oz	1 cup	
WORCESTERSHIRE SAUCE	4-1/4 oz	1/2 cup	
BUN,HAMBURGER	9-1/2 lbs	100 each	

Method

1. Drain beef chunks, break up into 3/4 to 1 inch pieces. Reserve 1 gallon beef juices.
2. Add reserved beef juices, tomato paste, onions, peppers, brown sugar, vinegar, Worcestershire sauce, chili powder and black pepper. Bring to a boil. Cover; reduce heat; simmer 15 minutes stirring occasionally.
3. Stir beef chunks gently into sauce. Cover; reduce heat; simmer 5 minutes. CCP: Internal temperature must reach 165 F. or higher for 15 seconds.
4. Ladle beef barbecue on the bottom half of the bun. Cover with top half of bun. CCP: Hold for service at 140 F. or higher.

SANDWICHES No.N 024 01

BARBECUED BEEF SANDWICH (DICED BEEF)

Yield 100   **Portion** 1 Sandwich

Calories	Carbohydrates	Protein	Fat	Cholesterol	Sodium	Calcium
346 cal	30 g	27 g	13 g	66 mg	456 mg	77 mg

Ingredient

	Weight	Measure	Issue
BEEF,DICED,LEAN,RAW	30 lbs		
WATER	12-1/2 lbs	1 gal 2 qts	
ONIONS,FRESH,CHOPPED	1-3/4 lbs	1 qts 1 cup	2 lbs
PEPPERS,GREEN,FRESH,CHOPPED	1 lbs	3 cup	1-1/4 lbs
TOMATO PASTE,CANNED	4-1/3 lbs	1 qts 3-1/2 cup	
CHILI POWDER,DARK,GROUND	1/2 oz	2 tbsp	
PEPPER,BLACK,GROUND	1/8 oz	1/8 tsp	
SUGAR,BROWN,PACKED	10-7/8 oz	2-1/8 cup	
VINEGAR,DISTILLED	8-1/3 oz	1 cup	
WORCESTERSHIRE SAUCE	4-1/4 oz	1/2 cup	
BUN,HAMBURGER	9-1/2 lbs	100 each	

Method

1 Cook beef cubes in a steam jacketed kettle or stock pot 15 minutes, uncovered, stirring constantly.
2 Add 1-1/2 gallon of water. Bring to a boil. Cover; reduce heat; simmer 1 hour or until tender.
3 Add beef broth, tomato paste, onions, peppers, brown sugar, vinegar, Worcestershire sauce, chili powder and black pepper. Bring to a boil. Cover; reduce heat; simmer 15 minutes stirring occasionally. CCP: Internal temperature must reach 165 F. or higher for 15 seconds.
4 Ladle beef barbecue on the bottom half of the bun. Cover with top half of bun. CCP: Hold for service at 140 F. or higher.

SANDWICHES No.N 025 00

MONTE CRISTO SANDWICH

Yield 100   **Portion**   1 Sandwich

Calories	Carbohydrates	Protein	Fat	Cholesterol	Sodium	Calcium
439 cal	27 g	26 g	25 g	154 mg	905 mg	360 mg

Ingredient

Ingredient	Weight	Measure	Issue
HAM,COOKED,BONELESS	6-1/4 lbs		
TURKEY,BNLS,WHITE AND DARK MEAT	6-1/4 lbs		
CHEESE,SWISS,SLICED	6-1/4 lbs	100 sl	
BREAD,WHITE	11 lbs	200 sl	
MILK,NONFAT,DRY	2-2/3 oz	1-1/8 cup	
WATER	2-7/8 lbs	1 qts 1-1/2 cup	
EGGS,WHOLE,FROZEN	5 lbs	2 qts 1-3/8 cup	
OIL,SALAD	1-7/8 lbs	1 qts	

Method

1. Slice ham and turkey into 1 ounce thin slices.
2. Place 1 slice each ham, turkey and cheese on 1 slice bread; top with second slice of bread.
3. Reconstitute milk; add eggs. Blend well.
4. Dip each side of sandwich into egg and milk mixture; drain.
5. Grill each sandwich on well-greased griddle at 350 F. for about 2-1/2 minutes on each side or until golden brown and cheese is melted.
6. Serve hot. CCP: Hold for service at 140 F. or higher.

SANDWICHES No.N 026 00

ITALIAN VEAL CUTLET SUBMARINE

Yield 100   **Portion** 1 Sandwich

Calories	Carbohydrates	Protein	Fat	Cholesterol	Sodium	Calcium
496 cal	37 g	31 g	24 g	103 mg	833 mg	175 mg

Ingredient   Weight   Measure   Issue

Ingredient	Weight	Measure	Issue
VEAL,STEAKS,BREADED	37-1/2 lbs		
SAUCE,PIZZA,CANNED	21-1/2 lbs	2 gal 7/8 qts	
ROLL,FRENCH	8-3/8 lbs	100 each	
CHEESE,PIZZA BLEND,SHREDDED	1-1/2 lbs	1 qts 2-1/4 cup	

Method

1. Place veal steaks on sheet pans. Bake at 425 F. for 20 minutes. Turn steaks. Bake 15 minutes or until thoroughly heated and browned. Cut in half lengthwise. CCP: Internal temperature must reach 145 F. or higher for 15 seconds.
2. Bring pizza sauce to a boil.
3. Split French rolls almost through. Spread 1 ounce (2 tbsp) sauce on bottom half of each roll. Add 2 steak halves; ladle 1/4 cup sauce over steak halves.
4. Sprinkle about 1 tablespoon of cheese over each sandwich; close top. CCP: Hold for service at 140 F. or higher.

SANDWICHES No.N 027 00

BARBECUED BEEF SANDWICH (SLOPPY JOE)

Yield 100   Portion   1 Sandwich

Calories	Carbohydrates	Protein	Fat	Cholesterol	Sodium	Calcium
352 cal	36 g	24 g	12 g	66 mg	881 mg	82 mg

Ingredient	Weight	Measure	Issue
BEEF,GROUND,BULK,RAW,90% LEAN	18-3/4 lbs		
ONIONS,FRESH,CHOPPED	5-1/4 lbs	3 qts 3 cup	5-7/8 lbs
CATSUP	9-1/2 lbs	1 gal 1/2 qts	
MUSTARD,DRY	2-1/4 oz	1/4 cup 2 tbsp	
SALT	3/4 oz	1 tbsp	
SUGAR,BROWN,PACKED	1-1/4 oz	1/4 cup 1/3 tbsp	
VINEGAR,DISTILLED	1 lbs	2 cup	
WATER	2 lbs	3-3/4 cup	
ROLL,SANDWICH BUNS,SPLIT	9-1/2 lbs	100 each	

Method

1. Cook beef until beef loses its pink color, stirring to break apart. Drain or skim off excess fat.
2. Combine onions, catsup, mustard, salt, brown sugar, vinegar, and water. Add to beef.
3. Cover; simmer 35 minutes. Stir occasionally to prevent scorching. CCP: Internal temperature must reach 155 F. or higher for 15 seconds.
4. Place 1/2 cup, or a No. 8 scoop of hot mixture on bottom half of bun. Top with second half.
5. CCP: Hold for service at 140 F. or higher.

SANDWICHES No.N 027 01

BARBECUED PORK SANDWICH (PORK BUTT)

Yield 100 **Portion** 1 Sandwich

Calories	Carbohydrates	Protein	Fat	Cholesterol	Sodium	Calcium
328 cal	32 g	19 g	14 g	53 mg	624 mg	89 mg

Ingredient	**Weight**	**Measure**	**Issue**
PORK,COOKED,DICED	13-1/2 lbs		
ONIONS,FRESH,CHOPPED	3-3/8 lbs	2 qts 1-5/8 cup	3-3/4 lbs
CATSUP	6-1/3 lbs	3 qts	
MUSTARD,DRY	2-1/4 oz	1/4 cup 2 tbsp	
SUGAR,BROWN,PACKED	1 oz	3 tbsp	
WATER	2-1/3 lbs	1 qts 1/2 cup	
VINEGAR,DISTILLED	9-3/8 oz	1-1/8 cup	
ROLL,SANDWICH BUNS,SPLIT	9-1/2 lbs	100 each	

Method

1. Combine onions, catsup, mustard, brown sugar, water and vinegar. Add to pork.
2. Cover; simmer 35 minutes. Stir occasionally to prevent scorching. CCP: Internal temperature must reach 145 F. or higher for 15 seconds.
3. Place 1/2 cup or No. 8 scoop of hot mixture on bottom half bun. Top with second half.
4. CCP: Hold for service at 140 F. or higher.

SANDWICHES No.N 027 02

BARBECUED PORK SANDWICH (FROZEN BARBECUED PORK)

Yield 100 **Portion** 1 Sandwich

Calories	Carbohydrates	Protein	Fat	Cholesterol	Sodium	Calcium
246 cal	30 g	14 g	8 g	27 mg	674 mg	70 mg

Ingredient

Ingredient	Weight	Measure	Issue
PORK W/BARBECUE SAUCE,COOKED,FROZEN	16-1/2 lbs	1 gal 2-2/3 qts	
ROLL,SANDWICH BUNS,SPLIT	9-1/2 lbs	100 each	

Method

1. Heat pork according to manufacturer's directions.
2. Place 1/2 cup or a No. 8 scoop of hot meat on bottom half of bun. Top with second half of bun. CCP: Internal temperature must reach 145 F. or higher for 15 seconds.
3. CCP: Hold for service at 140 F. or higher.

SANDWICHES No.N 027 03

TURKEY BARBECUE SANDWICH

Yield 100 **Portion** 1 Sandwich

Calories	Carbohydrates	Protein	Fat	Cholesterol	Sodium	Calcium
303 cal	36 g	21 g	9 g	56 mg	918 mg	95 mg

Ingredient	Weight	Measure	Issue
TURKEY,GROUND,90% LEAN,RAW	20 lbs		
ONIONS,FRESH,CHOPPED	5-1/4 lbs	3 qts 3 cup	5-7/8 lbs
CATSUP	9-1/2 lbs	1 gal 1/2 qts	
MUSTARD,DRY	2-1/4 oz	1/4 cup 2 tbsp	
SALT	3/4 oz	1 tbsp	
SUGAR,BROWN,PACKED	1-1/4 oz	1/4 cup 1/3 tbsp	
VINEGAR,DISTILLED	1 lbs	2 cup	
WATER	2 lbs	3-3/4 cup	
ROLL,SANDWICH BUNS,SPLIT	9-1/2 lbs	100 each	

Method

1. Cook turkey until it loses its pink color, stirring to break apart. Skim off excess fat.
2. Combine onions, catsup, mustard, salt, brown sugar, water and vinegar. Add to meat.
3. Cover; simmer 35 minutes. Stir occasionally to prevent scorching. CCP: Internal temperature must reach 165 F. or higher for 15 seconds.
4. Place 1/2 cup or No. 8 scoop of hot mixture on bottom half bun. Top with second half.
5. CCP: Hold for service at 140 F. or higher.

SANDWICHES No. N 028 00

ITALIAN PEPPER BEEF SANDWICH

Yield 100   **Portion** 1 Sandwich

Calories	Carbohydrates	Protein	Fat	Cholesterol	Sodium	Calcium
520 cal	63 g	35 g	13 g	70 mg	987 mg	99 mg

Ingredient	Weight	Measure	Issue
BEEF, OVEN ROAST, PRE COOKED	18 lbs		
OIL, OLIVE	7-5/8 oz	1 cup	
PEPPERS, GREEN, FRESH, CHOPPED	12-1/2 lbs	2 gal 1-1/2 qts	15-1/4 lbs
NATURAL PAN GRAVY (AU JUS)		3 qts	
GARLIC POWDER	1/8 oz	1/8 tsp	
OREGANO, CRUSHED	1/8 oz	1/4 tsp	
BREAD, FRENCH	25 lbs		

Method

1. Slice beef thin, about 16 slices per pound.
2. Cut slices in half lengthwise to form strips.
3. Saute peppers 5 minutes on 400 F. griddle.
4. Prepare Natural Pan Gravy, Recipe No. O 018 00. Add garlic powder and oregano. Simmer 10 minutes. CCP: Hold at 140 F. or higher for use in Step 6.
5. Slice bread lengthwise so that bottom is thicker than top. Slice loaves crosswise into equal pieces to yield proper amount of portions.
6. Place 5 to 6 beef strips, about 2-3/4 ounces on bottom half of bread.
7. Top beef with 8 to 10 pepper strips.
8. Pour 1 tablespoon hot gravy over peppers. Cover with top half of bread.
9. Serve hot. CCP: Hold for service at 140 F. or higher.

SANDWICHES No. N 030 00

SIMMERED FRANKFURTER ON ROLL

Yield 100   Portion   1 Each

Calories	Carbohydrates	Protein	Fat	Cholesterol	Sodium	Calcium
262 cal	22 g	9 g	15 g	22 mg	732 mg	65 mg

Ingredient	Weight	Measure	Issue
FRANKFURTERS	9-2/3 lbs	100 each	
WATER	10-1/2 lbs	1 gal 1 qts	
BUN, HOTDOG	9-3/8 lbs	100 each	

Method

1. Pierce each frankfurter and cover with water in steam-jacketed kettle or stock pot; bring to a boil; reduce heat; simmer 10 minutes. CCP: Internal temperature must reach 145 F. or higher for 15 seconds.
2. Drain, leaving enough water to cover bottom of steam-jacketed kettle or stock pot. Keep hot until served. CCP: Hold for service at 140 F. or higher.
3. Serve hot on rolls. Notes: Frankfurters may be grilled at 350 F., turning frequently until thoroughly heated or slightly browned.

SANDWICHES No.N 030 01

GRILLED POLISH SAUSAGE SANDWICH

Yield 100   Portion   1 Sandwich

Calories	Carbohydrates	Protein	Fat	Cholesterol	Sodium	Calcium
325 cal	23 g	12 g	20 g	43 mg	785 mg	67 mg

Ingredient	Weight	Measure	Issue
SAUSAGE,POLISH,PORK,RAW	18-3/4 lbs		
ROLL,HOT DOG	9-1/2 lbs	100 each	

Method

1. Cut sausage in 3 ounce pieces, then in half, lengthwise. Grill on 375 F. griddle until thoroughly cooked and browned. Turn frequently to ensure even browning. CCP: Internal temperature must reach 145 F. or higher for 15 seconds.
2. Place 2 pieces sausage in each hot roll. Hold for service at 140 F. or higher.

SANDWICHES No.N 030 02

SIMMERED KNOCKWURST ON ROLL

Yield 100   Portion 1 Each

Calories	Carbohydrates	Protein	Fat	Cholesterol	Sodium	Calcium
385 cal	23 g	14 g	26 g	49 mg	1101 mg	70 mg

Ingredient   Weight   Measure   Issue

KNOCKWURST,3 OZ   18-3/4 lbs   100 each
WATER   10-1/2 lbs   1 gal 1 qts
ROLL,HOT DOG   9-1/2 lbs   100 each

Method

1. Pierce each Knockwurst; cover with water in steam-jacketed kettle or stock pot. Cover. Bring to a boil; reduce heat; simmer 10 minutes. Drain. CCP: Internal temperature must reach 145 F. or higher for 15 seconds.
2. Serve on hot rolls. CCP: Hold for service at 140 F. or higher.

SANDWICHES No.N 030 03

SIMMERED QUARTER POUND FRANKFURTER

Yield 100   Portion   1 Each

Calories	Carbohydrates	Protein	Fat	Cholesterol	Sodium	Calcium
462 cal	21 g	17 g	34 g	69 mg	1396 mg	58 mg

Ingredient	Weight	Measure	Issue
FRANKFURTERS,BEEF	25 lbs		
WATER	10-1/2 lbs	1 gal 1 qts	
ROLL,FRENCH	8-3/8 lbs	100 each	

Method

1 Pierce each frankfurter and cover with water in steam-jacketed kettle or stock pot; bring to a boil; reduce heat; simmer 10 minutes. Drain. CCP: Internal temperature must reach 145 F. or higher for 15 seconds.
2 Serve hot on French rolls. CCP: Hold for service at 140 F. or higher.

SANDWICHES No.N 030 05

GRILLED FRANKFURTER WITH FRIED PEPPERS AND ONIONS

Yield 100   Portion 1 Each

Calories	Carbohydrates	Protein	Fat	Cholesterol	Sodium	Calcium
274 cal	25 g	9 g	15 g	22 mg	731 mg	69 mg

Ingredient | Weight | Measure | Issue

Ingredient	Weight	Measure	Issue
PEPPERS,GREEN,FRESH,MEDIUM,SLICED,THIN	3-1/8 lbs	2 qts 1-1/2 cup	3-3/4 lbs
ONIONS,FRESH,SLICED	4-1/2 lbs	1 gal 1/2 qts	5-1/8 lbs
PEPPER,BLACK,GROUND	1/8 oz	1/8 tsp	
GARLIC POWDER	1/8 oz	1/4 tsp	
FRANKFURTERS	9-2/3 lbs	100 each	
BUN,HOTDOG	9-3/8 lbs	100 each	

Method

1. Saute peppers and sliced onions on lightly greased 350 F. griddle until tender. Sprinkle with black pepper and garlic powder. CCP: Hold at 140 F. or higher.
2. Grill frankfurters slowly on greased 350 F. griddle turning frequently, until thoroughly heated and browned. CCP: Internal temperature must reach 145 F. or higher for 15 seconds.
3. Top with 1 tablespoon peppers and 1 tablespoon onions. Serve hot on rolls. CCP: Hold for service at 140 F. or higher.

SANDWICHES No.N 031 00

MONTE CARLO SANDWICH (OPEN-FACED TURKEY AND HAM)

Yield 100   Portion   1 Sandwich

Calories	Carbohydrates	Protein	Fat	Cholesterol	Sodium	Calcium
511 cal	27 g	20 g	36 g	131 mg	1543 mg	48 mg

Ingredient	Weight	Measure	Issue
THOUSAND ISLAND DRESSING		1 gal 2 qts	
OVEN FRIED BACON	1-1/2 kg	200 unit	
HAM,COOKED,BONELESS	6-1/4 lbs		
TURKEY,BNLS,WHITE AND DARK MEAT	6-1/4 lbs		
BREAD,PUMPERNICKEL	5-3/4 lbs	100 sl	
LETTUCE,ICEBERG,FRESH	4 lbs		4-1/3 lbs
TOMATOES,FRESH,SLICED	11-1/8 lbs	1 gal 3 qts	11-1/3 lbs
EGG,HARD COOKED,SLICED	2-2/3 lbs	24 Eggs	

Method

1 Prepare bacon according to Recipe No. L 002 00 or L 002 02.
2 Slice ham and turkey into thin slices, about 16 slices per pound.
3 Place lettuce leaf, 2 slices tomato, 1 slice ham, 1 slice turkey, 2 egg slices, and 2 strips bacon on 1 slice bread.
4 CCP: Hold for service at 41 F. or lower.
5 Just before serving, pour about 1/4 cup Thousand Island Dressing on top.

SANDWICHES No.N 032 00

FISHWICH

Yield 100 **Portion** 1 Sandwich

Calories	Carbohydrates	Protein	Fat	Cholesterol	Sodium	Calcium
498 cal	49 g	20 g	24 g	118 mg	940 mg	81 mg

Ingredient | Weight | Measure | Issue

Ingredient	Weight	Measure	Issue
FISH,PORTIONS,BREADED,FRZ	25 lbs		
TARTAR SAUCE		1 qts 3 cup	
ROLL,SANDWICH BUNS,SPLIT	9-1/2 lbs	100 each	

Method

1. Fry fish portions about 3 minutes in 350 F. deep fat or until lightly browned. CCP: Internal temperature must reach 145 F. or higher for 15 seconds. Portions will rise to the surface when done.
2. Drain well in basket or absorbent paper.
3. Place 1 fish portion on bottom half of bun. Spread 1 tablespoon tartar sauce on top half of bun. Cover with top half of bun.
4. Serve hot. CCP: Hold for service at 140 F. or higher. Notes: In Step 1, fish may be baked at 375 F. for 35 minutes or until browned. CCP: Internal temperature must reach 145 F. or higher for 15 seconds.

SANDWICHES No.N 032 01

CHEESE FISHWICH

Yield 100 **Portion** 1 Sandwich

Calories	Carbohydrates	Protein	Fat	Cholesterol	Sodium	Calcium
472 cal	40 g	22 g	25 g	72 mg	815 mg	248 mg

Ingredient | Weight | Measure | Issue

Ingredient	Weight	Measure	Issue
FISH,PORTIONS,BREADED,FRZ	25 lbs		
TARTAR SAUCE		1 qts 3 cup	
ROLL,SANDWICH BUNS,SPLIT	9-1/2 lbs	100 each	
CHEESE,AMERICAN,SLICED	3-1/8 lbs	50 sl	

Method

1. Fry fish portions about 3 minutes in 350 F. deep fat or until lightly browned. CCP: Internal temperature must reach 145 F. or higher for 15 seconds. Portions will rise to the surface when done.
2. Drain well in basket or absorbent paper.
3. Place 1 fish portion on bottom half of bun. Cut cheese slices in half. Place 1/2 slice cheese on bottom half of bun. Spread 1 tablespoon tartar sauce on top half of bun. Cover with top half of bun.
4. Serve hot. CCP: Hold for service at 140 F. or higher.

SANDWICHES No. N 032 03

FISHWICH (BATTER DIPPED)

Yield 100 **Portion** 1 Sandwich

Calories	Carbohydrates	Protein	Fat	Cholesterol	Sodium	Calcium
496 cal	49 g	20 g	24 g	118 mg	936 mg	81 mg

Ingredient	Weight	Measure	Issue
FISH, BATTER DIPPED, FROZEN	25 lbs		
TARTAR SAUCE		1 qts 3 cup	
ROLL, SANDWICH BUNS, SPLIT	9-1/2 lbs	100 each	

Method

1. Fry fish portions 3 to 5 minutes or until golden brown. CCP: Internal temperature must reach 145 F. or higher for 15 seconds. Portions will rise to the surface when done.
2. Drain well in basket or absorbent paper.
3. Place 1 fish portion on bottom half of bun. Spread 1 tablespoon tartar sauce on top half of bun. Cover with top half of bun.
4. Serve hot. CCP: Hold for service at 140 F. or higher.

SANDWICHES No. N 033 00

HOT ROAST TURKEY SANDWICH

Yield 100   **Portion** 1 Sandwich

Calories	Carbohydrates	Protein	Fat	Cholesterol	Sodium	Calcium
376 cal	34 g	23 g	16 g	54 mg	1720 mg	97 mg

Ingredient	Weight	Measure	Issue
TURKEY, BNLS, WHITE AND DARK MEAT	21 lbs		
BREAD, WHITE	11 lbs	200 sl	
CHICKEN OR TURKEY GRAVY		3 gal 1-1/2 qts	

Method

1. Slice turkey into thin slices, 16 to 22 per pound.
2. Place 3 to 4 slices turkey on 1 slice of bread; top with second slice of bread.
3. Prepare Chicken or Turkey Gravy, Recipe No. O 016 02. Pour about 1/2 cup, one Size 2 ladle, hot gravy over sandwich. CCP: Hold for service at 140 F. or higher.

SANDWICHES No.N 034 00

CORN DOG

Yield 100 **Portion** 1 Sandwich

Calories	Carbohydrates	Protein	Fat	Cholesterol	Sodium	Calcium
258 cal	20 g	8 g	16 g	35 mg	674 mg	49 mg

Ingredient	Weight	Measure	Issue
FRANKFURTERS	10 lbs		
FLOUR,WHEAT,GENERAL PURPOSE	3-1/3 lbs	3 qts	
CORN MEAL	1-2/3 lbs	1 qts 1-1/2 cup	
BAKING POWDER	1-1/3 oz	2-2/3 tbsp	
SALT	1 oz	1 tbsp	
SUGAR,GRANULATED	3-1/2 oz	1/2 cup	
MUSTARD,DRY	3-1/8 oz	1/2 cup	
MILK,NONFAT,DRY	3-5/8 oz	1-1/2 cup	
WATER	3-7/8 lbs	1 qts 3-1/2 cup	
EGGS,WHOLE,FROZEN	9-5/8 oz	1-1/8 cup	
OIL,SALAD	5-3/4 oz	3/4 cup	

Method

1 Insert 1 stirring stick lengthwise into each thawed frankfurter. Dry surface of frankfurter with paper towel.
2 Combine flour, cornmeal, baking powder, salt, sugar, mustard flour and milk.
3 Add water, eggs and salad oil or melted shortening to dry ingredients. Blend well.
4 Dip frankfurters in cornmeal mixture; allow excess batter to drain slightly; fry 2 to 4 minutes or until golden brown in 375 F. deep fat. CCP: Internal temperature must reach 145 F. or higher for 15 seconds.
5 Drain on absorbent paper.
6 Serve hot. CCP: Hold for service at 140 F. or higher.

Notes

1 18-3/4 lbs frozen corn dogs may also be used.

SANDWICHES No.N 034 01

CORN DOG (CORN BREAD MIX)

Yield 100 **Portion** 1 Sandwich

Calories	Carbohydrates	Protein	Fat	Cholesterol	Sodium	Calcium
256 cal	19 g	7 g	17 g	23 mg	792 mg	24 mg

Ingredient

Ingredient	Weight	Measure	Issue
FRANKFURTERS	10 lbs		
CORN BREAD MIX	5-5/8 lbs	1 gal 1/8 qts	
MUSTARD,DRY	3-1/8 oz	1/2 cup	

Method

1. Insert 1 stirring stick lengthwise into each thawed frankfurter. Dry surface of frankfurter with paper towel.
2. Combine canned cornbread mix and mustard flour. Prepare corn bread batter according to instructions on container.
3. Dip frankfurters in cornmeal mixture; allow excess batter to drain slightly; fry 2 to 4 minutes or until golden brown in 375 F. deep fat. CCP: Internal temperature must reach 145 F. or higher for 15 seconds.
4. Drain on absorbent paper.
5. Serve hot. CCP: Hold for service at 140 F. or higher.

SANDWICHES No.N 035 00

HOT ROAST BEEF SANDWICH (OVEN ROAST)

Yield 100　　　　　　　　　　　　　　　　**Portion**　1 Sandwich

Calories	Carbohydrates	Protein	Fat	Cholesterol	Sodium	Calcium
453 cal	35 g	31 g	20 g	74 mg	1022 mg	67 mg

Ingredient　　　　　　　　　　　　　　　Weight　　　　Measure　　　　Issue

BEEF,OVEN ROAST,PRE COOKED　　　　18-3/4 lbs
BREAD,WHITE,SLICE　　　　　　　　　11 lbs　　　　　200 sl
BROWN GRAVY　　　　　　　　　　　　　　　　　　3 gal

Method

1. Slice beef into thin slices, 16 to 22 per pound.
2. Place 3 to 4 slices beef on 1 slice of bread; top with second slice of bread.
3. Prepare Brown Gravy, Recipe No. O 016 00. CCP: Internal temperature must reach 165 F. or higher for 15 seconds. Pour about 1/2 cup, 4 ounces, or one Size 2 ladle of hot gravy over each sandwich. CCP: Hold for service at 140 F. or higher.

SANDWICHES No.N 035 01

HOT ROAST BEEF SANDWICH (PRECOOKED ROAST BEEF)

Yield 100   Portion   1 Sandwich

Calories	Carbohydrates	Protein	Fat	Cholesterol	Sodium	Calcium
453 cal	35 g	31 g	20 g	74 mg	1022 mg	67 mg

Ingredient | Weight | Measure | Issue

BEEF,OVEN ROAST,PRE COOKED   18-3/4 lbs
BREAD,WHITE   11 lbs   200 sl
BROWN GRAVY   3 gal

Method

1 Slice beef into thin slices, 16-22 per pound.
2 Place 3 to 4 pieces beef on 1 slice of bread; top with second slice of bread.
3 Prepare 2 recipes of brown gravy. Pour 1/2 cup hot gravy over each sandwich. CCP: Hold for service at 140 F. or higher.

SANDWICHES No.N 036 00

HOT ROAST PORK SANDWICH

Yield 100   **Portion** 1 Sandwich

Calories	Carbohydrates	Protein	Fat	Cholesterol	Sodium	Calcium
485 cal	35 g	29 g	25 g	71 mg	1016 mg	77 mg

Ingredient — Weight — Measure — Issue

Ingredient	Weight	Measure	Issue
PORK,LOIN,BONELESS,COOKED	18-3/4 lbs		
BREAD,WHITE	11 lbs	200 sl	
BROWN GRAVY		3 gal	

Method

1. Slice cooked pork into thin slices, 16 to 22 slices per pound.
2. Place 3 to 4 slices pork on 1 slice of bread. Top with second slice bread.
3. Prepare 2 recipes Brown Gravy, Recipe No. O 016 00 using pork drippings. CCP: Internal temperature must reach 165 F. or higher for 15 seconds. Pour about 1/2 cup, one Size 2 ladle hot gravy over each sandwich. CCP: Hold for service at 140 F. or higher.

SANDWICHES No.N 036 01

HOT ROAST PORK SANDWICH (FRESH HAM)

Yield 100   Portion   1 Sandwich

Calories	Carbohydrates	Protein	Fat	Cholesterol	Sodium	Calcium
425 cal	35 g	25 g	20 g	51 mg	2242 mg	67 mg

Ingredient

Ingredient	Weight	Measure	Issue
HAM,COOKED,BONELESS	18-3/4 lbs		
BREAD,WHITE,SLICE	11 lbs	200 sl	
BROWN GRAVY		3 gal	

Method

1 Slice into thin slices, about 16 to 22 slices per pound.
2 Place 3 to 4 slices pork on one side of bread. Top with second slice of bread.
3 Prepare Brown Gravy, Recipe No. O 016 00 using pork drippings. Pour about 1/2 cup, one Size 2 ladle hot gravy over sandwich. CCP: Hold for service at 140 F. or higher.

SANDWICHES No.N 037 00

GRILLED HAM, EGG AND CHEESE SANDWICH

Yield 100   **Portion** 1 Sandwich

Calories	Carbohydrates	Protein	Fat	Cholesterol	Sodium	Calcium
344 cal	23 g	21 g	18 g	237 mg	847 mg	263 mg

Ingredient	**Weight**	**Measure**	**Issue**
HAM,COOKED,BONELESS	6-1/4 lbs		
COOKING SPRAY,NONSTICK	2 oz	1/4 cup 1/3 tbsp	
EGGS,WHOLE,FROZEN	10 lbs	1 gal 2/3 qts	
CHEESE,AMERICAN,SLICED	6-1/4 lbs	100 sl	
BUN,HAMBURGER	9-1/2 lbs	100 each	

Method

1. Slice ham into 1 ounce slices.
2. Lightly spray griddle with non-stick cooking spray. Place thawed eggs into a bowl; ladle individual 1/4 cup portions of beaten eggs on 325 F. greased griddle. Cook until firm. CCP: Internal temperature must reach 145 F. or higher for 15 seconds. Place 1 slice hot ham on top of egg.
3. Place 1 slice cheese on top of 1 slice of ham. Continue to cook until cheese melts.
4. Serve hot on buns. CCP: Hold for service at 140 F. or higher.

SANDWICHES No.N 037 01

GRILLED BACON, EGG, AND CHEESE SANDWICH

Yield 100   **Portion** 1 Sandwich

Calories	Carbohydrates	Protein	Fat	Cholesterol	Sodium	Calcium
389 cal	23 g	20 g	24 g	236 mg	729 mg	263 mg

Ingredient	Weight	Measure	Issue
GRILLED BACON	1-1/2 kg	200 unit	
COOKING SPRAY,NONSTICK	2 oz	1/4 cup 1/3 tbsp	
EGGS,WHOLE,FROZEN	10 lbs	1 gal 2/3 qts	
CHEESE,AMERICAN,SLICED	6-1/4 lbs	100 sl	
BUN,HAMBURGER	9-1/2 lbs	100 each	

Method

1. Grill bacon according to instructions on Recipe L 002 02.
2. Lightly spray griddle with non-stick cooking spray. Place thawed eggs into a bowl; ladle individual 1/4 cup portions of beaten eggs on 325 F. greased griddle. Cook until firm. CCP: Internal temperature must reach 145 F. or higher for 15 seconds. Place 2 slices bacon on top of each egg.
3. Place 1 slice cheese on top of each sandwich. Continue to cook until cheese melts.
4. Serve on hot buns. CCP: Hold at 140 F. or higher for service.

SANDWICHES No.N 037 02

GRILLED HAM AND EGG SANDWICH

Yield 100   **Portion**   1 Sandwich

Calories	Carbohydrates	Protein	Fat	Cholesterol	Sodium	Calcium
237 cal	22 g	14 g	9 g	210 mg	663 mg	88 mg

Ingredient   Weight   Measure   Issue

Ingredient	Weight	Measure
HAM,COOKED,BONELESS	6-1/4 lbs	
EGGS,WHOLE,FROZEN	10 lbs	1 gal 2/3 qts
COOKING SPRAY,NONSTICK	2 oz	1/4 cup 1/3 tbsp
BUN,HAMBURGER,TOASTED	9-1/2 lbs	100 each

Method

1 Slice ham into 1 ounce slices.
2 Lightly spray griddle with non-stick cooking spray. Place thawed eggs into a bowl; ladle individual 1/4 cup portions of beaten eggs on 325 F. greased griddle. Cook until firm. CCP: Internal temperature must reach 145 F. or higher for 15 seconds.
3 Serve hot on buns. CCP: Hold at 140 F. or higher for service.

SANDWICHES No.N 037 03

GRILLED SAUSAGE, EGG, AND CHEESE SANDWICH

Yield 100   **Portion**   1 Sandwich

Calories	Carbohydrates	Protein	Fat	Cholesterol	Sodium	Calcium
448 cal	23 g	23 g	29 g	256 mg	1003 mg	274 mg

Ingredient	Weight	Measure	Issue
GRILLED SAUSAGE PATTIES (PREFORMED)		100 each	
EGGS,WHOLE,FROZEN	10 lbs	1 gal 2/3 qts	
COOKING SPRAY,NONSTICK	2 oz	1/4 cup 1/3 tbsp	
CHEESE,AMERICAN,SLICED	6-1/4 lbs	100 sl	
BUN,HAMBURGER	9-1/2 lbs	100 each	

Method

1 Grill sausages according to instructions on Recipe L 089 02.
2 Lightly spray griddle with non-stick cooking spray. Place thawed eggs into a bowl; ladle individual 1/4 cup portions of beaten eggs on 325 F. greased griddle. Cook until firm. CCP: Internal temperature must reach 145 F. or higher for 15 seconds. Place 1 sausage patty on top of each egg.
3 Place 1 slice cheese on top of each patty; continue to cook until cheese melts.
4 Serve hot on buns. CCP: Hold at 140 F. or higher for service.

SANDWICHES No.N 038 00

MOROCCAN POCKETS

Yield 100 **Portion** 1 Serving

Calories	Carbohydrates	Protein	Fat	Cholesterol	Sodium	Calcium
429 cal	66 g	24 g	9 g	46 mg	587 mg	208 mg

Ingredient	Weight	Measure	Issue
YOGURT,PLAIN,LOWFAT	13-1/2 lbs	1 gal 2-1/4 qts	
PARSLEY,FRESH,BUNCH,CHOPPED	6-1/3 oz	3 cup	6-2/3 oz
BEEF,GROUND,BULK,RAW,90% LEAN	12 lbs		
TOMATOES,CANNED,CRUSHED,INCL LIQUIDS	13-1/4 lbs	1 gal 2 qts	
BARLEY,UNCOOKED	4-3/8 lbs	2 qts 2 cup	
ONIONS,FRESH,CHOPPED	3-1/2 lbs	2 qts 2 cup	3-7/8 lbs
PEPPERS,GREEN,FRESH,CHOPPED	3-1/4 lbs	2 qts 2 cup	4 lbs
CHILI POWDER,DARK,GROUND	5-1/4 oz	1-1/4 cup	
SALT	1-1/2 oz	2-1/3 tbsp	
OREGANO,CRUSHED	1-1/4 oz	1/2 cup	
GARLIC POWDER	1/2 oz	1 tbsp	
CUMIN,GROUND	1/4 oz	1 tbsp	
PEPPER,BLACK,GROUND	1/4 oz	1 tbsp	
WATER	7-1/3 lbs	3 qts 2 cup	
RAISINS	3-1/4 lbs	2 qts 2 cup	
BREAD,PITA,WHITE,8-INCH	10-1/2 lbs	50 each	

Method

1 Combine yogurt and parsley. CCP: Refrigerate at 41 F. or lower for use in Step 5.
2 Cook beef until it loses its pink color, stirring beef to break apart. Drain or skim off fat.
3 Add tomatoes, barley, onions, green peppers, chili powder, salt, oregano, garlic, cumin, pepper, and water. Stir well. Bring to a boil; reduce heat. Cover; simmer 40 to 45 minutes or until barley is tender and most of liquid is absorbed. CCP: Internal temperature must reach 155 F. or higher for 15 seconds.
4 Stir in raisins. Simmer 5 minutes. Meat mixture is done when all moisture has been absorbed and product holds together.
5 Cut pita bread in halves to make 2 pockets. Place about 5-1/2 ounces of meat mixture in each pocket. Top with yogurt topping just before serving. CCP: Hold for service at 140 F. or higher.

SANDWICHES No. N 039 00

CHEESE DELI SANDWICH

Yield 100   **Portion** 1 Sandwich

Calories	Carbohydrates	Protein	Fat	Cholesterol	Sodium	Calcium
434 cal	35 g	19 g	25 g	57 mg	756 mg	422 mg

Ingredient

Ingredient	Weight	Measure	Issue
BREAD,WHEAT,SLICED	12-1/2 lbs	200 sl	
SALAD DRESSING,MAYONNAISE TYPE	2 lbs	1 qts	
CHEESE,AMERICAN,SLICED	12-1/2 lbs	200 sl	
LETTUCE,ICEBERG,FRESH	4 lbs		4-1/3 lbs
TOMATOES,FRESH,SLICED	11-1/2 lbs	1 gal 3-1/4 qts	11-3/4 lbs
ONIONS,FRESH,SLICED	1-1/2 lbs	1 qts 2 cup	1-2/3 lbs
ALFALFA SPROUTS,FRESH,RAW	7-5/8 oz	1 qts 2-1/2 cup	

Method

1 Spread 1 slice of bread with salad dressing.
2 Add 2 slices of cheese. Top with lettuce. May also top with 2 slices of tomatoes, sliced onions, and 1 tablespoon of alfalfa sprouts. Top with second slice of bread.
3 Cut each sandwich in half. CCP: Hold for service at 41 F. or lower.

SANDWICHES No.N 040 00

TACO BURGER

Yield 100   **Portion** 1 Sandwich

Calories	Carbohydrates	Protein	Fat	Cholesterol	Sodium	Calcium
348 cal	27 g	24 g	16 g	70 mg	580 mg	168 mg

Ingredient	Weight	Measure	Issue
BEEF,GROUND,BULK,RAW,90% LEAN	16 lbs		
FLOUR,WHEAT,GENERAL PURPOSE	10-1/4 oz	2-3/8 cup	
WATER,WARM	7-1/3 lbs	3 qts 2 cup	
TOMATO PASTE,CANNED	1-1/3 lbs	2-1/4 cup	
CHILI POWDER,DARK,GROUND	8-1/2 oz	2 cup	
CUMIN,GROUND	1-1/8 oz	1/4 cup 1-2/3 tbsp	
SALT	1-1/4 oz	2 tbsp	
PEPPER,RED,CRUSHED	1/8 oz	1 tbsp	
ROLL,SANDWICH BUNS,SPLIT	9-1/2 lbs	100 each	
CHEESE,AMERICAN,SLICED	3-1/8 lbs	50 sl	
LETTUCE,ICEBERG,FRESH,SHREDDED	4 lbs	2 gal 1/4 qts	4-1/3 lbs

Method

1 Cook beef in steam-jacketed kettle or stock pot until beef loses its pink color, stirring to break apart. Drain or skim off excess fat.
2 Sprinkle flour over cooked beef. Stir well. Cook about 5 minutes or until flour is absorbed, stirring occasionally.
3 Combine water, tomato paste, chili powder, cumin, salt and red pepper; mix well. Bring to a boil; simmer 2 to 3 minutes or until thoroughly heated.
4 Combine sauce with beef mixture; mix well. Simmer 2 to 3 minutes. CCP: Internal temperature must reach 155 F. or higher for 15 seconds.
5 On bottom half of bun, place 1/2 slice cheese, 1/3 cup meat mixture, and 2-1/2 tablespoons lettuce. Cover with top half of bun. Serve hot. CCP: Hold for service at 140 F. or higher.

SANDWICHES No. N 041 00

CHILI DOG WITH CHEESE AND ONIONS

Yield 100   Portion 1 Sandwich

Calories	Carbohydrates	Protein	Fat	Cholesterol	Sodium	Calcium
342 cal	26 g	14 g	20 g	36 mg	948 mg	145 mg

Ingredient	Weight	Measure	Issue
FRANKFURTERS	9-2/3 lbs	100 each	
WATER	10-1/2 lbs	1 gal 1 qts	
CHILI CON CARNE,CANNED,NO BEANS	6-3/4 lbs	3 qts	
BUN,HOTDOG	9-3/8 lbs	100 each	
CHEESE,AMERICAN,SHREDDED	2-1/2 lbs	2 qts 2 cup	
ONIONS,FRESH,CHOPPED	1 lbs	3 cup	1-1/8 lbs

Method

1. Pierce each frankfurter; cover with water in steam-jacketed kettle or stock pot; bring to a boil; reduce heat; simmer 10 minutes. Drain. CCP: Internal temperature must reach 145 F. or higher for 15 seconds.
2. Keep hot until served. CCP: Hold for service at 140 F. or higher.
3. Thoroughly heat chili. CCP: Internal temperature must reach 165 F. or higher for 15 seconds.
4. Place frankfurter in roll.
5. Place 1 ounce hot chili over each frankfurter.
6. Place 2 tablespoons cheese and 1/2 teaspoon onions on top of chili.
7. CCP: Hold for service at 140 F. or higher.

SANDWICHES No.N 041 01

CHILI DOG (CANNED CHILI CON CARNE)

Yield 100 **Portion** 1 Sandwich

Calories	Carbohydrates	Protein	Fat	Cholesterol	Sodium	Calcium
297 cal	26 g	11 g	16 g	25 mg	874 mg	74 mg

Ingredient / Weight / Measure / Issue

Ingredient	Weight	Measure	Issue
FRANKFURTERS	9-2/3 lbs	100 each	
WATER	10-1/2 lbs	1 gal 1 qts	
CHILI CON CARNE,CANNED,NO BEANS	6-3/4 lbs	3 qts	
BUN,HOTDOG	9-3/8 lbs	100 each	

Method

1. Pierce each frankfurter; cover with water in steam-jacketed kettle or stock pot; bring to a boil; reduce heat; simmer 10 minutes. Drain. CCP: Internal temperaturemust reach 145 F. or higher for 15 seconds.
2. Thoroughly heat chili. CCP: Internal temperature must reach 165 F. or higher for 15 seconds.
3. Place frankfurter in roll.
4. Place 1 ounce hot chili over each frankfurter.
5. CCP: Hold for service at 140 F.or higher.

SANDWICHES No. N 041 02

CHILI DOG

Yield 100   **Portion** 1 Sandwich

Calories	Carbohydrates	Protein	Fat	Cholesterol	Sodium	Calcium
350 cal	25 g	15 g	21 g	46 mg	890 mg	144 mg

Ingredient	Weight	Measure	Issue
FRANKFURTERS	9-2/3 lbs	100 each	
WATER	10-1/2 lbs	1 gal 1 qts	
CHILI (WITHOUT BEANS)		3 qts 1/2 cup	
BUN, HOTDOG	9-3/8 lbs	100 each	
CHEESE, AMERICAN, SHREDDED	2-1/2 lbs	2 qts 2 cup	
ONIONS, FRESH, CHOPPED	1 lbs	3 cup	1-1/8 lbs

Method

1. Pierce each frankfurter; cover with water in steam-jacketed kettle or stock pot; bring to a boil; reduce heat; simmer 10 minutes. Drain. CCP: Internal temperature must reach 145 F. or higher for 15 seconds.
2. Prepare 1/8 recipe Chili, Recipe No. L 170 00, per 100 portions.
3. Place frankfurter in roll.
4. Place 1 ounce hot chili over each frankfurter.
5. Place 2 tablespoons cheese and 1/2 teaspoon onions on top of chili.
6. CCP: Hold for service at 140 F. or higher.

SANDWICHES No.N 042 00

BEEF FAJITA PITA

Yield 100   Portion   1/2 Pita

Calories	Carbohydrates	Protein	Fat	Cholesterol	Sodium	Calcium
389 cal	37 g	35 g	11 g	87 mg	881 mg	78 mg

Ingredient	Weight	Measure	Issue
JUICE,LIME	1-1/2 lbs	3 cup	
SALT	3 oz	1/4 cup 1 tbsp	
GARLIC POWDER	2-3/8 oz	1/2 cup	
ONION POWDER	1-1/8 oz	1/4 cup 1 tbsp	
PEPPER,BLACK,GROUND	3/4 oz	3-1/3 tbsp	
CUMIN,GROUND	1/3 oz	1 tbsp	
PEPPER,RED,GROUND	1/4 oz	1 tbsp	
TOMATOES,CANNED,CRUSHED,DRAINED	7-1/4 lbs	1 #10cn	
BEEF,FAJITA STRIPS	30-3/8 lbs		
BREAD,PITA,WHITE,8-INCH	10-1/2 lbs	50 each	
COOKING SPRAY,NONSTICK	1-1/2 oz	3 tbsp	
ONIONS,FRESH,CHOPPED	7 lbs	1 gal 1 qts	7-7/8 lbs
PEPPERS,GREEN,FRESH,MEDIUM,SLICED,THIN	7-7/8 lbs	1 gal 2 qts	9-5/8 lbs
COOKING SPRAY,NONSTICK	1-1/2 oz	3 tbsp	
SAUCE,SALSA	7 lbs	3 qts 1 cup	

Method

1 Combine lime juice, salt, garlic powder, onion powder, black pepper, cumin, tomatoes and red pepper. Stir to blend well.
2 Pour mixture over beef strips. Mix thoroughly to evenly distribute seasonings around all surfaces of beef. Cover. CCP: Marinate under refrigeration at 41 F. or lower for 45 minutes for use in Step 5.
3 Cut each pita in half forming 2 pockets. Cover; set aside for use in Step 6.
4 Lightly spray griddle with cooking spray. Grill onions and peppers 6 to 8 minutes while tossing intermittently; lightly spray with cooking spray as needed. CCP: Hold for service at 140 F. or higher for use in Step 6.
5 Lightly spray griddle with cooking spray. Grill beef strips 3 to 4 minutes or until lightly browned while tossing intermittently; lightly spray with cooking spray as needed. CCP: Internal temperature must reach 155 F. or higher for 15 seconds.
6 Place 6 to 7 cooked fajita strips (3 oz), 3 tbsp onion/sweet pepper mixture into each pita pocket. If desired, top each pocket with 2 tbsp salsa. Batch preparation techniques should be utilized. Pitas may be served with guacamole or sour cream.

SANDWICHES No.N 042 01

CHICKEN FAJITA PITA

Yield 100　　　　　　　　　　　　　　　　　　　　**Portion**　1/2 Pita

Calories	Carbohydrates	Protein	Fat	Cholesterol	Sodium	Calcium
293 cal	33 g	29 g	4 g	65 mg	758 mg	68 mg

Ingredient	Weight	Measure	Issue
JUICE,LIME	1-1/4 lbs	2-1/2 cup	
SALT	2-1/3 oz	1/4 cup	
GARLIC POWDER	2 oz	1/4 cup 3 tbsp	
ONION POWDER	7/8 oz	1/4 cup	
PEPPER,BLACK,GROUND	5/8 oz	2-2/3 tbsp	
CUMIN,GROUND	1/4 oz	1 tbsp	
PEPPER,RED,GROUND	1/8 oz	3/8 tsp	
CHICKEN,FAJITA STRIPS	23 lbs		
BREAD,PITA,WHITE,8-INCH	10-1/2 lbs	50 each	
COOKING SPRAY,NONSTICK	1-1/2 oz	3 tbsp	
ONIONS,GREEN,FRESH,CHOPPED	4-3/8 lbs	1 gal 1 qts	4-7/8 lbs
PEPPERS,GREEN,FRESH,CHOPPED	7-7/8 lbs	1 gal 2 qts	9-5/8 lbs
COOKING SPRAY,NONSTICK	1-1/2 oz	3 tbsp	
SAUCE,SALSA	7 lbs	3 qts 1 cup	

Method

1. Combine lime juice, salt, garlic powder, onion powder, black pepper, cumin and red pepper. Stir to blend well.
2. Pour mixture over chicken strips. Mix thoroughly. Cover. CCP: Marinate under refrigeration at 41 F. or lower for 45 minutes for use in Step 5.
3. Cut pita bread in half forming 2 pockets. Cover; set aside for use in Step 6.
4. Lightly spray griddle with cooking spray. Grill onions and peppers 6 to 8 minutes while tossing intermittently; lightly spray with cooking spray as needed. CCP: Hold for service at 140 F. or higher.
5. Lightly spray griddle with cooking spray. Grill chicken strips 5 to 7 minutes or until lightly browned while tossing intermittently; lightly spray with cooking spray as needed. CCP: Internal temperature must reach 165 F. or higher for 15 seconds.
6. Place 6 to 7 cooked fajita strips (3 oz) 3 tbsp onion/sweet pepper mixture into each pita pocket. If desired, top each pocket with 2 tbsp salsa. Batch preparation techniques should be utilized. Pitas may be served with guacamole or sour cream.

SANDWICHES No.N 043 00

CHICKEN PITA POCKET SANDWICH

Yield 100   **Portion** 1 Sandwich

Calories	Carbohydrates	Protein	Fat	Cholesterol	Sodium	Calcium
294 cal	32 g	30 g	4 g	67 mg	337 mg	119 mg

Ingredient	**Weight**	**Measure**	**Issue**
YOGURT,PLAIN,LOWFAT	6-1/2 lbs	3 qts	
CUCUMBER,FRESH,CHOPPED	4-1/4 lbs	1 gal <1/16th qts	5 lbs
DILL WEED,DRIED	1/2 oz	1/4 cup 1 tbsp	
GARLIC POWDER	1/2 oz	1 tbsp	
TOMATOES,FRESH,SLICED	6-1/2 lbs	1 gal 1/8 qts	6-5/8 lbs
LETTUCE,ICEBERG,FRESH,SHREDDED	5-1/8 lbs	2 gal 2-5/8 qts	5-1/2 lbs
ONIONS,FRESH,SLICED	2 lbs	1 qts 3-7/8 cup	2-1/4 lbs
BREAD,PITA,WHITE,8-INCH	10-1/2 lbs	50 each	
COOKING SPRAY,NONSTICK	1-1/2 oz	3 tbsp	
CHICKEN,FAJITA STRIPS	23 lbs		

Method

1 Combine yogurt, cucumbers, dillweed and garlic powder. Mix well; cover. CCP: Refrigerate at 41 F. or lower for use in Step 6.
2 Slice tomatoes, shred lettuce and separate onion slices into rings; cover.
3 Cut pita bread in half forming 2 pockets.
4 Lightly spray griddle with cooking spray.
5 Grill chicken strips 5 to 7 minutes or until lightly browned while tossing intermittently; lightly spray with cooking spray as needed. CCP: Internal temperature must reach 165 F. or higher for 15 seconds.
6 Place 1/3 cup shredded lettuce, 1 tomato slice and 4 to 6 onion rings into each pita pocket. Place 6 to 7 cooked fajita strips (2-3/4 oz) into each pita pocket. If desired, top each pocket with about 3 tbsp yogurt-cucumber sauce. CCP: Hold for service at 140 F. or higher.

SANDWICHES No.N 044 00

GRILLED CHICKEN BREAST SANDWICH

Yield 100   **Portion** 1 Sandwich

Calories	Carbohydrates	Protein	Fat	Cholesterol	Sodium	Calcium
363 cal	26 g	36 g	12 g	92 mg	386 mg	83 mg

Ingredient	Weight	Measure	Issue
CHICKEN,BREAST,BNLS/SKNLS,5 OZ	31-1/4 lbs		
COOKING SPRAY,NONSTICK	1-2/3 oz	3-1/3 tbsp	
ROLL,SANDWICH BUNS,SPLIT	9-1/2 lbs	100 each	
SALAD DRESSING,MAYONNAISE TYPE	2-1/3 lbs	1 qts 5/8 cup	
LETTUCE,ICEBERG,FRESH,LEAF	4 lbs		4-1/3 lbs
ONIONS,FRESH,RED,SLICED	3 lbs	2 qts 3-7/8 cup	3-1/3 lbs
TOMATOES,FRESH,SLICED	6-1/2 lbs	1 gal 1/8 qts	6-5/8 lbs

Method

1. Wash chicken breasts thoroughly under cold running water. Drain well. Remove excess fat.
2. Lightly spray griddle with cooking spray. Grill breasts 5 minutes; lightly spray with cooking spray; turn; grill second side 4 minutes. CCP: Internal temperature must reach 165 F. or higher for 15 seconds.
3. Place 1 chicken breast on the bottom half of bun. Spread 2 tsp of salad dressing on top half of bun.
4. Place lettuce leaf, onion slice and tomato slice over chicken breast. Cover with top half of bun. CCP: Hold for service at 140 F. or higher.

SANDWICHES No.N 045 00

CHICKEN CAESAR ROLLUP SANDWICH

Yield 100   Portion   1 Sandwich

Calories	Carbohydrates	Protein	Fat	Cholesterol	Sodium	Calcium
316 cal	41 g	26 g	6 g	52 mg	718 mg	181 mg

Ingredient   Weight   Measure   Issue

Ingredient	Weight	Measure	Issue
CHICKEN,BREAST,BNLS/SKNLS,5 OZ	17 lbs		
COOKING SPRAY,NONSTICK	1-1/2 oz	3 tbsp	
SALAD DRESSING,CAESAR,FAT FREE	6-7/8 lbs	3 qts 1 cup	
CHEESE,PARMESAN,GRATED	1-1/8 lbs	1 qts 1 cup	
LETTUCE,ROMAINE,FRESH	8 lbs	4 gal 1/4 qts	8-1/2 lbs
TOMATOES,FRESH,CHOPPED	12 lbs	1 gal 3-5/8 qts	12-1/4 lbs
TORTILLAS,WHEAT,10 INCH	12-3/8 lbs	100 each	

Method

1 Wash chicken breasts thoroughly under cold running water. Drain well. Remove excess fat. Cut breasts into 1/2 inch cubes.
2 Lightly spray grill with cooking spray.
3 Grill chicken cubes 3 to 5 minutes while tossing intermittently; lightly spray with cooking spray as needed. Grill until lightly browned. CCP: Internal temperature must reach 165 F. or higher is reached for 15 seconds.
4 Combine chicken, caesar dressing and parmesan cheese; cover. CCP: Refrigerate at 41 F. or lower for use in Step 8.
5 Cut romaine into 1/2-inch strips. Toss romaine and tomatoes together.
6 Wrap tortillas in foil; place in warm oven (150 F.) or in a warmer 15 minutes or until warm and pliable.
7 Place about 3/4 cup romaine mixture on warmed tortilla.
8 Distribute 1/4 cup chicken cubes over romaine mixture.
9 Roll up tortilla; wrap with parchment, wax paper, or foil. CCP: Hold for service at 41 F. or lower.

Notes

1 In Step 4, 13 lb frozen, cooked, diced, thawed (RTU) chicken may be used. Omit Steps 1 through 3. Follow Steps 4 through 9.

SANDWICHES No.N 046 00

GARDEN VEGETABLE WRAP

Yield 100   Portion   1 Sandwich

Calories	Carbohydrates	Protein	Fat	Cholesterol	Sodium	Calcium
212 cal	44 g	6 g	2 g	0 mg	403 mg	116 mg

Ingredient	Weight	Measure	Issue
TORTILLAS,WHEAT,10 INCH	12-3/8 lbs	100 each	
LETTUCE,LEAF,FRESH,HEAD	5 lbs		7-3/4 lbs
TOMATOES,FRESH,CHOPPED	6 lbs	3 qts 3-1/8 cup	6-1/8 lbs
CARROTS,FRESH,GRATED	3-1/8 lbs	3 qts 7/8 cup	3-3/4 lbs
PEPPERS,GREEN,FRESH,CHOPPED	3-1/8 lbs	2 qts 1-1/2 cup	3-3/4 lbs
MUSHROOMS,FRESH,WHOLE,SLICED	3-1/8 lbs	1 gal 1-1/8 qts	3-3/8 lbs
SQUASH,FRESH,SUMMER,DICED	3-1/8 lbs	3 qts 1/2 cup	3-1/4 lbs
CUCUMBER,FRESH,CHOPPED	3-1/8 lbs	2 qts 3-7/8 cup	3-3/4 lbs
ONIONS,GREEN,FRESH,SLICED	6 oz	1-3/4 cup	6-2/3 oz
SALAD DRESSING,CREAMY GARLIC,FAT FREE	5-5/8 lbs	2 qts 1 cup	

Method

1 Wrap tortillas in foil; place in warm oven, about 150 F. or warmer for 15 minutes or until warm and pliable.
2 Cut lettuce into 1/2-inch strips.
3 Toss lettuce, tomatoes, carrots, peppers, mushrooms, squash, cucumbers, green onions and dressing.
4 Place 5 ounces, (about 1 cup), vegetable mixture on warmed tortilla.
5 Roll up tortilla; wrap with parchment, wax paper, or foil. CCP: Hold for service at 41 F. or lower Batch preparation methods should be used to prevent the lettuce from wilting and the tortillas from getting soggy.

SANDWICHES No.N 047 00

ROAST BEEF & CHEESE ROLLUP SANDWICH

Yield 100   **Portion** 1 Sandwich

Calories	Carbohydrates	Protein	Fat	Cholesterol	Sodium	Calcium
335 cal	35 g	29 g	9 g	54 mg	1020 mg	226 mg

Ingredient	**Weight**	**Measure**	**Issue**
SALSA		1 gal 3-7/8 qts	
TORTILLAS,WHEAT,10 INCH	12-3/8 lbs	100 each	
LETTUCE,LEAF,FRESH,HEAD	7-7/8 lbs		
BEEF,OVEN ROAST,PRE COOKED	12-1/2 lbs		
CHEESE,MONTEREY JACK,REDUCED FAT,SHREDDED	5-1/2 lbs	1 gal 1-1/2 qts	

Method

1. Prepare 2 recipes of salsa (Recipe No. O 007 01) for use in Step 8.
2. Wrap tortillas in foil; place in warm oven (150 F.) or warmer for 15 minutes or until warm and pliable.
3. Cut lettuce into 1/2-inch strips.
4. Slice beef thin, about 16 slices per pound.
5. Place 2 ounces (2 slices) roast beef on warmed tortilla.
6. Evenly distribute 1 ounce (2 tablespoons) of shredded cheese over beef.
7. Combine lettuce with salsa.
8. Distribute about 3 ounces (3/4 cup) salsa and lettuce over beef and cheese.
9. Roll up tortilla; wrap with parchment, wax paper, or foil. CCP: Hold for service at 41 F. or lower. Batch preparation methods should be used to prevent the lettuce from wilting and the tortillas from getting soggy.

SANDWICHES No.N 047 01

HOT ROAST BEEF & CHEESE ROLLUP SANDWICH

Yield 100   Portion   1 Sandwich

Calories	Carbohydrates	Protein	Fat	Cholesterol	Sodium	Calcium
335 cal	35 g	29 g	9 g	54 mg	1020 mg	226 mg

Ingredient	Weight	Measure	Issue
SALSA		1 gal 3-7/8 qts	
BEEF,OVEN ROAST,PRE COOKED	12-1/2 lbs		
TORTILLAS,WHEAT,10 INCH	12-3/8 lbs	100 each	
CHEESE,MONTEREY JACK,REDUCED FAT,SHREDDED	5-1/2 lbs	1 gal 1-1/2 qts	

Method

1 Prepare 2 recipes of salsa (Recipe No. O 007 01) for use in Step 8.
2 Slice beef thin, about 16 slices per pound.
3 Place 2 ounces (2 slices) of roast beef on tortilla.
4 Evenly distribute 1 ounce (2 tablespoons) of shredded cheese over beef.
5 Distribute about 2 ounces (1/4 cup) salsa over beef and cheese.
6 Roll up tortilla; wrap with foil sheet. Place 20 roll-ups on each sheet pan.
7 Using a convection oven, bake at 325 F. for 20 minutes or until cheese is melted on high fan, closed vent. CCP: Hold for service at 140 F. or higher.

SANDWICHES No.N 048 00

JAMAICAN JERK CHICKEN SANDWICH

Yield 100   **Portion** 1 Sandwich

Calories	Carbohydrates	Protein	Fat	Cholesterol	Sodium	Calcium
387 cal	41 g	38 g	7 g	88 mg	623 mg	91 mg

Ingredient	Weight	Measure	Issue
TROPICAL FRUIT SALSA		2 gal 3/4 qts	
PEPPER,BLACK,GROUND	2-3/8 oz	1/2 cup 2-2/3 tbsp	
ONION POWDER	2-1/2 oz	1/2 cup 2-2/3 tbsp	
SALT	2-1/8 oz	3-1/3 tbsp	
PEPPER,RED,GROUND	2 oz	1/2 cup 2-2/3 tbsp	
NUTMEG,GROUND	1-1/3 oz	1/4 cup 1-2/3 tbsp	
ALLSPICE,GROUND	1-1/8 oz	1/4 cup 1-2/3 tbsp	
PEPPER,RED,CRUSHED	3/8 oz	1/4 cup 1-2/3 tbsp	
THYME,GROUND	3/4 oz	1/4 cup 1-2/3 tbsp	
JUICE,LIME	1 lbs	2 cup	
CHICKEN,BREAST,BNLS/SKNLS,5 OZ	31-1/4 lbs		
COOKING SPRAY,NONSTICK	1-1/2 oz	3 tbsp	
ROLL,KAISER	12-5/8 lbs	100 each	

Method

1. Prepare Tropical Fruit Salsa, Recipe No. O 030 00. Cover. CCP: Refrigerate product at 41F or lower for use in Step 8.
2. Combine black pepper, onion powder, salt, ground red pepper, nutmeg, allspice, crushed red pepper, and thyme. Stir until well blended.
3. Add lime juice to spices. Mix until smooth paste is formed.
4. Add jerk paste to chicken. Evenly coat chicken with paste.
5. Place chicken breasts on lightly sprayed sheet pans. Lightly spray breasts with cooking spray.
6. Using a convection oven, bake 10 to 12 minutes at 325 F. on high fan, closed vent. CCP: Internal temperature must reach 165 F. or higher for 15 seconds.
7. Place chicken breast on bottom half of roll. CCP: Hold for service at 140 F. or higher. Cover with top half. Serve with 1/4 cup Tropical Fruit Salsa or Pineapple Salsa.

SANDWICHES No. N 049 00

MEXICAN BEEF WRAP

Yield 100   Portion   1 Each

Calories	Carbohydrates	Protein	Fat	Cholesterol	Sodium	Calcium
399 cal	40 g	30 g	14 g	74 mg	734 mg	177 mg

Ingredient	Weight	Measure	Issue
BEEF,GROUND,BULK,RAW,90% LEAN	20 lbs		
TOMATOES,CANNED,DICED,DRAINED	8-1/4 lbs	3 qts 3 cup	
CORN,FROZEN,WHOLE KERNEL	5-3/8 lbs	3 qts 3 cup	
PEPPERS,GREEN,FRESH,CHOPPED	2-3/4 lbs	2 qts 1/4 cup	3-1/3 lbs
ONIONS,FRESH,CHOPPED	2-7/8 lbs	2 qts 1/4 cup	3-1/4 lbs
TOMATO PASTE,CANNED	1-3/4 lbs	3 cup	
VINEGAR,DISTILLED	1-1/8 lbs	2-1/4 cup	
SALT	2-1/8 oz	3-1/3 tbsp	
CHILI POWDER,DARK,GROUND	1-3/4 oz	1/4 cup 3 tbsp	
GARLIC POWDER	1 oz	3-1/3 tbsp	
CUMIN,GROUND	3/4 oz	3-1/3 tbsp	
PEPPER,BLACK,GROUND	3/4 oz	3-1/3 tbsp	
PEPPER,RED,GROUND	1/8 oz	1/3 tsp	
TORTILLAS,WHEAT,10 INCH	12-3/8 lbs	100 each	
CHEESE,MONTEREY JACK,REDUCED FAT,SHREDDED	3-1/8 lbs	3 qts 1/2 cup	

Method

1. In a steam-jacketed kettle, cook beef until it loses its pink color.
2. Add tomatoes, corn, peppers, onions, tomato paste, vinegar, salt, chili powder, garlic powder, cumin, black pepper, and red pepper to beef. Stir well.
3. Bring to a boil; reduce heat; simmer, uncovered for 35 to 40 minutes or until sauce has reduced and meat mixture is a moderately dry, packable consistency, stirring occasionally. CCP: Internal temperature must reach 155 F. or higher for 15 seconds. Remove to serving pans. CCP: Hold for service at 140 F. or higher.
4. Wrap tortillas in foil; place in warm oven, about 150 F. or warmer for 15 minutes or until warm and pliable.
5. Place 4-1/4 ounces or 1/2 cup beef mixture in the center of the warmed tortilla.
6. Evenly distribute 1/2 ounce or 1 tablespoon cheese over beef.
7. Fold in sides of tortilla, roll up burrito style; wrap with parchment, wax, or foil. CCP: Hold for service at 140 F. or higher. Batch preparation methods should be used to prevent tortillas from getting soggy.

SANDWICHES No.N 049 01

MEXICAN TURKEY WRAP

Yield 100   Portion 1 Each

Calories	Carbohydrates	Protein	Fat	Cholesterol	Sodium	Calcium
329 cal	41 g	28 g	6 g	52 mg	509 mg	159 mg

Ingredient	Weight	Measure	Issue
TOMATOES,CANNED,DICED,INCL LIQUIDS	6-7/8 lbs	3 qts	
CORN,FROZEN,WHOLE KERNEL	4 lbs	2 qts 3 cup	
PEPPERS,GREEN,FRESH,CHOPPED	2-1/2 lbs	1 qts 3-5/8 cup	3 lbs
ONIONS,FRESH,CHOPPED	2-1/4 lbs	1 qts 2-3/8 cup	2-1/2 lbs
TOMATO PASTE,CANNED	1-1/2 lbs	2-1/2 cup	
VINEGAR,DISTILLED	13-7/8 oz	1-5/8 cup	
SEASONING, SANTE FE	6-7/8 oz	2 cup	
TURKEY,BREAST,COOKED,DICED	13 lbs		
TORTILLAS,FLOUR,10 INCH	12-3/8 lbs		
CHEESE,MONTEREY JACK,REDUCED FAT,SHREDDED	3-1/8 lbs	3 qts 1/2 cup	

Method

1 Add tomatoes, corn, peppers, onions, tomato paste, vinegar, and Sante Fe Style seasoning to steam jacketed kettle or stockpot. Stir.
2 Bring to a boil; reduce heat; simmer, covered, 5 to 7 minutes stirring frequently.
3 Add turkey to sauce/vegetable mixture; stir well. Bring to a simmer; cover; simmer 5 to 7 minutes stirring frequently to prevent sticking. CCP: Temperature must register 165 F. or higher for 15 seconds. Remove from heat. CCP: Hold at 140 F. or higher for use in Step 5.
4 Wrap tortillas in foil; place in warm oven (about 150 F.) or in a warmer 15 minutes or until warm and pliable.
5 Place 1/2 cup, 1-No. 8 scoop of turkey filling in the center of each warmed tortilla.
6 Evenly distribute 2 tablespoon shredded cheese over turkey filling.
7 Fold up front of tortilla to cover filling; fold in sides of tortilla; roll tightly to the back of tortilla like a burrito. Wrap with parchment, wax paper or foil. CCP: Serve immediately or hold for service at 140 F. or higher.

SANDWICHES No.N 050 00

CRUNCHY VEGETABLE BURRITO

Yield 100   Portion   1 Burrito

Calories	Carbohydrates	Protein	Fat	Cholesterol	Sodium	Calcium
280 cal	50 g	14 g	3 g	4 mg	753 mg	234 mg

Ingredient	Weight	Measure	Issue
YOGURT,PLAIN,NONFAT	6-1/4 lbs	2 qts 3-1/2 cup	
SALAD DRESSING,RANCH,FAT FREE	4-1/4 lbs	2 qts	
GARLIC POWDER	1/2 oz	1 tbsp	
CHILI POWDER,DARK,GROUND	1/2 oz	1 tbsp	
CUMIN,GROUND	1/4 oz	1 tbsp	
BEANS,KIDNEY,DARK RED,CANNED,DRAINED	9-1/8 lbs	1 gal 1-7/8 qts	
SWEET POTATOES,FRESH,PARED,SHREDDED	4-1/2 lbs	3 qts 3-3/8 cup	5-5/8 lbs
TOMATOES,FRESH,CHOPPED	4-1/2 lbs	2 qts 3-3/8 cup	4-5/8 lbs
BROCCOLI,FRESH,FLORETS	3-1/2 lbs	1 gal 1/2 qts	5-3/4 lbs
ONIONS,GREEN,FRESH,SLICED	1 lbs	1 qts 1/2 cup	1-1/8 lbs
PEPPERS,JALAPENOS,CANNED,DRAINED,CHOPPED	7-1/4 oz	1-1/2 cup	
TORTILLAS,WHEAT,10 INCH	12-3/8 lbs	100 each	
CHEESE,MONTEREY JACK,REDUCED FAT,SHREDDED	3-1/4 lbs	3 qts 1 cup	

Method

1 Combine yogurt, ranch dressing, garlic powder, chili powder, and cumin. Blend well. CCP: Refrigerate at 41 F. or lower for use in Step 3.
2 Combine kidney beans, sweet potatoes, tomatoes, broccoli, green onions, and jalapeno peppers.
3 Toss vegetables with dressing until well coated. CCP: Refrigerate at 41 F. or lower for use in Step 5.
4 Wrap tortillas in foil; place in warm oven, about 150 F., or in a warmer for 15 minutes or until warm and pliable.
5 Place 5-1/2 ounces (about 2/3 cup) vegetable mixture on warm tortilla. Top with 1/2 ounce (2 tablespoons) cheese. Spread evenly in center of tortilla. Fold up sides of tortilla; fold up front of tortilla to cover filling; roll tightly to back of tortilla like a burrito; wrap with parchment, wax paper or foil.
6 CCP: Hold for service at 41 F. or lower.

SANDWICHES No.N 051 00

VEGETARIAN HEARTY BURGER

Yield 100   **Portion** 1 Burger

Calories	Carbohydrates	Protein	Fat	Cholesterol	Sodium	Calcium
409 cal	52 g	21 g	13 g	11 mg	647 mg	244 mg

Ingredient	**Weight**	**Measure**	**Issue**
EGG WHITES	7-1/2 lbs	3 qts 2 cup	
CHEESE,MOZZARELLA,PART SKIM,SHREDDED	4-1/2 lbs	1 gal 1/2 qts	
ONIONS,FRESH,GRATED	2-7/8 lbs	2 qts 1/4 cup	3-1/4 lbs
SOY SAUCE	1 lbs	1-1/2 cup	
CEREAL,OATMEAL,ROLLED	8-5/8 lbs	1 gal 2-1/4 qts	
WALNUTS,SHELLED,CHOPPED	1-5/8 lbs	1 qts 2 cup	
GARLIC POWDER	2-3/8 oz	1/2 cup	
SAGE,GROUND	1/4 oz	1/4 cup 1/3 tbsp	
COOKING SPRAY,NONSTICK	2 oz	1/4 cup 1/3 tbsp	
ROLL,SANDWICH BUNS,SPLIT	9-1/2 lbs	100 each	

Method

1. Place egg whites, cheese, onions, and soy sauce in mixer bowl. Using a dough hook, mix on low speed 1 minute or until well blended.
2. Add oats, walnuts, garlic powder, and sage; mix on low speed 1 minute. Scrape down sides; continue mixing 30 seconds, or until well blended. Refrigerate mixture at least one hour to allow mixture to absorb moisture. CCP: Refrigerate at 41 F. or lower.
3. Shape 3-1/2 ounce balls. Place 20 balls on each sheet pan. Cover with parchment paper; flatten into burgers by pressing down with another sheet pan to a thickness of 1/2-inch. Mixture will be very moist and fragile.
4. Grill burgers on lightly sprayed griddle at 400 F. for 6 minutes or bake on lightly sprayed sheet pans in a convection oven at 350 F. for 15 to 20 minutes on high fan, open vent or until golden brown. CCP: Internal temperature must reach 145 F. or higher for 15 seconds.
5. Serve on buns. CCP: Hold for service at 140 F. or higher.

SANDWICHES No.N 052 00

CAJUN CHICKEN SANDWICH

Yield 100   Portion 6 Ounces

Calories	Carbohydrates	Protein	Fat	Cholesterol	Sodium	Calcium
389 cal	41 g	39 g	7 g	88 mg	600 mg	96 mg

Ingredient	**Weight**	**Measure**	**Issue**
TROPICAL FRUIT SALSA		2 gal 3/4 qts	
CHICKEN,BREAST,BNLS/SKNLS,5 OZ	22-3/4 lbs	100 each	
PEPPER,BLACK,GROUND	3-5/8 oz	1 cup	
SALT	1-7/8 oz	3 tbsp	
PAPRIKA,GROUND	1-1/3 oz	1/4 cup 1-2/3 tbsp	
FENNEL,GROUND	1-1/8 oz	1/4 cup 1-2/3 tbsp	
MUSTARD,DRY	2-1/8 oz	1/4 cup 1-2/3 tbsp	
THYME,GROUND	3/4 oz	1/4 cup 1-2/3 tbsp	
PEPPER,RED,GROUND	1/2 oz	3 tbsp	
SAGE,GROUND	3/8 oz	1/4 cup 1-2/3 tbsp	
GARLIC POWDER	1-5/8 oz	1/4 cup 1-2/3 tbsp	
JUICE,LEMON	1-1/8 lbs	2 cup	
COOKING SPRAY,NONSTICK	2 oz	1/4 cup 1/3 tbsp	
ROLL,KAISER	12-5/8 lbs	100 each	

Method

1 Prepare 1 recipe Tropical Fruit Salsa (O 030 00); cover. CCP: Refrigerate at 41 F. or lower for use in Step 8.
2 Wash chicken breasts thoroughly under cold running water. Drain well. Remove excess fat.
3 Combine black pepper, salt, garlic powder, paprika, fennel, mustard flour, thyme, red pepper and sage. Stir until well blended.
4 Add lemon juice to spices. Mix until smooth paste is formed.
5 Add cajun paste to chicken. Mix well to evenly distribute paste mixture.
6 Lightly spray each sheet pan and chicken breast with non-stick cooking spray. Place 25 chicken breasts on each sheet pan.
7 Using a convection oven, bake at 325 F. 10 to 12 minutes on high fan, closed vent. CCP: Internal temperature must reach 165 F. or higher for 15 seconds.
8 Place chicken breast on bottom half of roll. Cover with top half. Serve with 1/4 cup of Tropical Fruit Salsa. CCP: Hold for service at 140 F. or higher.

SANDWICHES No.N 502 00

BAKED TURKEY MELT

Yield 100   **Portion** 1 Sandwich

Calories	Carbohydrates	Protein	Fat	Cholesterol	Sodium	Calcium
363 cal	32 g	30 g	13 g	80 mg	622 mg	194 mg

Ingredient	Weight	Measure	Issue
TURKEY,GROUND,90% LEAN,RAW	25-1/2 lbs		
ONIONS,FRESH,CHOPPED	2-7/8 lbs	2 qts	3-1/8 lbs
PARSLEY,FRESH,BUNCH,CHOPPED	3-1/2 oz	1-5/8 cup	3-3/4 oz
BREADCRUMBS,DRY,GROUND,FINE	3-1/8 lbs	3 qts 1 cup	
SALT	1-1/4 oz	2 tbsp	
GARLIC POWDER	1-1/4 oz	1/4 cup 1/3 tbsp	
PEPPER,WHITE,GROUND	1/2 oz	2 tbsp	
WORCESTERSHIRE SAUCE	8-1/2 oz	1 cup	
MUSTARD,DRY	3/4 oz	2 tbsp	
CHEESE,MOZZARELLA,PART SKIM	3 lbs		
ROLL,SANDWICH BUNS,SPLIT	9-1/2 lbs	100 each	
TOMATOES,FRESH,SLICED	2 lbs	1 qts 1 cup	2 lbs
PEPPERS,GREEN,FRESH,MEDIUM,SLICED,THIN	2 lbs	1 qts 2-1/8 cup	2-1/2 lbs

Method

1. Combine turkey, bread crumbs, onions, parsley, salt, garlic powder, pepper, Worcestershire sauce, and mustard. Mix thoroughly.
2. Shape into patties 1/2-inch thick, weighing 5 ounces.
3. Place turkey on sheet pans. Using a convection oven, bake at 325 F. for 20 to 25 minutes on high fan, open vent. CCP: Internal temperature must reach 165 F. or higher for 15 seconds.
4. Place a 1/2 ounce slice of low fat mozzarella cheese on top of each patty and melt in oven. Serve patty on a hamburger bun. CCP: Hold for service at 140 F. or higher.
5. Garnish with slice of fresh green pepper or tomato (optional).

GUIDELINES FOR PREPARING SAUCES AND GRAVIES

Sauces and gravies are thickened liquids or stocks. They are served with meat, fish, poultry, vegetables, and desserts to add flavor and garnish.

A. INGREDIENTS USED IN PREPARATION OF SAUCES AND GRAVIES:
1. *Liquids* - Vegetable juice, fruit juice, milk, meat or poultry stock maybe used. Liquid and browned particles from meat drippings should be added to stock for flavor and color. In large quantity preparation, liquid should be just below boiling point when thickening agent is added because most starches thicken immediately in 180° F. to 190° F. liquids.
NOTE: It is important to keep the temperature of the reconstituted nonfat dry milk to just below the boiling point because the proteins in milk tend to coagulate at boiling temperature and give the sauce a rough texture.
2. *Thickening Agents*
    a. Roux is a French word for a mixture of flour and fat, cooked to eliminate the raw, uncooked taste of flour.
       (1) Blonde or Light Roux - A smooth mixture of melted fat and flour that must be cooked to eliminate the raw, uncooked taste of flour but should not be browned.
       (2) Brown Roux - A browned mixture of fat and flour. Flour is added to hot fat and cooked over low heat until a golden brown color is formed, about 10 minutes with continuous stirring to prevent scorching. Roux may also be browned in 350° F. to 375° F. oven (about 30 minutes).
    b. Slurry - A lump-free mixture made by dissolving cornstarch into cold water and/or other cold liquids.

3. *Fats* - Fat gives flavor, body, and a finish to sauces and gravies. Fat is also valuable because it separates the starch granules and decreases the chance of lumping. Whenever possible, use fat from meat or poultry drippings. Separate clear fat from meat or poultry drippings to use in roux. Butter, margarine, or shortening may also be used. DO NOT use meat juice; it causes lumps.

B. METHODS USED IN PREPARATION OF SAUCES AND GRAVIES:

Sauces and gravies should be cooked in a heavy saucepan, double boiler, steam-jacketed kettle or stock pot.

1. *Combining Sauces and Gravies - Roux and Paste - When* sauces or gravies are prepared in more than 1/2 gallon volumes, it is preferable to add the near-boiling liquid slowly to the roux or paste while stirring with wire whip. Follow recipe directions, cook ingredients at low heat, stirring constantly until mixture is smooth, thickened, and no longer has a starchy taste.

2. *Prevention of skin on surface of sauce or gravy* - Cover with lid immediately, or spread a thin film of melted butter, margarine, or shortening over surface. Whip thoroughly before serving.

3. *Reheating Sauces or Gravies* - Cold sauces and gravies will scorch easily over direct heat. If possible, reheat in double boiler, steam-jacketed kettle, or over hot water, stirring occasionally.

4. *Adjustments* - If sauce or gravy is too thin, sprinkle a small amount of potato granules into hot mixture, stirring constantly. A mixture of cornstarch and cold water may also be added, stirring constantly until mixture no longer has a starchy taste. If sauce or gravy is too thick, thin with a small amount of hot liquid.

O. SAUCES, GRAVIES AND DRESSINGS No. 0 (1)

INDEX

Card No.		Card No.	
O 001 00	White Sauce	O 016 00	Brown Gravy
O 001 01	Cheese Sauce	O 016 02	Chicken or Turkey Gravy
O 002 00	Barbecue Sauce	O 016 03	Chili Gravy
O 003 00	Cherry Sauce (for Meat)	O 016 04	Giblet Gravy
O 004 00	Marinara Sauce	O 016 05	Mushroom Gravy
O 004 01	Marinara Sauce with Clams	O 016 06	Onion Gravy
O 005 00	Creole Sauce	O 016 07	Quick Onion Gravy
O 005 01	Spanish Sauce	O 016 08	Vegetable Gravy
O 005 02	Cajun Creole Sauce	O 016 09	Onion and Mushroom Gravy
O 006 00	Mustard Sauce	O 017 00	Cream Gravy
O 007 00	Taco Sauce	O 017 01	Cream Onion Gravy
O 007 01	Salsa	O 018 00	Natural Pan Gravy (Au Jus)
O 008 00	Sweet and Sour Sauce	O 019 00	Tomato Gravy
O 009 00	Pineapple Sauce	O 020 00	Corn Bread Dressing
O 009 01	Raisin Sauce	O 021 00	Bread Dressing
O 010 00	Szechwan Sauce	O 021 01	Apple Bread Dressing
O 011 00	Seafood Cocktail Sauce	O 021 02	Sausage Bread Dressing
O 012 00	Pizza Sauce	O 021 03	Oyster Bread Dressing
O 012 01	Pizza Sauce (Canned)	O 022 00	Chinese Mustard Sauce
O 013 00	Tartar Sauce	O 023 00	Horseradish Sauce
O 014 00	Teriyaki Sauce	O 024 00	Yogurt-Cucumber Sauce
O 015 00	Tomato Sauce	O 025 00	Herbed Mayonnaise

O. SAUCES, GRAVIES AND DRESSINGS No. 0 (1)

Card No...

O 026 00	Oriental Sweet and Sour Sauce
O 027 00	Dill Sauce
O 028 00	Horseradish Dijon Sauce
O 029 00	Honey Mustard Sauce
O 030 00	Tropical Fruit Salsa
O 030 01	Pineapple Salsa
O 030 02	Tropical Fruit Salsa (Canned)
O 031 00	Shrimp Sauce
O 801 00	Dressing, Traditional Mix
O 801 01	Dressing, Cornbread Mix

SAUCES, GRAVIES, AND DRESSINGS   No.O 001 00

WHITE SAUCE

Yield 100   Portion   1 Ounce

Calories	Carbohydrates	Protein	Fat	Cholesterol	Sodium	Calcium
41 cal	3 g	1 g	3 g	8 mg	110 mg	27 mg

Ingredient	Weight	Measure	Issue
BUTTER,MELTED	12 oz	1-1/2 cup	
FLOUR,WHEAT,GENERAL PURPOSE	8-7/8 oz	2 cup	
MILK,NONFAT,DRY	7-1/4 oz	3 cup	
WATER,WARM	7-7/8 lbs	3 qts 3 cup	
SALT	5/8 oz	1 tbsp	

Method

1. Blend butter or margarine and flour together using wire whip to form a roux; stir until smooth.
2. Reconstitute milk; heat to just below boiling. DO NOT BOIL.
3. Add milk gradually to roux stirring constantly.
4. Add salt. Simmer 10 to 15 minutes or until thickened. Stir as necessary. CCP: Internal temperature must reach 145 F. or higher for 15 seconds. Hold for service at 140 F. or higher.

SAUCES, GRAVIES, AND DRESSINGS No.O 001 01

CHEESE SAUCE

Yield 100   **Portion** 1 Ounce

Calories	Carbohydrates	Protein	Fat	Cholesterol	Sodium	Calcium
61 cal	3 g	2 g	4 g	13 mg	142 mg	64 mg

Ingredient | Weight | Measure | Issue

Ingredient	Weight	Measure	Issue
BUTTER,MELTED	12 oz	1-1/2 cup	
FLOUR,WHEAT,GENERAL PURPOSE	8-7/8 oz	2 cup	
MILK,NONFAT,DRY	7-1/4 oz	3 cup	
WATER,WARM	7-7/8 lbs	3 qts 3 cup	
SALT	5/8 oz	1 tbsp	
CHEESE,CHEDDAR,SHREDDED	1-1/8 lbs	1 qts 1/2 cup	

Method

1. Blend butter or margarine and flour together using wire whip to form a roux; stir until smooth.
2. Reconstitute milk; heat to just below boiling. DO NOT BOIL.
3. Add milk gradually to roux stirring constantly.
4. Add salt. Simmer 10 to 15 minutes or until thickened. Stir as necessary.
5. Add shredded American or Cheddar Cheese. Stir until blended CCP: Internal temperature must reach 145 F. or higher for 15 seconds. Hold for service at 140 F. or higher.

SAUCES. GRAVIES. AND DRESSINGS No.O 002 00

BARBECUE SAUCE

Yield 100 **Portion** 1/4 Cup

Calories	Carbohydrates	Protein	Fat	Cholesterol	Sodium	Calcium
56 cal	14 g	1 g	0 g	0 mg	569 mg	19 mg

Ingredient	**Weight**	**Measure**	**Issue**
VINEGAR,DISTILLED	1-1/3 lbs	2-1/2 cup	
TOMATO PASTE,CANNED	3-1/2 lbs	1 qts 2 cup	
CATSUP	3-2/3 lbs	1 qts 3 cup	
WATER	3-2/3 lbs	1 qts 3 cup	
SUGAR,BROWN,PACKED	1 lbs	3-1/4 cup	
SALT	1-7/8 oz	3 tbsp	
MUSTARD,PREPARED	8-7/8 oz	1 cup	
PEPPER,RED,GROUND	1/4 oz	1 tbsp	
ONIONS,FRESH,CHOPPED	1 lbs	2-7/8 cup	1-1/8 lbs
CELERY,FRESH,CHOPPED	1 lbs	3-3/4 cup	1-3/8 lbs
GARLIC POWDER	1-5/8 oz	1/4 cup 1-2/3 tbsp	
CHILI POWDER,DARK,GROUND	1/4 oz	1 tbsp	
LIQUID SMOKE	1-7/8 oz	3 tbsp	

Method

1. Combine vinegar, tomato paste, catsup, water, sugar, salt, mustard, red pepper, onions, celery, garlic, chili powder, and liquid smoke (optional).
2. Bring to a boil; reduce heat; cover and simmer for 40 minutes or until sauce is blended. CCP: Internal temperature must reach 145 F. or higher for 15 seconds. Hold for service at 140 F. or higher.

SAUCES, GRAVIES, AND DRESSINGS No.O 003 00

CHERRY SAUCE (FOR MEAT)

Yield 100   **Portion** 3 Tablespoons

Calories	Carbohydrates	Protein	Fat	Cholesterol	Sodium	Calcium
69 cal	17 g	0 g	0 g	1 mg	6 mg	4 mg

Ingredient	**Weight**	**Measure**	**Issue**
CHERRIES,CANNED,RED,TART,WATER PACK,INCL LIQUIDS	6-1/2 lbs	3 qts	
CORNSTARCH	4-1/2 oz	1 cup	
SUGAR,GRANULATED	2-2/3 lbs	1 qts 2 cup	
WATER	1 lbs	2 cup	
RESERVED LIQUID	3-1/8 lbs	1 qts 2 cup	
BUTTER	2 oz	1/4 cup 1/3 tbsp	
FOOD COLOR,RED	1/8 oz	1/8 tsp	
JUICE,LEMON	4-1/3 oz	1/2 cup	

Method

1. Drain cherries; reserve juice for use in Step 3; reserve cherries for use in Step 4.
2. Combine cornstarch and sugar in mixer bowl; add water and stir until smooth.
3. Add water to reserved juice to make recipe amount. Bring to boil and add cornstarch-sugar mixture stirring constantly. Cook 10 minutes or until thick and clear. Remove from heat. CCP: Internal temperature must reach 145 F. or higher for 15 seconds.
4. Add cherries, butter or margarine, food coloring and lemon juice. Mix well.
5. Serve hot or cold. CCP: Hold for service at 140 F. or higher.

SAUCES, GRAVIES, AND DRESSINGS No.O 004 00

MARINARA SAUCE

Yield 100   **Portion** 3/4 Cup

Calories	Carbohydrates	Protein	Fat	Cholesterol	Sodium	Calcium
93 cal	21 g	4 g	1 g	0 mg	891 mg	66 mg

Ingredient	Weight	Measure	Issue
GARLIC POWDER	7/8 oz	3 tbsp	
ONIONS,FRESH,CHOPPED	3-1/8 lbs	2 qts 1 cup	3-1/2 lbs
SHORTENING,VEGETABLE,MELTED	1-3/4 oz	1/4 cup 1/3 tbsp	
TOMATOES,CANNED,CRUSHED,INCL LIQUIDS	26-1/2 lbs	3 gal	
TOMATO PASTE,CANNED	10 lbs	1 gal 1/3 qts	
WATER	8-1/3 lbs	1 gal	
BAY LEAF,WHOLE,DRIED	1/4 oz	6 lf	
OREGANO,CRUSHED	1/3 oz	2 tbsp	
BASIL,DRIED,CRUSHED	1/3 oz	2 tbsp	
SALT	3-3/8 oz	1/4 cup 1-2/3 tbsp	
SUGAR,GRANULATED	5-1/4 oz	3/4 cup	
THYME,GROUND	1/3 oz	2 tbsp	

Method

1. Saute garlic and onions in shortening, salad oil, or olive oil until tender.
2. Combine sauteed onions and garlic with tomatoes, tomato paste, water, bay leaves, oregano, basil, salt, sugar and thyme. Mix well.
3. Bring to a boil; reduce heat and simmer 1 hour or until thickened, stirring occasionally. Remove bay leaves before serving. CCP: Internal temperature must reach 145 F. or higher for 15 seconds. Hold for service at 140 F. or higher.

SAUCES, GRAVIES, AND DRESSINGS No.O 004 01

MARINARA SAUCE WITH CLAMS

Yield 100   **Portion** 3/4 Cup

Calories	Carbohydrates	Protein	Fat	Cholesterol	Sodium	Calcium
95 cal	21 g	4 g	1 g	2 mg	1013 mg	74 mg

Ingredient	**Weight**	**Measure**	**Issue**
GARLIC POWDER	7/8 oz	3 tbsp	
ONIONS,FRESH,CHOPPED	3-1/8 lbs	2 qts 1 cup	3-1/2 lbs
OIL,OLIVE	1-7/8 oz	1/4 cup 1/3 tbsp	
CLAMS,CANNED,CHOPPED	12-1/2 lbs	1 gal 1-7/8 qts	
TOMATOES,CANNED,CRUSHED,INCL LIQUIDS	26-1/2 lbs	3 gal	
TOMATO PASTE,CANNED	10 lbs	1 gal 1/3 qts	
WATER	8-1/3 lbs	1 gal	
BAY LEAF,WHOLE,DRIED	1/4 oz	6 lf	
OREGANO,CRUSHED	1/3 oz	2 tbsp	
BASIL,SWEET,WHOLE,CRUSHED	1/3 oz	2 tbsp	
SALT	3-3/8 oz	1/4 cup 1-2/3 tbsp	
SUGAR,GRANULATED	5-1/4 oz	3/4 cup	
THYME,GROUND	1/3 oz	2 tbsp	

Method

1 Saute garlic and onions in salad oil or olive oil until tender.
2 Drain clams and reserve clam liquid. CCP: Refrigerate clams at 41 F. or lower for use in Step 3. Add water to clam liquid to equal 1 gallon per 100 portions. Combine clam liquid with sauteed onions, garlic, tomatoes, tomato paste, water, bay leaves, oregano, basil, salt, sugar and thyme. Mix well.
3 Bring to a boil; reduce heat and simmer for 1 hour or until thickened, stirring occasionally. Add clams. Stir and simmer about 5 minutes, stirring constantly. DO NOT OVERCOOK. CCP: Internal temperature must reach 145 F. or higher for 15 seconds. Hold for service at 140 F. or higher. Remove bay leaves before serving.

SAUCES, GRAVIES, AND DRESSINGS   No.O 005 00

CREOLE SAUCE

Yield 100   Portion   1/3 Cup

Calories	Carbohydrates	Protein	Fat	Cholesterol	Sodium	Calcium
43 cal	8 g	1 g	1 g	0 mg	212 mg	28 mg

Ingredient

Ingredient	Weight	Measure	Issue
ONIONS,FRESH,CHOPPED	1-1/2 lbs	1 qts 1/4 cup	1-2/3 lbs
PEPPERS,GREEN,FRESH,CHOPPED	1-1/2 lbs	1 qts 1/2 cup	1-7/8 lbs
CELERY,FRESH,CHOPPED	1-1/2 lbs	1 qts 1-5/8 cup	2 lbs
SHORTENING,VEGETABLE,MELTED	3-5/8 oz	1/2 cup	
TOMATOES,CANNED,CRUSHED,INCL LIQUIDS	14-7/8 lbs	1 gal 2-3/4 qts	
SALT	1 oz	1 tbsp	
PEPPER,BLACK,GROUND	1/4 oz	1 tbsp	
SUGAR,GRANULATED	1-3/4 oz	1/4 cup 1/3 tbsp	
WORCESTERSHIRE SAUCE	1 oz	2 tbsp	
FLOUR,WHEAT,GENERAL PURPOSE	4-3/8 oz	1 cup	
WATER	8-1/3 oz	1 cup	

Method

1. Saute onions, peppers and celery in shortening, salad or olive oil for 10 minutes or until tender.
2. Add tomatoes, salt, pepper, sugar, and Worcestershire sauce to vegetables. Bring to a boil; reduce heat; cover and simmer for 10 minutes.
3. Blend flour and water to make a smooth paste; add to sauce. Stir to combine. Simmer for 5 minutes or until thickened, stirring constantly. CCP: Internal temperature must reach 145 F. or higher for 15 seconds. Hold for service at 140 F. or higher.

SAUCES, GRAVIES, AND DRESSINGS No.O 005 01

SPANISH SAUCE

Yield 100 **Portion** 1/3 Cup

Calories	Carbohydrates	Protein	Fat	Cholesterol	Sodium	Calcium
45 cal	8 g	2 g	1 g	0 mg	247 mg	29 mg

Ingredient	Weight	Measure	Issue
ONIONS,FRESH,CHOPPED	1-5/8 lbs	1 qts 5/8 cup	1-3/4 lbs
PEPPERS,GREEN,FRESH,CHOPPED	1-1/2 lbs	1 qts 1/2 cup	1-7/8 lbs
CELERY,FRESH,CHOPPED	1-1/4 lbs	1 qts 3/4 cup	1-3/4 lbs
SHORTENING,VEGETABLE,MELTED	3-5/8 oz	1/2 cup	
TOMATOES,CANNED,CRUSHED,INCL LIQUIDS	14-7/8 lbs	1 gal 2-3/4 qts	
SALT	1 oz	1 tbsp	
PEPPER,BLACK,GROUND	1/4 oz	1 tbsp	
SUGAR,GRANULATED	1-3/4 oz	1/4 cup 1/3 tbsp	
WORCESTERSHIRE SAUCE	1 oz	2 tbsp	
HOT SAUCE	<1/16th oz	<1/16th tsp	
BAY LEAF,WHOLE,DRIED	<1/16th oz	1 lf	
CHILI POWDER,DARK,GROUND	1/4 oz	1 tbsp	
GARLIC POWDER	1/8 oz	1/8 tsp	
MUSHROOMS,CANNED,SLICED,DRAINED	1-3/4 lbs	1 qts 1-1/4 cup	
FLOUR,WHEAT,GENERAL PURPOSE	4-3/8 oz	1 cup	
WATER	8-1/3 oz	1 cup	

Method

1. Saute onions, peppers and celery in shortening, salad or olive oil for 10 minutes or until tender.
2. Add tomatoes, salt, pepper, sugar, Worcestershire sauce, hot sauce, bay leaf, chili powder, garlic, and canned sliced drained mushrooms to vegetables. Bring to a boil; reduce heat; cover and simmer for 10 minutes.
3. Blend flour and water to make a smooth paste; add to sauce. Stir to combine. Simmer for 5 minutes or until thickened, stirring constantly.
4. Remove bay leaves. CCP: Internal temperature must reach 145 F. or higher for 15 seconds. Hold at 140 F. or higher for service.

SAUCES, GRAVIES, AND DRESSINGS No. O 005 02

CAJUN CREOLE SAUCE

Yield 100   Portion 1/3 Cup

Calories	Carbohydrates	Protein	Fat	Cholesterol	Sodium	Calcium
45 cal	8 g	2 g	1 g	0 mg	212 mg	35 mg

Ingredient	Weight	Measure	Issue
ONIONS,FRESH,CHOPPED	1-1/2 lbs	1 qts 1/4 cup	1-2/3 lbs
PEPPERS,GREEN,FRESH,CHOPPED	1-1/2 lbs	1 qts 1/2 cup	1-7/8 lbs
CELERY,FRESH,CHOPPED	1-1/2 lbs	1 qts 1-5/8 cup	2 lbs
SHORTENING,VEGETABLE,MELTED	3-5/8 oz	1/2 cup	
TOMATOES,CANNED,CRUSHED,INCL LIQUIDS	14-7/8 lbs	1 gal 2-3/4 qts	
SALT	1 oz	1 tbsp	
PEPPER,BLACK,GROUND	1/3 oz	1 tbsp	
PEPPER,RED,GROUND	1/8 oz	1/3 tsp	
OREGANO,CRUSHED	3/8 oz	2-2/3 tbsp	
BASIL,SWEET,WHOLE,CRUSHED	3/8 oz	2-2/3 tbsp	
THYME,GROUND	3/8 oz	2-2/3 tbsp	
GARLIC POWDER	1/3 oz	1 tbsp	
PAPRIKA,GROUND	1/4 oz	1 tbsp	
SUGAR,GRANULATED	1-3/4 oz	1/4 cup 1/3 tbsp	
WORCESTERSHIRE SAUCE	1 oz	2 tbsp	
FLOUR,WHEAT,GENERAL PURPOSE	4-3/8 oz	1 cup	
WATER	8-1/3 oz	1 cup	

Method

1. Saute onions, peppers and celery in shortening, salad or olive oil for 10 minutes or until tender.
2. Add tomatoes, salt, black pepper, red pepper, oregano, basil, thyme, garlic powder, paprika, sugar, and Worcestershire sauce to vegetables. Bring to a boil; reduce heat; cover and simmer for 10 minutes.
3. Blend flour and water to make a smooth paste; add to sauce. Stir to combine. Simmer for 5 minutes or until thickened, stirring constantly. CCP: Internal temperature must reach 145 F. or higher for 15 seconds. Hold at 140 F. or higher for service.

SAUCES, GRAVIES, AND DRESSINGS No.O 006 00

MUSTARD SAUCE

Yield 100   **Portion** 2 Tablespoons

Calories	Carbohydrates	Protein	Fat	Cholesterol	Sodium	Calcium
19 cal	3 g	0 g	1 g	1 mg	221 mg	6 mg

Ingredient	**Weight**	**Measure**	**Issue**
CHICKEN BROTH		2 qts 3 cup	
PEPPER,BLACK,GROUND	<1/16th oz	1/8 tsp	
CORNSTARCH	7-7/8 oz	1-3/4 cup	
SUGAR,GRANULATED	1-1/3 oz	3 tbsp	
WATER,COLD	8-1/3 oz	1 cup	
MUSTARD,PREPARED	8-7/8 oz	1 cup	
HORSERADISH,PREPARED	6-1/3 oz	3/4 cup	
VINEGAR,DISTILLED	2-1/8 oz	1/4 cup 1/3 tbsp	
BUTTER	2 oz	1/4 cup 1/3 tbsp	

Method

1. Prepare broth according to directions.
2. Combine pepper, cornstarch, sugar, and water to make a smooth paste. Stir gradually into hot stock. Cook until smooth and thickened, stirring constantly.
3. Add mustard, horseradish, vinegar and butter or margarine; stir until smooth. CCP: Internal temperature must reach 145 F. or higher for 15 seconds. Hold for service at 140 F. or higher.

SAUCES, GRAVIES, AND DRESSINGS   No.O 007 00

TACO SAUCE

Yield 100   Portion   2 Tablespoons

Calories	Carbohydrates	Protein	Fat	Cholesterol	Sodium	Calcium
12 cal	3 g	0 g	0 g	0 mg	225 mg	11 mg

Ingredient   Weight   Measure   Issue

Ingredient	Weight	Measure	Issue
TOMATOES,CANNED,CRUSHED,INCL LIQUIDS	6-5/8 lbs	3 qts	
ONIONS,FRESH,CHOPPED	8-1/2 oz	1-1/2 cup	9-3/8 oz
PEPPERS,JALAPENOS,CANNED,CHOPPED	4-3/4 oz	1 cup	
SALT	1-1/2 oz	2-1/3 tbsp	
SUGAR,GRANULATED	7/8 oz	2 tbsp	

Method

1. Combine tomatoes, onions, peppers, salt, and sugar; blend well.
2. Cover and refrigerate at 41 F. or lower at least 1 hour before serving.

SAUCES, GRAVIES, AND DRESSINGS No. O 007 01

SALSA

Yield 100   **Portion** 2 Tablespoons

Calories	Carbohydrates	Protein	Fat	Cholesterol	Sodium	Calcium
13 cal	3 g	1 g	0 g	0 mg	228 mg	12 mg

Ingredient	Weight	Measure	Issue
TOMATOES,CANNED,DICED,DRAINED	7-1/8 lbs	3 qts 1 cup	
ONIONS,FRESH,CHOPPED	8-1/2 oz	1-1/2 cup	9-3/8 oz
PEPPERS,JALAPENOS,CANNED,CHOPPED	4-3/4 oz	1 cup	
SALT	1-1/2 oz	2-1/3 tbsp	
SUGAR,GRANULATED	7/8 oz	2 tbsp	

Method

1 Combine coarsely chopped canned tomatoes or finely chopped fresh tomatoes with onions, peppers, salt, and sugar. Blend well.

2 Cover and refrigerate at 41 F. or lower at least 1 hour before serving.

SAUCES, GRAVIES, AND DRESSINGS No.O 008 00

SWEET AND SOUR SAUCE

Yield 100 Portion 2 Tablespoons

Calories	Carbohydrates	Protein	Fat	Cholesterol	Sodium	Calcium
94 cal	25 g	0 g	0 g	0 mg	46 mg	8 mg

Ingredient	Weight	Measure	Issue
JAM, PEACH	8-1/2 lbs	3 qts	
VINEGAR, DISTILLED	12-1/2 oz	1-1/2 cup	
WATER	4-1/8 oz	1/2 cup	
SOY SAUCE	1-7/8 oz	3 tbsp	

Method

1 Combine jam, vinegar, water, and soy sauce; optional. Blend well.

SAUCES, GRAVIES, AND DRESSINGS No.O 009 00

PINEAPPLE SAUCE

Yield 100   Portion 1/4 Cup

Calories	Carbohydrates	Protein	Fat	Cholesterol	Sodium	Calcium
71 cal	18 g	0 g	0 g	0 mg	2 mg	5 mg

Ingredient   **Weight**   **Measure**   **Issue**

WATER,BOILING   4-1/8 lbs   2 qts
SUGAR,GRANULATED   2-2/3 lbs   1 qts 2 cup
CORNSTARCH   5-5/8 oz   1-1/4 cup
WATER,COLD   2-1/8 lbs   1 qts
PINEAPPLE,CANNED,CRUSHED,JUICE PACK,INCL LIQUIDS   6-5/8 lbs   3 qts
NUTMEG,GROUND   1/8 oz   1/4 tsp
JUICE,LEMON   4-1/3 oz   1/2 cup
LEMON RIND,GRATED   3/8 oz   2 tbsp

Method

1 Combine sugar and boiling water; stir until dissolved.
2 Blend cornstarch and cold water to make a smooth paste. Add paste to hot water, stirring constantly. Cook for 10 minutes or until thick and clear, stirring constantly.
3 Add pineapple, nutmeg, lemon juice, and rind; mix and return to a boil. Reduce heat; cover and simmer for about 5 minutes. CCP: Internal temperature must reach 145 F. or higher for 15 seconds.
4 Serve hot. CCP: Hold for service at 140 F. or higher.

SAUCES, GRAVIES, AND DRESSINGS   No. O 009 01

RAISIN SAUCE

Yield 100   Portion   3 Tablespoons

Calories	Carbohydrates	Protein	Fat	Cholesterol	Sodium	Calcium
49 cal	13 g	0 g	0 g	0 mg	4 mg	10 mg

Ingredient	Weight	Measure	Issue
SUGAR,BROWN,PACKED	1 lbs	3-1/4 cup	
WATER,BOILING	6-1/4 lbs	3 qts	
RAISINS	1-7/8 lbs	1 qts 2 cup	
CORNSTARCH	4-1/2 oz	1 cup	
WATER	2-1/8 lbs	1 qts	
CINNAMON,GROUND	1/8 oz	1/8 tsp	
CLOVES,GROUND	<1/16th oz	1/8 tsp	
JUICE,LEMON	3-1/4 oz	1/4 cup 2-1/3 tbsp	

Method

1. Combine packed brown sugar and boiling water. Stir until sugar is dissolved.
2. Add raisins and bring to a boil.
3. Blend cornstarch and cold water to make a smooth paste.
4. Add ground cinnamon and ground cloves. Blend well.
5. Slowly add cornstarch mixture to boiling raisin mixture, stirring constantly.
6. Bring to a boil; cook for 5 minutes or until thick and clear, stirring constantly. Remove from heat. CCP: Internal temperature must reach 145 F. or higher for 15 seconds.
7. Add lemon juice and stir well. Serve hot. CCP: Hold for service at 140 F. or higher.

SAUCES, GRAVIES, AND DRESSINGS No.O 010 00

SZECHWAN SAUCE

Yield 100   **Portion** 1/3 Cup

Calories	Carbohydrates	Protein	Fat	Cholesterol	Sodium	Calcium
130 cal	13 g	1 g	9 g	0 mg	569 mg	5 mg

Ingredient	**Weight**	**Measure**	**Issue**
WATER	4-2/3 lbs	2 qts 1 cup	
OIL, SALAD	1-7/8 lbs	1 qts	
VINEGAR, DISTILLED	2-1/8 lbs	1 qts	
SUGAR, GRANULATED	1-3/4 lbs	1 qts	
SOY SAUCE	1-7/8 lbs	3 cup	
CATSUP	1-5/8 lbs	3 cup	
PEPPER, RED, CRUSHED	2/3 oz	1/2 cup	
CORNSTARCH	6-3/4 oz	1-1/2 cup	
WATER, COOL	2-1/8 lbs	1 qts	

Method

1. Combine water, salad oil, vinegar, sugar, soy sauce, catsup, and pepper in steam jacketed kettle or stock-pot; bring to a boil. Reduce heat and simmer for 5 minutes.
2. Combine water and cornstarch. Blend until smooth. Add to mixture slowly while stirring. Bring to a boil; reduce heat and simmer for 3 minutes. CCP: Internal temperature must reach 145 F. or higher for 15 seconds. Hold at 140 F. or higher for service.

Notes

1. This sauce is peppery hot.

SAUCES, GRAVIES, AND DRESSINGS   No.O 011 00

SEAFOOD COCKTAIL SAUCE

Yield 100   Portion   2 Tablespoons

Calories	Carbohydrates	Protein	Fat	Cholesterol	Sodium	Calcium
32 cal	8 g	0 g	0 g	0 mg	357 mg	7 mg

Ingredient

Ingredient	Weight	Measure	Issue
CATSUP	6-1/3 lbs	3 qts	
HORSERADISH, PREPARED	12-2/3 oz	1-1/2 cup	
HOT SAUCE	1/2 oz	1 tbsp	

Method

1 Combine catsup, thawed horseradish, and hot sauce; blend well.

2 Cover and refrigerate at 41 F. or lower.

SAUCES, GRAVIES, AND DRESSINGS   No.O 012 00

PIZZA SAUCE

Yield 100 **Portion** 2-1/2 Tablespoons

Calories	Carbohydrates	Protein	Fat	Cholesterol	Sodium	Calcium
23 cal	5 g	1 g	0 g	0 mg	221 mg	20 mg

Ingredient	Weight	Measure	Issue
OIL,SALAD	1/2 oz	1 tbsp	
ONIONS,FRESH,CHOPPED	12-2/3 oz	2-1/4 cup	14-1/8 oz
TOMATOES,CANNED,CRUSHED,INCL LIQUIDS	8-1/4 lbs	3 qts 3 cup	
TOMATO PASTE,CANNED	1-1/2 lbs	2-5/8 cup	
SUGAR,GRANULATED	1-3/4 oz	1/4 cup 1/3 tbsp	
SALT	1 oz	1 tbsp	
PEPPER,BLACK,GROUND	1/8 oz	1/8 tsp	
BASIL,DRIED,CRUSHED	1/3 oz	2 tbsp	
BAY LEAF,WHOLE,DRIED	1/8 oz	3 each	
GARLIC POWDER	1/8 oz	1/8 tsp	
OREGANO,CRUSHED	1/3 oz	2 tbsp	

Method

1 Saute onions in shortening, salad or olive oil until tender.
2 Add tomatoes, tomato paste, sugar, salt, pepper, basil, bay leaves, garlic, and oregano. Bring to a boil; reduce heat and simmer for 1 hour. Remove bay leaves. CCP: Internal temperature must reach 145 F. or higher for 15 seconds. Hold for service at 140 F. or higher.

SAUCES, GRAVIES, AND DRESSINGS No. O 012 01

PIZZA SAUCE (CANNED)

Yield 100　　　　　　　　　　　　　　　　　　　　Portion 2 Tablespoons

Calories	Carbohydrates	Protein	Fat	Cholesterol	Sodium	Calcium
20 cal	3 g	1 g	0 g	1 mg	67 mg	23 mg

Ingredient	Weight	Measure	Issue
SAUCE,PIZZA,CANNED	8 lbs	3 qts 1-1/4 cup	
BASIL,DRIED,CRUSHED	1/3 oz	2 tbsp	
OREGANO,CRUSHED	1/3 oz	2 tbsp	
GARLIC POWDER	1/4 oz	1/3 tsp	
PEPPER,BLACK,GROUND	1/8 oz	1/3 tsp	

Method

1 Heat canned pizza sauce to simmer. CCP: Internal temperature must reach 145 F. or higher for 15 seconds. Hold for service at 140 F. or higher.

2 If desired, crushed basil, crushed oregano, garlic powder, and black pepper may be added to the pizza sauce.

SAUCES, GRAVIES, AND DRESSINGS No. O 013 00

TARTAR SAUCE

Yield 100   Portion   2 Tablespoons

Calories	Carbohydrates	Protein	Fat	Cholesterol	Sodium	Calcium
103 cal	6 g	0 g	9 g	6 mg	189 mg	1 mg

Ingredient	Weight	Measure	Issue
SALAD DRESSING,MAYONNAISE TYPE	4 lbs	2 qts	
PICKLE RELISH,SWEET	2-1/8 lbs	1 qts	
PARSLEY,FRESH,BUNCH,CHOPPED	1/2 oz	1/4 cup 1/3 tbsp	1/2 oz
PIMIENTO,CANNED,DRAINED,CHOPPED	5-1/8 oz	3/4 cup	
ONIONS,FRESH,CHOPPED	2-7/8 oz	1/2 cup	3-1/8 oz
PAPRIKA,GROUND	<1/16th oz	1/8 tsp	
PEPPER,BLACK,GROUND	<1/16th oz	<1/16th tsp	

Method

1 Combine salad dressing, relish, parsley, pimientos, onions, paprika, and pepper.
2 Cover and refrigerate to chill. Keep refrigerated until ready to serve. CCP: Hold for service at 41 F. or lower.

SAUCES, GRAVIES, AND DRESSINGS No.O 014 00

TERIYAKI SAUCE

Yield 100 **Portion** 2-1/2 Ounces

Calories	Carbohydrates	Protein	Fat	Cholesterol	Sodium	Calcium
89 cal	9 g	4 g	4 g	0 mg	1934 mg	15 mg

Ingredient	Weight	Measure	Issue
SOY SAUCE	7-5/8 lbs	3 qts	
OIL,SALAD	1 lbs	2 cup	
JUICE,PINEAPPLE,CANNED,UNSWEETENED	3-1/3 lbs	1 qts 2 cup	
WATER	4-1/8 lbs	2 qts	
GARLIC POWDER	1/4 oz	3/8 tsp	
GINGER,GROUND	1-1/8 oz	1/4 cup 2-1/3 tbsp	
SUGAR,BROWN,PACKED	1 lbs	3-1/4 cup	
JUICE,LEMON	6-1/2 oz	3/4 cup	
VINEGAR,DISTILLED	8-1/3 oz	1 cup	
ONIONS,FRESH,CHOPPED	12-2/3 oz	2-1/4 cup	14-1/8 oz

Method

1 Combine soy sauce, salad oil, pineapple juice, and water.
2 Add garlic, ginger, brown sugar, lemon juice, vinegar, and onions. Stir to mix well.
3 Pour sauce over meat; cover and refrigerate. Marinate meat 2 hours before cooking. Drain well.

SAUCES, GRAVIES, AND DRESSINGS No.O 015 00

TOMATO SAUCE

Yield 100   **Portion**   1/4 Cup

Calories	Carbohydrates	Protein	Fat	Cholesterol	Sodium	Calcium
47 cal	7 g	1 g	2 g	0 mg	263 mg	9 mg

Ingredient

Ingredient	Weight	Measure	Issue
ONIONS,FRESH,CHOPPED	1-1/4 lbs	3-1/2 cup	1-3/8 lbs
SHORTENING,VEGETABLE,MELTED	7-1/4 oz	1 cup	
FLOUR,WHEAT,GENERAL PURPOSE	7-3/4 oz	1-3/4 cup	
WATER	10-1/2 lbs	1 gal 1 qts	
TOMATO PASTE,CANNED	4 lbs	1 qts 3 cup	
SUGAR,GRANULATED	3-1/2 oz	1/2 cup	
SALT	1 oz	1 tbsp	
PEPPER,BLACK,GROUND	1/8 oz	1/8 tsp	
PEPPER,RED,GROUND	<1/16th oz	1/8 tsp	
GARLIC POWDER	<1/16th oz	<1/16th tsp	

Method

1. Saute onions in shortening, salad or olive oil in steam jacketed kettle or stock pot for 5 minutes or until onions are tender.
2. Add flour to sauteed mixture; stir until well blended. Cook for 5 minutes.
3. Combine water, tomato paste, sugar, salt, pepper, red pepper, and garlic powder. Add to flour and onion mixture.
4. Bring to a boil; reduce heat and simmer for 15 minutes. CCP: Internal temperature must reach 145 F. or higher for 15 seconds. Hold for service at 140 F. or higher.

SAUCES, GRAVIES, AND DRESSINGS   No. O 016 00

BROWN GRAVY

Yield 100 **Portion** 1/4 Cup

Calories	Carbohydrates	Protein	Fat	Cholesterol	Sodium	Calcium
73 cal	5 g	1 g	5 g	0 mg	363 mg	3 mg

Ingredient	Weight	Measure	Issue
SHORTENING	1-1/8 lbs	2-1/2 cup	
FLOUR,WHEAT,GENERAL PURPOSE	1-3/8 lbs	1 qts 1 cup	
BEEF BROTH		1 gal 2 qts	
PEPPER,BLACK,GROUND	<1/16th oz	1/8 tsp	

Method

1. Sprinkle flour evenly over drippings and shortening in bottom of pan. Scrape and use brown particles remaining in pan.
2. Cook at low heat on top of range in a steam-jacketed kettle or in 375 F. oven for 30 minutes until flour is a rich brown color. Stir frequently to avoid over-browning.
3. Add stock to roux, stirring constantly. Bring to a boil; reduce heat; simmer 10 minutes or until thickened, stirring constantly. CCP: Internal temperature must reach 165 F. or higher for 15 seconds.
4. Add pepper. Stir to blend. CCP: Hold at 140 F. or higher for service.

SAUCES, GRAVIES, AND DRESSINGS No.O 016 02

CHICKEN OR TURKEY GRAVY

Yield 100 **Portion** 1/4 Cup

Calories	Carbohydrates	Protein	Fat	Cholesterol	Sodium	Calcium
47 cal	3 g	1 g	3 g	0 mg	414 mg	6 mg

Ingredient | Weight | Measure | Issue

Ingredient	Weight	Measure	Issue
SHORTENING	10-7/8 oz	1-1/2 cup	
FLOUR,WHEAT,GENERAL PURPOSE	13-1/4 oz	3 cup	
CHICKEN BROTH		1 gal 2-1/4 qts	
PEPPER,BLACK,GROUND	<1/16th oz	1/8 tsp	

Method

1. Combine melted shortening or salad oil and sifted general purpose flour. Blend together until smooth and cook at low heat for 2 minutes.
2. Prepare broth according to directions. Add broth to roux, stirring constantly. Bring to a boil; reduce heat and simmer 10 minutes or until thickened, stirring constantly. CCP: Internal temperature must reach 165 F. or higher for 15 seconds.
3. Add pepper. Stir to blend. CCP: Hold at 140 F. or higher for service.

SAUCES. GRAVIES. AND DRESSINGS No.O 016 03

CHILI GRAVY

Yield 100 **Portion** 1/4 Cup

Calories	Carbohydrates	Protein	Fat	Cholesterol	Sodium	Calcium
54 cal	5 g	1 g	4 g	0 mg	421 mg	9 mg

Ingredient

Ingredient	Weight	Measure	Issue
SHORTENING	10-7/8 oz	1-1/2 cup	
FLOUR,WHEAT,GENERAL PURPOSE	13-1/4 oz	3 cup	
TOMATO PASTE,CANNED	1-1/2 lbs	2-1/2 cup	
CHILI POWDER,DARK,GROUND	2-3/8 oz	1/2 cup 1 tbsp	
CUMIN,GROUND	5/8 oz	3 tbsp	
BEEF BROTH		1 gal 2 qts	
PEPPER,BLACK,GROUND	<1/16th oz	1/8 tsp	

Method

1. Use melted shortening or salad oil and sifted general purpose flour. Blend together until smooth and cook at low heat for 20 minutes.
2. Add canned tomato paste, chili powder, and ground cumin; blend well.
3. Prepare broth according to directions. Add broth to roux, stirring constantly. Bring to a boil; reduce heat and simmer for 10 minutes or until thickened, stirring constantly. CCP: Internal temperature must reach 165 F. or higher for 15 seconds.
4. Add pepper. Stir to blend. Hold for service at 140 F. or higher.

SAUCES, GRAVIES, AND DRESSINGS No.O 016 04

GIBLET GRAVY

Yield 100   **Portion** 1/4 Cup

Calories	Carbohydrates	Protein	Fat	Cholesterol	Sodium	Calcium
87 cal	5 g	3 g	6 g	32 mg	402 mg	7 mg

Ingredient   Weight   Measure   Issue

Ingredient	Weight	Measure	Issue
CHICKEN,GIBLETS,FROZEN	3 lbs	1 qts 1-5/8 cup	
FLOUR,WHEAT,GENERAL PURPOSE	1-3/8 lbs	1 qts 1 cup	
SHORTENING,VEGETABLE,MELTED	1-1/8 lbs	2-1/2 cup	
CHICKEN BROTH		1 gal 2 qts	
PEPPER,BLACK,GROUND	<1/16th oz	1/8 tsp	

Method

1. Wash and clean giblets.
2. Cover with water; bring to a boil; reduce heat and simmer for 1 hour or until tender. CCP: Internal temperature must reach 165 F. or higher for 15 seconds.
3. Drain; reserve liquid for use as part of stock for chicken gravy or turkey gravy.
4. Sprinkle flour evenly over shortening in bottom of pan. Cook at low heat on top of range, in a steam-jacketed kettle or in 375 F. oven 30 minutes until flour is a rich brown color. Stir frequently to avoid overbrowning.
5. Use reserved liquid from giblets when preparing chicken broth from mix. Add stock to roux, stirring constantly. Bring to a boil; reduce heat; simmer 10 minutes or until thickened, stirring constantly.
6. Chop giblets coarsely; add to thickened chicken or turkey gravy.
7. Add pepper. Stir to blend. CCP: Hold for service at 140 F. or higher.

SAUCES. GRAVIES. AND DRESSINGS No.O 016 05

MUSHROOM GRAVY

Yield 100   Portion 1/4 Cup

Calories	Carbohydrates	Protein	Fat	Cholesterol	Sodium	Calcium
77 cal	6 g	1 g	6 g	0 mg	404 mg	4 mg

Ingredient	Weight	Measure	Issue
SHORTENING	1-1/8 lbs	2-1/2 cup	
FLOUR,WHEAT,GENERAL PURPOSE	1-3/8 lbs	1 qts 1 cup	
BEEF BROTH		1 gal 2 qts	
MUSHROOMS,CANNED,DRAINED	2 lbs	1 qts 1-3/4 cup	
MARGARINE	1 oz	2 tbsp	
PEPPER,BLACK,GROUND	<1/16th oz	1/8 tsp	

Method

1. Combine melted shortening and flour. Blend together until smooth and cook on low heat for 2 minutes.
2. Prepare broth according to directions. Add broth to roux, stirring constantly. Bring to a boil; reduce heat and simmer for 10 minutes or until thickened, stirring constantly. CCP: Internal temperature must reach 165 F. or higher for 15 seconds.
3. Saute drained canned mushrooms in butter or margarine; drain well. Add to gravy.
4. Add pepper. Stir to blend. CCP: Hold at 140 F. or higher for service.

SAUCES, GRAVIES, AND DRESSINGS No. O 016 06

ONION GRAVY

Yield 100   **Portion** 1/4 Cup

Calories	Carbohydrates	Protein	Fat	Cholesterol	Sodium	Calcium
84 cal	6 g	1 g	6 g	0 mg	363 mg	5 mg

Ingredient	Weight	Measure	Issue
SHORTENING	1-1/8 lbs	2-1/2 cup	
FLOUR,WHEAT,GENERAL PURPOSE	1-3/8 lbs	1 qts 1 cup	
BEEF BROTH		1 gal 2 qts	
ONIONS,FRESH,SLICED	2-1/4 lbs	2 qts 1 cup	2-1/2 lbs
SHORTENING	2-3/4 oz	1/4 cup 2-1/3 tbsp	
PEPPER,BLACK,GROUND	<1/16th oz	1/8 tsp	

Method

1. Combine melted shortening and flour. Blend together until smooth and cook on low heat for 2 minutes.
2. Prepare stock according to package directions. Add stock to roux, stirring constantly. Bring to a boil; reduce heat; simmer 10 minutes or until thickened, stirring constantly. CCP: Internal temperature must reach 165 F. or higher for 15 seconds.
3. Saute thinly sliced, fresh onions in melted shortening or salad oil until onions are tender. Drain and add to gravy.
4. Add pepper. Stir to blend. CCP: Hold at 140 F. or higher for service.

SAUCES, GRAVIES, AND DRESSINGS No. O 016 07

QUICK ONION GRAVY

Yield 100 **Portion** 1/4 Cup

Calories	Carbohydrates	Protein	Fat	Cholesterol	Sodium	Calcium
78 cal	7 g	1 g	5 g	0 mg	307 mg	7 mg

Ingredient	**Weight**	**Measure**	**Issue**
SHORTENING	1-1/8 lbs	2-1/2 cup	
FLOUR,WHEAT,GENERAL PURPOSE	1-3/8 lbs	1 qts 1 cup	
SOUP,DEHYDRATED,ONION	12 oz	2-5/8 cup	
WATER,BOILING	13 lbs	1 gal 2-1/4 qts	

Method

1. Combine melted shortening and flour. Blend together until smooth and cook on low heat for 2 minutes.
2. Use boiling water combined with canned, dehydrated onion soup; simmer for 10 minutes.
3. Add soup mixture to roux, stirring constantly. Bring to a boil; reduce heat and simmer for 10 minutes or until thickened, stirring constantly. CCP: Internal temperature must reach 165 F. or higher for 15 seconds. Hold at 140 F. or higher for service.

SAUCES, GRAVIES, AND DRESSINGS No.O 016 08

VEGETABLE GRAVY

Yield 100   Portion   1/4 Cup

Calories	Carbohydrates	Protein	Fat	Cholesterol	Sodium	Calcium
85 cal	7 g	1 g	6 g	0 mg	365 mg	6 mg

Ingredient	Weight	Measure	Issue
SHORTENING	1-1/8 lbs	2-1/2 cup	
FLOUR,WHEAT,GENERAL PURPOSE	1-3/8 lbs	1 qts 1 cup	
BEEF BROTH		1 gal 2 qts	
CARROTS,FRESH,CHOPPED	15 oz	3-3/8 cup	1-1/8 lbs
ONIONS,FRESH,CHOPPED	12 oz	2-1/8 cup	13-1/3 oz
SHORTENING	1-3/4 oz	1/4 cup 1/3 tbsp	
PEAS,GREEN,FROZEN	1 lbs	3-1/8 cup	
PEPPER,BLACK,GROUND	<1/16th oz	1/8 tsp	

Method

1. Combine melted shortening and flour. Blend together until smooth and cook on low heat for 2 minutes.
2. Prepare broth according to directions. Add broth to roux, stirring constantly. Bring to a boil. Reduce heat; simmer 10 minutes or until thickened, stirring constantly. CCP: Internal temperature must reach 165 F. or higher for 15 seconds.
3. Saute diced fresh carrots and chopped onions in melted shortening or salad oil until tender.
4. Add onions, carrots, and frozen peas to boiling stock. Reduce heat and simmer for 10 minutes or until thickened, stirring constantly.
5. Add pepper. Stir to blend. CCP: Hold at 140 F. or higher for service.

SAUCES, GRAVIES, AND DRESSINGS No. O 016 09

ONION AND MUSHROOM GRAVY

Yield 100   Portion 1/4 Cup

Calories	Carbohydrates	Protein	Fat	Cholesterol	Sodium	Calcium
81 cal	6 g	1 g	6 g	0 mg	380 mg	5 mg

Ingredient	Weight	Measure	Issue
SHORTENING	1-1/8 lbs	2-1/2 cup	
FLOUR,WHEAT,GENERAL PURPOSE	1-3/8 lbs	1 qts 1 cup	
BEEF BROTH		1 gal 2 qts	
MUSHROOMS,CANNED,DRAINED	14 oz	2-1/2 cup	
ONIONS,FRESH,SLICED	1-1/8 lbs	1 qts 1/2 cup	1-1/4 lbs
SHORTENING	1-3/4 oz	1/4 cup 1/3 tbsp	
PEPPER,BLACK,GROUND	<1/16th oz	1/8 tsp	

Method

1. Combine melted shortening and flour. Blend together until smooth and cook on low heat for 2 minutes.
2. Prepare broth according to directions. Add broth to roux, stirring constantly. Bring to a boil; reduce heat and simmer for 10 minutes or until thickened, stirring constantly. CCP: Internal temperature must reach 165 F. or higher for 15 seconds.
3. Saute drained canned mushrooms, and thinly sliced dry onions in melted shortening or salad oil until onions are tender.
4. Add mushrooms and onions to thickened gravy.
5. Add pepper. Stir to blend. CCP: Hold at 140 F. or higher for service.

SAUCES. GRAVIES. AND DRESSINGS No.O 017 00

CREAM GRAVY

Yield 100   Portion   1/4 Cup

Calories	Carbohydrates	Protein	Fat	Cholesterol	Sodium	Calcium
53 cal	5 g	2 g	3 g	1 mg	230 mg	44 mg

Ingredient	Weight	Measure	Issue
MILK,NONFAT,DRY	12 oz	1 qts 1 cup	
WATER,WARM	12-1/2 lbs	1 gal 2 qts	
SHORTENING	10-7/8 oz	1-1/2 cup	
FLOUR,WHEAT,GENERAL PURPOSE	13-1/4 oz	3 cup	
SALT	1-7/8 oz	3 tbsp	
PEPPER,BLACK,GROUND	1/8 oz	3/8 tsp	

Method

1 Reconstitute milk; heat to just below boiling. DO NOT BOIL. Set aside for use in Step 3.
2 Add flour to shortening (and drippings) in roasting pan. Use brown particles remaining in pan. Cook about 5 minutes until light brown, stirring until smooth.
3 Add hot milk from Step 1, stirring constantly.
4 Bring to a simmer and simmer 5 minutes until thickened. CCP: Internal temperature must reach 145 F. or higher for 15 seconds.
5 Add salt and pepper. CCP: Hold for service at 140 F. or higher.

SAUCES. GRAVIES. AND DRESSINGS No.O 017 01

CREAM ONION GRAVY

Yield 100　　　　　　　　　　　　　　　　　　**Portion**　1/4 Cup

Calories	Carbohydrates	Protein	Fat	Cholesterol	Sodium	Calcium
59 cal	6 g	2 g	3 g	1 mg	230 mg	47 mg

Ingredient	Weight	Measure	Issue
MILK,NONFAT,DRY	12 oz	1 qts 1 cup	
WATER,WARM	12-1/2 lbs	1 gal 2 qts	
SHORTENING	10-7/8 oz	1-1/2 cup	
ONIONS,FRESH,CHOPPED	3-1/8 lbs	2 qts 1 cup	3-1/2 lbs
FLOUR,WHEAT,GENERAL PURPOSE	13-1/4 oz	3 cup	
SALT	1-7/8 oz	3 tbsp	
PEPPER,BLACK,GROUND	1/8 oz	3/8 tsp	

Method

1. Reconstitute milk; heat to just below boiling. DO NOT BOIL. Set aside for use in Step 3.
2. Saute chopped fresh onions in shortening and (fat drippings) until tender. Add flour and blend together.
3. Add hot milk from Step 1, stirring constantly.
4. Bring to a simmer and simmer 5 minutes until thickened. CCP: Internal temperature must reach 145 F. or higher for 15 seconds.
5. Add salt and pepper. CCP: Hold for service at 140 F. or higher.

SAUCES, GRAVIES, AND DRESSINGS   No.O 018 00

NATURAL PAN GRAVY (AU JUS)

Yield 100   Portion   2 Tablespoons

Calories	Carbohydrates	Protein	Fat	Cholesterol	Sodium	Calcium
3 cal	0 g	0 g	0 g	0 mg	251 mg	2 mg

Ingredient	Weight	Measure	Issue
BEEF BROTH		3 qts	
SALT	5/8 oz	1 tbsp	
PEPPER,BLACK,GROUND	1/4 oz	1 tbsp	

Method

1 Prepare broth according to directions.
2 Add salt and pepper. CCP: Internal temperature must reach 165 F. or higher for 15 seconds. Hold at 140 F. or higher for service.

SAUCES, GRAVIES, AND DRESSINGS No.O 019 00

TOMATO GRAVY

Yield 100   **Portion**   1/4 Cup

Calories	Carbohydrates	Protein	Fat	Cholesterol	Sodium	Calcium
50 cal	4 g	1 g	3 g	0 mg	315 mg	5 mg

Ingredient | Weight | Measure | Issue

Ingredient	Weight	Measure	Issue
ONIONS,FRESH,CHOPPED	12-2/3 oz	2-1/4 cup	14-1/8 oz
SHORTENING	10-7/8 oz	1-1/2 cup	
FLOUR,WHEAT,GENERAL PURPOSE	13-1/4 oz	3 cup	
BEEF BROTH		3 qts 3 cup	
JUICE,TOMATO,CANNED	5-1/3 lbs	2 qts 2 cup	
PEPPER,BLACK,GROUND	1/8 oz	3/8 tsp	

Method

1. Saute onions in drippings and shortening until tender.
2. Add flour to sauteed onions and stir until well blended.
3. Prepare broth according to package directions. CCP: Internal temperature must reach 145 F. or higher for 15 seconds.
4. Combine tomato juice and broth.
5. Add tomato juice to warm roux, stirring constantly. Bring to a boil; reduce heat and simmer for 5 minutes or until thickened.
6. Add pepper. CCP: Hold at 140 F. or higher for service.

SAUCES. GRAVIES. AND DRESSINGS No. O 020 00

CORN BREAD DRESSING

Yield 100   Portion   3-1/2 Ounces

Calories	Carbohydrates	Protein	Fat	Cholesterol	Sodium	Calcium
175 cal	25 g	5 g	6 g	35 mg	611 mg	97 mg

Ingredient	Weight	Measure	Issue
CELERY,FRESH,CHOPPED	3 lbs	2 qts 3-3/8 cup	4-1/8 lbs
ONIONS,FRESH,CHOPPED	3 lbs	2 qts 1/2 cup	3-1/3 lbs
COOKING SPRAY,NONSTICK	2 oz	1/4 cup 1/3 tbsp	
BREAD,WHITE,SLICED	3-3/8 lbs	2 gal 3 qts	
CORN BREAD		50 pc	
PEPPER,BLACK,GROUND	1/4 oz	1 tbsp	
SEASONING,POULTRY	1/2 oz	1/4 cup 1/3 tbsp	
CHICKEN BROTH		1 gal 1 qts	
EGGS,WHOLE,FROZEN	1 lbs	1-7/8 cup	
COOKING SPRAY,NONSTICK	2 oz	1/4 cup 1/3 tbsp	

Method

1 Stir cook celery and onions in a lightly sprayed steam jacketed kettle, about 10 minutes, stirring constantly.
2 Combine breads, pepper, and poultry seasoning. Toss lightly.
3 Pour cooked vegetables over bread mixture and toss lightly.
4 Prepare stock according to directions. CCP: Internal temperature must reach 165 F. or higher for 15 seconds.
5 Mix stock and eggs together and pour over bread and vegetable mixture. Mix lightly but thoroughly.
6 Place 1-3/4 gallon mixture into each sprayed pan.
7 Using a convection oven, bake 300 F. 1 hour or until top is lightly browned, on high fan, open vent.
8 Cut each pan 5 by 10. CCP: Hold for service at 140 F. or higher.

SAUCES, GRAVIES, AND DRESSINGS No. O 021 00

BREAD DRESSING

Yield 100   **Portion**   3-1/2 Ounces

Calories	Carbohydrates	Protein	Fat	Cholesterol	Sodium	Calcium
142 cal	24 g	4 g	3 g	1 mg	682 mg	63 mg

Ingredient

Ingredient	Weight	Measure	Issue
CELERY,FRESH,CHOPPED	2 lbs	1 qts 3-1/2 cup	2-3/4 lbs
ONIONS,FRESH,CHOPPED	2 lbs	1 qts 1-5/8 cup	2-1/4 lbs
COOKING SPRAY,NONSTICK	2 oz	1/4 cup 1/3 tbsp	
BREAD,WHITE,SLICED	10 lbs	8 gal 3/8 qts	
CHICKEN BROTH		1 gal 2-1/2 qts	
THYME,GROUND	1/3 oz	2 tbsp	
SEASONING,POULTRY	1/4 oz	2 tbsp	
PEPPER,BLACK,GROUND	1/4 oz	1 tbsp	
COOKING SPRAY,NONSTICK	2 oz	1/4 cup 1/3 tbsp	

Method

1. Stir cook celery and onions in a lightly sprayed steam jacketed kettle, about 10 minutes, stirring constantly.
2. Pour cooked vegetables over bread; toss lightly.
3. Prepare chicken broth according to package directions.
4. Combine stock, thyme, poultry seasoning, and pepper; add to bread mixture. Mix lightly. DO NOT OVERMIX.
5. Place 13 lb 1 oz (6-1/2 quart) mixture into each lightly sprayed pan.
6. Using a convection oven, bake at 325 F. 50 to 55 minutes or until top is lightly browned on low fan, open vent. CCP: Internal temperature must reach 165 F. for 15 seconds.
7. Cut each pan 5 by 10. CCP: Hold for service at 140 F. or higher.

SAUCES. GRAVIES. AND DRESSINGS No.O 021 01

APPLE BREAD DRESSING

Yield 100   **Portion**   3-1/2 Ounces

Calories	Carbohydrates	Protein	Fat	Cholesterol	Sodium	Calcium
151 cal	27 g	4 g	3 g	1 mg	517 mg	60 mg

Ingredient | Weight | Measure | Issue

Ingredient	Weight	Measure	Issue
CELERY,FRESH,CHOPPED	2 lbs	1 qts 3-1/2 cup	2-3/4 lbs
ONIONS,FRESH,CHOPPED	2 lbs	1 qts 1-5/8 cup	2-1/4 lbs
COOKING SPRAY,NONSTICK	2 oz	1/4 cup 1/3 tbsp	
BREAD,WHITE,SLICED	10 lbs	8 gal 3/8 qts	
APPLES,FRESH,PEELED,SLICED	4-3/4 lbs	1 gal 1/3 qts	6-1/8 lbs
CHICKEN BROTH		1 gal	
SEASONING,POULTRY	1/4 oz	2 tbsp	
PEPPER,BLACK,GROUND	1/4 oz	1 tbsp	
COOKING SPRAY,NONSTICK	2 oz	1/4 cup 1/3 tbsp	

Method

1. Stir cook celery and onions in a lightly sprayed steam jacketed kettle, about 10 minutes, stirring constantly.
2. Combined bread and apples. Pour cooked vegetables over bread and apples; toss lightly.
3. Combine stock, poultry seasoning, and pepper; add to bread mixture. Mix lightly. DO NOT OVERMIX.
4. Place 13 lb (6-3/4 quart) mixture into each lightly sprayed pan.
5. Using a convection oven, bake at 325 F. 1 hour on low fan, open vent. CCP: Internal temperature must reach 165 F. for 15 seconds.
6. Cut each pan 5 by 10.
7. CCP: Hold for service at 140 F. or higher.

SAUCES. GRAVIES. AND DRESSINGS No.O 021 02

SAUSAGE BREAD DRESSING

Yield 100   Portion   3-1/2 Ounces

Calories	Carbohydrates	Protein	Fat	Cholesterol	Sodium	Calcium
191 cal	24 g	7 g	7 g	12 mg	759 mg	66 mg

Ingredient

Ingredient	Weight	Measure	Issue
CELERY,FRESH,CHOPPED	2 lbs	1 qts 3-1/2 cup	2-3/4 lbs
ONIONS,FRESH,CHOPPED	2-1/8 lbs	1 qts 2 cup	2-1/3 lbs
COOKING SPRAY,NONSTICK	2 oz	1/4 cup 1/3 tbsp	
BREAD,WHITE,SLICED	10 lbs	8 gal 3/8 qts	
SAUSAGE,PORK,COOKED,DICED	3 lbs		
CHICKEN BROTH		1 gal 1 qts	
THYME,GROUND	1/3 oz	2 tbsp	
SEASONING,POULTRY	1/4 oz	2 tbsp	
PEPPER,BLACK,GROUND	1/4 oz	1 tbsp	
COOKING SPRAY,NONSTICK	2 oz	1/4 cup 1/3 tbsp	

Method

1 Lightly spray non-stick cooking spray in steam-jacketed kettle. Stir-cook celery and onions about 10 minutes, stirring constantly.
2 Combine bread and sausage. Pour cooked vegetables over bread and sausage; toss lightly.
3 Combine stock, thyme, poultry seasoning, and pepper; add to bread mixture. Mix lightly. DO NOT OVER MIX.
4 Place 13 lb 2 oz (6-3/4 quart) mixture into each lighly sprayed pan.
5 Using a convection oven, bake at 325 F. 1 hour on low fan, open vent. CCP: Internal temperature must reach 165 F. for 15 seconds.
6 Cut each pan 5 by 10.
7 Hold for service at 140 F. or higher.

SAUCES, GRAVIES, AND DRESSINGS No.O 021 03

OYSTER BREAD DRESSING

Yield 100   **Portion** 3-1/2 Ounces

Calories	Carbohydrates	Protein	Fat	Cholesterol	Sodium	Calcium
173 cal	26 g	8 g	4 g	22 mg	496 mg	63 mg

Ingredient	**Weight**	**Measure**	**Issue**
OYSTERS,FROZEN	6 lbs		
CELERY,FRESH,CHOPPED	2 lbs	1 qts 3-1/2 cup	2-3/4 lbs
ONIONS,FRESH,CHOPPED	2 lbs	1 qts 1-5/8 cup	2-1/4 lbs
COOKING SPRAY,NONSTICK	2 oz	1/4 cup 1/3 tbsp	
BREAD,WHITE,SLICED	10 lbs	8 gal 3/8 qts	
CHICKEN BROTH		3 qts	
RESERVED LIQUID	2-1/8 lbs	1 qts	
THYME,GROUND	1/3 oz	2 tbsp	
SEASONING,POULTRY	1/4 oz	2 tbsp	
PEPPER,BLACK,GROUND	1/4 oz	1 tbsp	
COOKING SPRAY,NONSTICK	2 oz	1/4 cup 1/3 tbsp	

Method

1. Thaw frozen oysters. Drain oysters; reserve and refrigerate liquid for use in Step 4. Chop oysters; reserve and refrigerate for use in Step 4. CCP: Refrigerate at 41 F. or lower.
2. Stir cook celery and onions in a lightly sprayed steam jacketed kettle, about 10 minutes, stirring constantly.
3. Pour cooked vegetables over bread; toss lightly.
4. Combine stock, oysters, reserved oyster liquid, thyme, poultry seasoning, and pepper; add to bread mixture. Mix lightly. DO NOT OVERMIX.
5. Lightly spray each pan with non-stick cooking spray. Place 12 lbs 9 oz (6-1/2 quart) mixture into each lightly sprayed pan.
6. Using a convection oven, bake at 325 F. 1 hour on low fan, open vent. CCP: Internal temperature must reach 165 F. or higher for 15 seconds.
7. Cut each pan 5 by 10.
8. CCP: Hold for service at 140 F. or higher.

SAUCES, GRAVIES, AND DRESSINGS No.O 022 00

CHINESE MUSTARD SAUCE

Yield 100 **Portion** 1 Teaspoon

Calories	Carbohydrates	Protein	Fat	Cholesterol	Sodium	Calcium
17 cal	1 g	1 g	1 g	0 mg	0 mg	19 mg

Ingredient	Weight	Measure	Issue
WATER	12-1/2 oz	1-1/2 cup	
MUSTARD,DRY	12-5/8 oz	2 cup	

Method

1 Add water gradually to mustard and blend until smooth.

SAUCES, GRAVIES, AND DRESSINGS No.O 023 00

HORSERADISH SAUCE

Yield 100   **Portion** 1 Tablespoon

Calories	Carbohydrates	Protein	Fat	Cholesterol	Sodium	Calcium
40 cal	2 g	0 g	3 g	3 mg	62 mg	12 mg

Ingredient | **Weight** | **Measure** | **Issue**

Ingredient	Weight	Measure	Issue
HORSERADISH,PREPARED	1-1/4 lbs	2-1/4 cup	
SALAD DRESSING,MAYONNAISE TYPE	1-1/2 lbs	3 cup	
MILK,NONFAT,DRY	2-3/8 oz	1 cup	
GARLIC POWDER	1/8 oz	1/8 tsp	
ONION POWDER	1/4 oz	1 tbsp	
SUGAR,GRANULATED	7/8 oz	2 tbsp	
PEPPER,WHITE,GROUND	<1/16th oz	1/8 tsp	
PEPPER,RED,GROUND	<1/16th oz	1/8 tsp	

Method

1. Combine horseradish, salad dressing, milk, garlic, onion powder, sugar, white pepper, and red pepper in mixer bowl. Blend on high speed for 1 minute.
2. Cover and refrigerate to chill. CCP: Hold for service at 41 F. or lower.

SAUCES. GRAVIES. AND DRESSINGS No.O 024 00

YOGURT-CUCUMBER SAUCE

Yield 100 **Portion** 3 Tablespoons

Calories	Carbohydrates	Protein	Fat	Cholesterol	Sodium	Calcium
22 cal	3 g	2 g	0 g	2 mg	21 mg	59 mg

Ingredient	**Weight**	**Measure**	**Issue**
YOGURT,PLAIN,LOWFAT	6-1/2 lbs	3 qts	
CUCUMBER,FRESH,CHOPPED	4-1/4 lbs	1 gal <1/16th qts	5 lbs
DILL WEED,DRIED	1/2 oz	1/4 cup 1 tbsp	
GARLIC POWDER	1/2 oz	1 tbsp	

Method

1 Combine yogurt, cucumbers, dill weed, and garlic powder. Mix well.
2 CCP: Refrigerate for service at 41 F. or lower.

SAUCES. GRAVIES. AND DRESSINGS No.O 025 00

HERBED MAYONNAISE

Yield 100   **Portion** 2 Tablespoons

Calories	Carbohydrates	Protein	Fat	Cholesterol	Sodium	Calcium
141 cal	4 g	0 g	14 g	10 mg	170 mg	4 mg

Ingredient	**Weight**	**Measure**	**Issue**
SALAD DRESSING,MAYONNAISE TYPE	6-1/8 lbs	3 qts 1/2 cup	
BASIL,DRIED,CRUSHED	1/3 oz	2 tbsp	
PEPPER,WHITE,GROUND	1/8 oz	1/3 tsp	
OREGANO,CRUSHED	1/3 oz	2 tbsp	
MARJORAM,SWEET,GROUND	<1/16th oz	1/3 tsp	

Method

1. Combine salad dressing, basil, pepper, and marjoram in mixer bowl. Blend well at medium speed; about 1 minute.
2. CCP: Refrigerate for service at 41 F. or lower.

SAUCES, GRAVIES, AND DRESSINGS No.O 026 00

ORIENTAL SWEET AND SOUR SAUCE

Yield 100   **Portion** 2 Tablespoons

Calories	Carbohydrates	Protein	Fat	Cholesterol	Sodium	Calcium
35 cal	9 g	0 g	0 g	0 mg	41 mg	3 mg

Ingredient	Weight	Measure	Issue
JUICE,PINEAPPLE,CANNED,UNSWEETENED	3-1/8 lbs	1 qts 1-3/4 cup	
WATER	1-1/3 lbs	2-1/2 cup	
SUGAR,GRANULATED	1-1/8 lbs	2-1/2 cup	
VINEGAR,DISTILLED	14-5/8 oz	1-3/4 cup	
SOY SAUCE	2-1/2 oz	1/4 cup 1/3 tbsp	
GINGER,GROUND	1/8 oz	1/3 tsp	
WATER	1 lbs	2 cup	
CORNSTARCH	5-5/8 oz	1-1/4 cup	

Method

1. Combine pineapple juice, water, sugar, vinegar, soy sauce, and ginger. Bring to a boil and reduce heat.
2. Dissolve cornstarch in water; stir until smooth. Add to sauce, stirring constantly. Simmer until thick and clear, about 5 minutes. Serve hot or cold. CCP: To serve hot, hold for service at 140 F. or higher. CCP: To serve cold, hold for service at 41 F. or lower.

SAUCES. GRAVIES. AND DRESSINGS No.O 027 00

DILL SAUCE

Yield 100   **Portion**   2 Tablespoons

Calories	Carbohydrates	Protein	Fat	Cholesterol	Sodium	Calcium
31 cal	3 g	1 g	1 g	6 mg	22 mg	52 mg

Ingredient

Ingredient	Weight	Measure	Issue
SOUR CREAM,LOW FAT	3-1/2 lbs	1 qts 3 cup	
YOGURT,PLAIN,LOWFAT	3-1/4 lbs	1 qts 2 cup	
SUGAR,GRANULATED	1-3/4 oz	1/4 cup 1/3 tbsp	
DILL WEED,DRIED	5/8 oz	1/4 cup 1-2/3 tbsp	
GARLIC POWDER	1/4 oz	1/3 tsp	

Method

1. Combine sour cream, yogurt, sugar, dill weed, and garlic powder.
2. Using a wire whip, mix at medium speed for 1 minute or until well blended.
3. CCP: Refrigerate for service at 41 F. or lower.

SAUCES. GRAVIES. AND DRESSINGS No.O 028 00

HORSERADISH DIJON SAUCE

Yield 100 **Portion** 2 Tablespoons

Calories	Carbohydrates	Protein	Fat	Cholesterol	Sodium	Calcium
34 cal	2 g	1 g	2 g	9 mg	39 mg	38 mg

Ingredient	**Weight**	**Measure**	**Issue**
SOUR CREAM,LOW FAT	5-1/2 lbs	2 qts 3 cup	
HORSERADISH,PREPARED	1 lbs	2 cup	
MUSTARD,DIJON	2-1/8 oz	1/4 cup 1/3 tbsp	
GARLIC POWDER	1/3 oz	1 tbsp	

Method

1. Place sour cream, horseradish, mustard, and garlic powder in mixer bowl.
2. Using a wire whip, mix on medium speed for 1 minute or until well blended.
3. CCP: Refrigerate for service at 41 F. or lower.

SAUCES, GRAVIES, AND DRESSINGS No.O 029 00

HONEY MUSTARD SAUCE

Yield 100   **Portion**   2 Tablespoons

Calories	Carbohydrates	Protein	Fat	Cholesterol	Sodium	Calcium
74 cal	19 g	1 g	0 g	0 mg	169 mg	13 mg

Ingredient	Weight	Measure	Issue
HONEY	4-2/3 lbs	1 qts 2-1/4 cup	
MUSTARD,DIJON	3-1/3 lbs	1 qts 2-1/4 cup	

Method

1. Combine honey and mustard in mixer bowl.
2. Using a wire whip, mix on medium speed for 3 minutes or until well blended.
3. Whip or stir well before serving. CCP: Refrigerate at 41 F. or lower.

SAUCES, GRAVIES, AND DRESSINGS No.O 030 00

TROPICAL FRUIT SALSA

Yield 100 **Portion** 1/4 Cup

Calories	Carbohydrates	Protein	Fat	Cholesterol	Sodium	Calcium
26 cal	6 g	0 g	0 g	0 mg	1 mg	5 mg

Ingredient	Weight	Measure	Issue
PINEAPPLE,FRESH,DICED	5-1/2 lbs	1 gal	10-1/2 lbs
MANGO,FRESH,DICED-1/2 IN	3-1/8 lbs	2 qts 1/2 cup	4-1/2 lbs
PEPPERS,RED FRESH,DICED	1-1/8 lbs	3-1/2 cup	1-3/8 lbs
PEPPERS,GREEN,FRESH,CHOPPED	1 lbs	3 cup	1-1/4 lbs
ONIONS,RED,FRESH,CHOPPED	10-5/8 oz	1-7/8 cup	11-3/4 oz
JUICE,LIME	7 oz	3/4 cup 2 tbsp	
CILANTRO,DRY	1/8 oz	1 tbsp	

Method

1. Combine pineapple, mangoes, red and green peppers, red onion, lime juice, and cilantro. Mix lightly.
2. CCP: Refrigerate for service at 41 F. or lower.

SAUCES, GRAVIES, AND DRESSINGS No.O 030 01

PINEAPPLE SALSA

Yield 100   **Portion** 1/4 Cup

Calories	Carbohydrates	Protein	Fat	Cholesterol	Sodium	Calcium
27 cal	7 g	0 g	0 g	0 mg	2 mg	7 mg

Ingredient

Ingredient	Weight	Measure	Issue
PINEAPPLE,CANNED,CHUNKS,JUICE PACK,DRAINED	5 lbs	2 qts 3-1/2 cup	
PEACHES,CANNED,SLICED,JUICE PACK,DRAINED,CHOPPED	4-3/8 lbs	2 qts	
PEPPERS,RED FRESH,DICED	1-1/8 lbs	3-1/2 cup	1-3/8 lbs
PEPPERS,GREEN,FRESH,CHOPPED	1 lbs	3 cup	1-1/4 lbs
ONIONS,RED,FRESH,CHOPPED	11-1/4 oz	2 cup	12-1/2 oz
JUICE,LIME	3 oz	1/4 cup 2-1/3 tbsp	
CILANTRO,DRY	1/8 oz	1 tbsp	
RESERVED LIQUID	6-1/4 oz	3/4 cup	

Method

1 Drain fruit. Reserve pineapple juice. Combine pineapple, peaches, red and green peppers, red onion, pineapple juice, lime juice, and cilantro. Mix lightly.
2 CCP: Refrigerate for service at 41 F. or lower.

SAUCES. GRAVIES. AND DRESSINGS No.O 030 02

TROPICAL FRUIT SALSA (CANNED)

Yield 100 **Portion** 1/4 Cup

Calories	Carbohydrates	Protein	Fat	Cholesterol	Sodium	Calcium
50 cal	13 g	0 g	0 g	0 mg	2 mg	9 mg

Ingredient	Weight	Measure	Issue
FRUIT SALAD,TROPICAL,CANNED,HEAVY SYRUP,DRAINED	11-7/8 lbs	1 gal 1-1/4 qts	
PEPPERS,RED FRESH,DICED	1-1/8 lbs	3-1/2 cup	1-3/8 lbs
PEPPERS,GREEN,FRESH,CHOPPED	1 lbs	3 cup	1-1/4 lbs
ONIONS,RED,FRESH,CHOPPED	10-5/8 oz	1-7/8 cup	11-3/4 oz
JUICE,LIME	2-7/8 oz	1/4 cup 2 tbsp	
RESERVED LIQUID	6-1/4 oz	3/4 cup	
CILANTRO,DRY	1/8 oz	1 tbsp	

Method

1. Drain canned fruit salad and reserve juice. Coarsely chop fruit pieces. Add red and green peppers, red onion, reserved juice, lime juice, and cilantro. Mix lightly.
2. CCP: Refrigerate for service at 41 F. or lower.

SAUCES. GRAVIES. AND DRESSINGS No.O 031 00

SHRIMP SAUCE

Yield 100   **Portion**   3/4 Cup

Calories	Carbohydrates	Protein	Fat	Cholesterol	Sodium	Calcium
178 cal	10 g	19 g	6 g	130 mg	387 mg	188 mg

Ingredient	Weight	Measure	Issue
SHRIMP,FROZEN,RAW,PEELED,DEVEINED	18 lbs		
WATER	10-1/2 lbs	1 gal 1 qts	
MARGARINE,MELTED	1-1/8 lbs	2-3/8 cup	
FLOUR,WHEAT,GENERAL PURPOSE	1-2/3 lbs	1 qts 2 cup	
RESERVED LIQUID	27-1/8 lbs	3 gal 1 qts	
MILK,NONFAT,DRY	1-2/3 lbs	2 qts 3 cup	
GARLIC POWDER	3/4 oz	2-2/3 tbsp	
ONION POWDER	5/8 oz	2-2/3 tbsp	
SALT	5/8 oz	1 tbsp	
DILL WEED,DRIED	5/8 oz	1/4 cup 1-2/3 tbsp	
BASIL,DRIED,CRUSHED	7/8 oz	1/4 cup 1-2/3 tbsp	
PEPPER,WHITE,GROUND	1/3 oz	1 tbsp	
CHEESE,PARMESAN,GRATED	14-1/8 oz	1 qts	
PARSLEY,FRESH,BUNCH,CHOPPED	2-1/8 oz	1 cup	2-1/4 oz

Method

1 CCP: Thaw shrimp under constant refrigeration at 41 F. or lower. Thoroughly rinse under cold running water; drain.
2 Bring water to a boil in steam-jacketed kettle or stock pot. Add shrimp; simmer 2 to 3 minutes. DO NOT OVERCOOK. Drain immediately. Reserve liquid to reconstitute milk. Spread shrimp on sheet pans in single layer; cover loosely. Coarsely chop cooled shrimp. Refrigerate product at 41 F. or lower for use in Step 6.
3 Blend together margarine and flour to form roux; stir until smooth. Cook roux 5 to 7 minutes.
4 Reconstitute milk; add garlic powder, onion powder, salt, dill weed, basil, and pepper. Stir to thoroughly rehydrate herbs.
5 Bring reconstituted milk mixture to a simmer; gradually add roux, stirring constantly. Simmer for 8 to 10 minutes or until thickened.
6 Add shrimp; simmer for 1 minute while stirring. CCP: Internal temperature must reach 145 F. or higher for 15 seconds.
7 Add cheese and parsley; stir. Remove immediately to serving pans. CCP: Hold for service at 140 F. or higher.

INDEX

Card No.		Card No.	
P 001 00	Beef Rice Soup	P 009 03	Chicken Noodle Soup (Canned)
P 001 01	Beef Barley Soup	P 009 04	Chicken with Rice Soup (Canned)
P 001 02	Beef Noodle Soup	P 009 05	Manhattan Clam Chowder (Canned)
P 002 00	Chicken Rice Soup	P 009 06	Minestrone Soup (Canned)
P 002 01	Chicken Noodle Soup	P 009 07	Split Pea Soup with Ham (Canned)
P 003 00	Creole Soup	P 009 08	Tomato Soup (Canned)
P 004 00	Onion Soup	P 009 09	Vegetable Soup (Canned)
P 004 01	French Onion Soup	P 009 10	Vegetable with Beef Soup (Canned)
P 005 00	Tomato Bouillon	P 010 00	Chicken Gumbo Soup
P 006 00	Tomato Soup	P 010 01	Shrimp Gumbo
P 006 01	Tomato Rice Soup	P 011 00	Corn Chowder
P 007 00	Vegetable Soup	P 011 01	Chicken Corn Chowder
P 007 01	Minestrone Soup	P 012 00	Manhattan Clam Chowder
P 008 00	Navy Bean Soup	P 013 00	New England Fish Chowder
P 008 01	Bean Soup with Smoked, Cured Ham Hocks	P 013 01	New England Clam Chowder
P 008 02	Knickerbocker Soup (Bean, Tomato and Bacon)	P 014 00	Cream of Mushroom Soup
		P 014 01	Cream of Broccoli Soup
P 008 03	Old Fashioned Bean Soup	P 015 00	Cream of Potato Soup (Dehydrated Sliced Potatoes)
P 009 00	Beef with Vegetables and Barley Soup (Canned)	P 015 01	Cream of Potato Soup (Fresh White Potatoes)
P 009 01	Bean With Bacon Soup (Canned)	P 016 00	Cream of Potato Soup (Instant Potatoes)
P 009 02	Beef Noodle Soup (Canned)	P 017 00	Spanish Soup (Dehydrated Onion Soup)

P. SOUPS No. 0 (1)

Card No. ....

P 017 01	Onion Soup (Dehydrated Mix)
P 017 02	Mexican Onion Corn Soup (Dehydrated Mix)
P 018 00	Tomato Vegetable Soup (Dehydrated)
P 018 01	Beef Noodle Soup with Vegetables (Dehydrated)
P 018 02	Chicken Noodle Soup (Dehydrated)
P 018 03	Chicken Noodle Soup with Vegetables (Dehydrated)
P 019 00	Pepper Pot Soup
P 020 00	Chicken Vegetable (Mulligatawny) Soup
P 021 00	Zesty Bean Soup
P 021 01	Zesty Bean Soup (Dry Beans)
P 022 00	Chicken Mushroom Soup (Canned)
P 022 01	Doubly Good Chicken Soup (Canned)
P 022 02	Logging Soup (Canned)
P 022 03	Tomato Noodle Soup (Canned)
P 022 04	Vegetable Beef Supreme Soup (Canned)
P 023 00	Split Pea Soup with Ham
P 023 01	Puree Mongole
P 024 00	Cream of Broccoli Soup (Canned)
P 024 01	Cream of Chicken Soup (Canned)
P 024 02	Cream of Mushroom Soup (Canned)
P 025 00	Texas Tortilla Soup

Card No.

P 026 00	Tortellini Soup
P 027 00	Lentil Vegetable Soup
P 028 00	Curried Vegetable Soup
P 029 00	Turkey Vegetable Soup
P 500 00	Asian Stir Fry Soup
P 800 00	Carrot Soup
P 801 00	Velvet Corn Soup
P 802 00	Nutty Split Pea Soup
P 803 00	Egg Drop Soup
P 804 00	Midwestern Tomato Rice Soup

SOUPS No.P 001 00

BEEF RICE SOUP

Yield 100   Portion 1 Cup

Calories	Carbohydrates	Protein	Fat	Cholesterol	Sodium	Calcium
71 cal	11 g	4 g	2 g	4 mg	1702 mg	21 mg

Ingredient   Weight   Measure   Issue

BEEF,DICED,LEAN,RAW   1-1/2 lbs
BEEF BROTH   7 gal
CARROTS,FROZEN,SLICED   1 lbs   3-1/2 cup
CELERY,FRESH,CHOPPED   12-1/8 oz   2-7/8 cup   1 lbs
ONIONS,FRESH,CHOPPED   1 lbs   2-7/8 cup   1-1/8 lbs
PEPPER,BLACK,GROUND   1/8 oz   1/3 tsp
BAY LEAF,WHOLE,DRIED   1/8 oz   3 each
RICE,LONG GRAIN   2 lbs   1 qts 7/8 cup

Method

1 Cook beef in a steam jacketed kettle for 5 minutes. Dice beef into 1/2 inch pieces.
2 Prepare broth according to package directions.
3 Add beef, carrots, celery, onions, pepper and bay leaves to broth in a steam jacketed kettle or stock pot. Cover; bring to a boil.
4 Add rice. Cover; Simmer 20 to 25 minutes stirring occasionally until rice is tender. Remove bay leaves. CCP: Internal temperature must reach 165 F. or higher for 15 seconds. Hold for service at 140 F. or higher.

SOUPS No.P 001 01

BEEF BARLEY SOUP

Yield 100   Portion 1 Cup

Calories	Carbohydrates	Protein	Fat	Cholesterol	Sodium	Calcium
80 cal	13 g	4 g	2 g	4 mg	1703 mg	19 mg

Ingredient	Weight	Measure	Issue
BEEF,DICED,LEAN,RAW	1-1/2 lbs		
BEEF BROTH		7 gal	
CARROTS,FROZEN,SLICED	1 lbs	3-1/2 cup	
CELERY,FRESH,CHOPPED	12-1/8 oz	2-7/8 cup	1 lbs
ONIONS,FRESH,CHOPPED	1 lbs	2-7/8 cup	1-1/8 lbs
PEPPER,BLACK,GROUND	1/8 oz	1/3 tsp	
BAY LEAF,WHOLE,DRIED	1/8 oz	3 each	
BARLEY,UNCOOKED	2-2/3 lbs	1 qts 2 cup	

Method

1. Cook beef in a steam jacketed kettle for 5 minutes. Dice beef into 1/2 inch pieces.
2. Prepare beef broth according to package directions.
3. Add beef broth, beef, carrots, celery, onions, pepper and bay leaves to steam jacketed kettle or stock pot. Cover; bring to a boil.
4. Add barley. Cover; Simmer 25 to 30 stirring occasionally until barley is tender. Remove bay leaves. CCP: Internal temperature must reach 165 F. or higher for 15 seconds. Hold for service at 140 F. or higher.

SOUPS No.P 001 02

BEEF NOODLE SOUP

Yield 100 **Portion** 1 Cup

Calories	Carbohydrates	Protein	Fat	Cholesterol	Sodium	Calcium
55 cal	6 g	4 g	2 g	9 mg	1702 mg	17 mg

Ingredient	Weight	Measure	Issue
BEEF,DICED,LEAN,RAW	1-1/2 lbs		
BEEF BROTH		7 gal	
CARROTS,FROZEN,SLICED	1 lbs	3-1/2 cup	
CELERY,FRESH,CHOPPED	12-1/8 oz	2-7/8 cup	1 lbs
ONIONS,FRESH,CHOPPED	1 lbs	2-7/8 cup	1-1/8 lbs
PEPPER,BLACK,GROUND	1/8 oz	1/3 tsp	
BAY LEAF,WHOLE,DRIED	1/8 oz	3 each	
NOODLES,EGG	1 lbs	2 qts 3-7/8 cup	

Method

1. Cook beef in a steam jacketed kettle for 5 minutes. Dice beef into 1/2 inch pieces.
2. Prepare beef broth according to package directions.
3. Add beef broth, beef, carrots, celery, onions, pepper and bay leaves to steam jacketed kettle or stock pot. Cover; bring to a boil.
4. Add noodles. Stir; bring to a boil. Reduce heat; cover; simmer 15 to 20 minutes stirring occasionally until noodles and vegetables are tender. Remove bay leaves. CCP: Internal temperature must reach 165 F. or higher for 15 seconds. Hold for service at 140 F. or higher.

SOUPS No.P 002 00

CHICKEN RICE SOUP

Yield 100 **Portion** 1 Cup

Calories	Carbohydrates	Protein	Fat	Cholesterol	Sodium	Calcium
73 cal	9 g	4 g	2 g	7 mg	1997 mg	35 mg

Ingredient	Weight	Measure	Issue
CHICKEN BROTH		7 gal 2 qts	
CHICKEN,COOKED,DICED	1-1/2 lbs		
CARROTS,FROZEN,SLICED	1 lbs	3-1/2 cup	
CELERY,FRESH,CHOPPED	12-1/8 oz	2-7/8 cup	1 lbs
ONIONS, FROZEN	1 lbs	3-1/2 cup	
PEPPER,BLACK,GROUND	1/8 oz	1/3 tsp	
BAY LEAF,WHOLE,DRIED	1/8 oz	2 each	
RICE,LONG GRAIN	1-5/8 lbs	1 qts	

Method

1. Prepare broth according to package directions. Combine broth, diced chicken, carrots, celery, onions, pepper, and bay leaves in a steam jacketed kettle or stock pot. Cover; bring to a boil.
2. Add rice and stir. Cover; bring to a boil; reduce heat; simmer for 20 to 25 minutes until chicken is cooked and rice and vegetables are tender. Remove bay leaves.
3. CCP: Internal temperature must reach 165 F. or higher for 15 seconds. Hold for service at 140 F. or higher.

SOUPS No.P 002 01

CHICKEN NOODLE SOUP

Yield 100   Portion 1 Cup

Calories	Carbohydrates	Protein	Fat	Cholesterol	Sodium	Calcium
62 cal	6 g	4 g	2 g	7 mg	1997 mg	31 mg

Ingredient	Weight	Measure	Issue
CHICKEN BROTH		7 gal 2 qts	
CHICKEN,COOKED,DICED	1-1/2 lbs		
CARROTS,FROZEN,SLICED	1 lbs	3-1/2 cup	
CELERY,FRESH,CHOPPED	12-2/3 oz	3 cup	1-1/8 lbs
ONIONS, FROZEN	1 lbs	3-1/2 cup	
PEPPER,BLACK,GROUND	1/8 oz	1/3 tsp	
BAY LEAF,WHOLE,DRIED	1/8 oz	2 each	
SPAGHETTI NOODLES,DRY	1 lbs	1 qts 3/8 cup	

Method

1. Prepare chicken broth according to directions. Combine chicken broth, diced chicken, carrots, celery, onions, pepper, and bay leaves in a steam jacketed kettle or stock pot. Cover; bring to a boil.
2. Add noodles and stir. Cover; bring to a boil; reduce heat; simmer for 15 to 20 minutes, stirring occasionally until chicken is cooked and noodles and vegetables are tender. Remove bay leaves. CCP: Internal temperature must reach 165 F. or higher for 15 seconds. Hold for service at 140 F. or higher.

SOUPS No.P 003 00

CREOLE SOUP

Yield 100 **Portion** 1 Cup

Calories	Carbohydrates	Protein	Fat	Cholesterol	Sodium	Calcium
69 cal	10 g	3 g	2 g	1 mg	1535 mg	17 mg

Ingredient	Weight	Measure	Issue
ONIONS,FRESH,CHOPPED	2-1/8 lbs	1 qts 2 cup	2-1/3 lbs
PEPPERS,GREEN,FRESH,CHOPPED	2 lbs	1 qts 2 cup	2-3/8 lbs
SHORTENING,VEGETABLE,MELTED	3-5/8 oz	1/2 cup	
BEEF BROTH		6 gal	
PEPPER,BLACK,GROUND	1/8 oz	1/3 tsp	
SPAGHETTI NOODLES,DRY	1-1/2 lbs	1 qts 2-1/2 cup	
TOMATO PASTE,CANNED	2-1/3 lbs	1 qts	

Method

1. Saute onions and peppers in salad oil, melted shortening or olive oil for 5 minutes in steam-jacketed kettle or stock pot. Stir occasionally.
2. Prepare stock according to directions.
3. Break spaghetti into 2-inch pieces. Add stock to sauteed peppers and onions. Add pepper, spaghetti, and tomato paste. Stir and bring to a boil; reduce heat and simmer 30 minutes. CCP: Internal temperature must reach 165 F. or higher for 15 seconds. Hold for service at 140 F. or higher.

SOUPS No.P 004 00

ONION SOUP

Yield 100   **Portion** 1 Cup

Calories	Carbohydrates	Protein	Fat	Cholesterol	Sodium	Calcium
107 cal	8 g	2 g	8 g	1 mg	1271 mg	19 mg

Ingredient	Weight	Measure	Issue
ONIONS,FRESH,SLICED	11-3/8 lbs	2 gal 3-1/4 qts	12-2/3 lbs
SHORTENING,VEGETABLE,MELTED	1-1/2 lbs	3-3/8 cup	
FLOUR,WHEAT,GENERAL PURPOSE	8-7/8 oz	2 cup	
PEPPER,BLACK,GROUND	1/8 oz	1/3 tsp	
BEEF BROTH		5 gal 1 qts	

Method

1. Saute onions in shortening or salad oil until lightly browned.
2. Blend flour and pepper with sauteed onions. Blend well. Prepare broth according to package directions. Add to onion mixture. Stir well. Simmer 15 minutes. CCP: Internal temperature must reach 165 F. or higher for 15 seconds. Hold for service at 140 F. or higher.

SOUPS No.P 004 01

FRENCH ONION SOUP

Yield 100   **Portion** 1 Cup

Calories	Carbohydrates	Protein	Fat	Cholesterol	Sodium	Calcium
163 cal	13 g	3 g	11 g	9 mg	1377 mg	51 mg

Ingredient	Weight	Measure	Issue
ONIONS,FRESH,SLICED	11-3/8 lbs	2 gal 3-1/4 qts	12-2/3 lbs
SHORTENING,VEGETABLE,MELTED	1-1/2 lbs	3-3/8 cup	
FLOUR,WHEAT,GENERAL PURPOSE	8-7/8 oz	2 cup	
PEPPER,BLACK,GROUND	1/8 oz	1/3 tsp	
WORCESTERSHIRE SAUCE	2-1/8 oz	1/4 cup 1/3 tbsp	
BEEF BROTH		5 gal 1 qts	
BREAD,WHITE,STALE,SLICED	2 lbs	1 gal 2-1/2 qts	
BUTTER,MELTED	12 oz	1-1/2 cup	
CHEESE,PARMESAN,GRATED	5-1/4 oz	1-1/2 cup	

Method

1 Saute onions in shortening or salad oil until lightly browned.
2 Blend flour, pepper and Worcestershire sauce with sauteed onions. Blend well. Prepare broth according to package directions. Add onion mixture; stir well. Simmer 15 minutes.
3 Prepare Parmesan Croutons. Trim crusts from bread; cut bread into 1/2-inch cubes. Place bread cubes on sheet pans. Brown lightly in 325 F. oven, 20 to 25 minutes or in 375 F. convection oven, 6 minutes on high fan, open vent. Melt butter or margarine; blend in grated Parmesan cheese. Pour mixture over lightly browned croutons in steam table pans; toss lightly.
4 Place 8 croutons in each soup bowl; pour soup over croutons. CCP: Internal temperature must reach 165 F. or higher for 15 seconds. Hold for service at 140 F. or higher.

Notes

1 In Step 1, 2 lbs bread will yield about 1 gallon lightly browned croutons.

SOUPS No. P 005 00

TOMATO BOUILLON

Yield 100   **Portion** 1 Cup

Calories	Carbohydrates	Protein	Fat	Cholesterol	Sodium	Calcium
35 cal	7 g	2 g	0 g	0 mg	974 mg	25 mg

Ingredient	Weight	Measure	Issue
CELERY,FRESH,CHOPPED	4 lbs	3 qts 3-1/8 cup	5-1/2 lbs
ONIONS,FRESH,CHOPPED	4 lbs	2 qts 3-3/8 cup	4-1/2 lbs
WATER,BOILING	16-3/4 lbs	2 gal	
BEEF BROTH		2 gal 2 qts	
JUICE,TOMATO,CANNED	21-3/8 lbs	2 gal 2 qts	
PEPPER,BLACK,GROUND	1/8 oz	1/8 tsp	

Method

1 Combine celery, onions and boiling water. Simmer 30 minutes; strain; discard vegetables; reserve broth for Step 3.
2 Prepare broth according to package directions.
3 Combine reserved vegetable broth, beef broth, tomato juice and pepper. CCP: Internal temperature must reach 165 F. or higher for 15 seconds. Hold for service at 140 F. or higher.

Notes

1 May be served with croutons. Prepare 1/2 recipe Croutons, Recipe No. D 016 00.

SOUPS No.P 006 00

TOMATO SOUP

Yield 100 Portion 1 Cup

Calories	Carbohydrates	Protein	Fat	Cholesterol	Sodium	Calcium
61 cal	12 g	2 g	1 g	0 mg	1028 mg	46 mg

Ingredient	Weight	Measure	Issue
ONIONS,FRESH,CHOPPED	2 lbs	1 qts 1-5/8 cup	2-1/4 lbs
CELERY,FRESH,CHOPPED	2 lbs	1 qts 3-1/2 cup	2-3/4 lbs
SHORTENING,VEGETABLE,MELTED	1-3/4 oz	1/4 cup 1/3 tbsp	
BEEF BROTH		3 gal	
BAY LEAF,WHOLE,DRIED	1/8 oz	3 each	
PEPPER,BLACK,GROUND	1/8 oz	1/3 tsp	
SUGAR,GRANULATED	5-1/4 oz	3/4 cup	
TOMATOES,CANNED,DICED,INCL LIQUIDS	29-7/8 lbs	3 gal 1 qts	

Method

1. Saute onions and celery in shortening or salad oil 5 minutes in steam-jacketed kettle or stock pot. Stir frequently.
2. Prepare broth according to package directions. Add to sauteed onions and celery.
3. Add bay leaves, pepper and sugar. Stir.
4. Cover; bring to a boil; reduce heat; simmer 10 minutes or until vegetables are tender.
5. Add tomatoes; mix well. Cover; bring to a boil; reduce heat; simmer for 5 minutes. Remove bay leaves. CCP: Internal temperature must reach 165 F. or higher for 15 seconds. Hold for service at 140 F. or higher.

SOUPS No.P 006 01

TOMATO RICE SOUP

Yield 100   **Portion** 1 Cup

Calories	Carbohydrates	Protein	Fat	Cholesterol	Sodium	Calcium
86 cal	17 g	2 g	1 g	0 mg	1089 mg	48 mg

Ingredient	Weight	Measure	Issue
ONIONS,FRESH,CHOPPED	2 lbs	1 qts 1-5/8 cup	2-1/4 lbs
CELERY,FRESH,CHOPPED	2 lbs	1 qts 3-1/2 cup	2-3/4 lbs
SHORTENING,VEGETABLE,MELTED	1-3/4 oz	1/4 cup 1/3 tbsp	
WATER,BOILING	2-1/8 lbs	1 qts	
BEEF BROTH		3 gal 1 qts	
BAY LEAF,WHOLE,DRIED	1/8 oz	3 each	
PEPPER,BLACK,GROUND	1/8 oz	1/3 tsp	
SUGAR,GRANULATED	5-1/4 oz	3/4 cup	
RICE,BROWN,LONG GRAIN,DRY	1-3/8 lbs	3-1/2 cup	
TOMATOES,CANNED,DICED,INCL LIQUIDS	29-7/8 lbs	3 gal 1 qts	

Method

1. Saute onions and celery in salad oil or shortening for 5 minutes in steam-jacketed kettle or stock pot. Stir frequently.
2. Prepare broth according to recipe directions. Add broth to sauteed onions and celery.
3. Add bay leaves, pepper and sugar. Stir.
4. Add rice. Cover, bring to a boil; reduce heat; simmer 25 minutes or until rice is tender.
5. Add tomatoes; mix well. Cover; bring to a boil; reduce heat; simmer 5 minutes. Remove bay leaves. CCP: Internal temperature must reach 165 F. or higher for 15 seconds. Hold for service at 140 F. or higher.

SOUPS No.P 007 00

VEGETABLE SOUP

Yield 100   Portion 1 Cup

Calories	Carbohydrates	Protein	Fat	Cholesterol	Sodium	Calcium
57 cal	10 g	3 g	1 g	1 mg	1278 mg	45 mg

Ingredient	Weight	Measure	Issue
CHICKEN BROTH		4 gal 2 qts	
TOMATOES,CANNED,CRUSHED,INCL LIQUIDS	13-1/4 lbs	1 gal 2 qts	
POTATOES,FRESH,PEELED,CUBED	3-1/8 lbs	2 qts 1-1/8 cup	3-7/8 lbs
CELERY,FRESH,CHOPPED	1-1/8 lbs	1 qts 1/4 cup	1-1/2 lbs
CARROTS,FRESH,CHOPPED	1-1/8 lbs	4 cup	1-3/8 lbs
CABBAGE,GREEN,FRESH,CHOPPED	1-1/8 lbs	1 qts 3-1/4 cup	1-3/8 lbs
ONIONS,FRESH,CHOPPED	2-1/8 lbs	1 qts 2 cup	2-1/3 lbs
PEPPERS,GREEN,FRESH,CHOPPED	7-1/8 oz	1-3/8 cup	8-2/3 oz
GARLIC POWDER	1/3 oz	1 tbsp	
PEPPER,BLACK,GROUND	1/8 oz	1/3 tsp	

Method

1. Prepare broth according to directions. Combine broth, tomatoes, potatoes, celery, carrots, cabbage, onions, peppers, garlic powder, and black pepper in a steam jacketed kettle or stock pot. Bring to a boil. Cover; simmer 30 minutes or until vegetables are tender.
2. CCP: Internal temperature must reach 165 F. or higher for 15 seconds. Hold for service at 140 F. or higher.

SOUPS No.P 007 01

MINESTRONE SOUP

Yield 100 Portion 1 Cup

Calories	Carbohydrates	Protein	Fat	Cholesterol	Sodium	Calcium
75 cal	14 g	3 g	1 g	1 mg	1157 mg	38 mg

Ingredient	Weight	Measure	Issue
CHICKEN BROTH		4 gal	
CELERY,FRESH,CHOPPED	1-1/8 lbs	1 qts 1/4 cup	1-1/2 lbs
CARROTS,FRESH,CHOPPED	1-1/8 lbs	4 cup	1-3/8 lbs
POTATOES,FRESH,PEELED,CUBED	2-1/8 lbs	1 qts 2-1/8 cup	2-5/8 lbs
CABBAGE,GREEN,FRESH,CHOPPED	1-1/8 lbs	1 qts 3-1/4 cup	1-3/8 lbs
ONIONS,FRESH,CHOPPED	2-1/8 lbs	1 qts 2 cup	2-1/3 lbs
PEPPERS,GREEN,FRESH,CHOPPED	7-1/8 oz	1-3/8 cup	8-2/3 oz
GARLIC POWDER	1/3 oz	1 tbsp	
PEPPER,BLACK,GROUND	1/8 oz	1/3 tsp	
TOMATOES,CANNED,CRUSHED,DRAINED	6-5/8 lbs	3 qts	
BEANS,KIDNEY,DARK RED,CANNED,DRAINED	2-1/3 lbs	1 qts 2 cup	
BEANS,GREEN,CANNED,DRAINED	1-1/4 lbs	1 qts	
MACARONI NOODLES,SHELLS,DRY	1-3/8 lbs	1 qts 2 cup	

Method

1. Prepare broth according to directions. Combine broth, celery, carrots, potatoes, cabbage, onions, peppers, garlic powder, and black pepper in a steam jacketed kettle or stock pot. Bring to a boil. Reduce heat; cover; simmer for 20 minutes.
2. Add tomatoes, kidney beans, and green beans. Bring to a boil.
3. Add macaroni. Bring to a boil; reduce heat; simmer 8 to 10 minutes or until macaroni is tender.
4. CCP: Internal temperature must reach 165 F. or higher for 15 seconds. Hold for service at 140 F. or higher.

SOUPS No.P 008 00

NAVY BEAN SOUP

Yield 100   Portion 1 Cup

Calories	Carbohydrates	Protein	Fat	Cholesterol	Sodium	Calcium
122 cal	23 g	8 g	0 g	1 mg	582 mg	78 mg

Ingredient	Weight	Measure	Issue
BEANS,WHITE,DRY	6-1/4 lbs	3 qts 2 cup	
WATER,COLD	16-3/4 lbs	2 gal	
HAM BROTH (FROM MIX)		5 gal	
CARROTS,FRESH,SHREDDED	1 lbs	1 qts 1/8 cup	1-1/4 lbs
ONIONS,FRESH,CHOPPED	2 lbs	1 qts 1-5/8 cup	2-1/4 lbs
PEPPER,BLACK,GROUND	1/8 oz	1/3 tsp	
FLOUR,WHEAT,GENERAL PURPOSE	13-1/4 oz	3 cup	
WATER,COLD	2-1/8 lbs	1 qts	

Method

1. Pick over beans, removing discolored beans and foreign matter. Wash thoroughly in cold water.
2. Cover with cold water; bring to a boil; boil 2 minutes. Turn off heat. Cover; let stand 1 hour.
3. Prepare broth according to package directions.
4. Add beans to stock; bring to a boil; cover; simmer 2 hours or until beans are tender.
5. Add carrots, onions and pepper to bean mixture. Simmer 30 minutes.
6. Blend flour and water to form a smooth paste. Stir into soup; cook 10 minutes. CCP: Internal temperature must reach 165 F. or higher for 15 seconds. Hold for service at 140 F. or higher.

SOUPS No.P 008 01

BEAN SOUP WITH SMOKED, CURED HAM HOCKS

Yield 100   **Portion** 1 Cup

Calories	Carbohydrates	Protein	Fat	Cholesterol	Sodium	Calcium
140 cal	23 g	9 g	2 g	4 mg	650 mg	79 mg

Ingredient	Weight	Measure	Issue
BEANS,WHITE,DRY	6-1/4 lbs	3 qts 2 cup	
WATER,COLD	16-3/4 lbs	2 gal	
HAM BROTH (FROM MIX)		5 gal	
PORK,HOCKS,(CURED & SMOKED),FROZEN	2-1/2 lbs		
CARROTS,FRESH,SHREDDED	1 lbs	1 qts 1/8 cup	1-1/4 lbs
ONIONS,FRESH,CHOPPED	2 lbs	1 qts 1-5/8 cup	2-1/4 lbs
PEPPER,BLACK,GROUND	1/8 oz	1/3 tsp	
FLOUR,WHEAT,GENERAL PURPOSE	13-1/4 oz	3 cup	
WATER,COLD	2-1/8 lbs	1 qts	

Method

1. Pick over beans, removing discolored beans and foreign matter. Wash thoroughly in cold water.
2. Cover with cold water; bring to a boil; boil 2 minutes. Turn off heat. Cover; let stand 1 hour.
3. Prepare stock according to recipe. Add to beans; bring to a boil; cover; simmer 2 hours or until beans are tender.
4. Place thawed, smoked, cured pork hocks in water to cover. Simmer 1 hour; remove from heat; cool. Remove lean meat; chop into small pieces.
5. Add carrots, onions, pepper and chopped ham hocks to bean mixture. Simmer 30 minutes.
6. Blend flour and water to form a smooth paste. Stir into soup; cook 10 minutes. CCP: Internal temperature must reach 165 F. or higher for 15 seconds. Hold for service at 140 F. or higher.

SOUPS No.P 008 02

KNICKERBOCKER SOUP (BEAN, TOMATO AND BACON)

Yield 100　　　　　　　　　　　　　　　　　　Portion 1 Cup

Calories	Carbohydrates	Protein	Fat	Cholesterol	Sodium	Calcium
143 cal	26 g	8 g	1 g	2 mg	446 mg	89 mg

Ingredient	Weight	Measure	Issue
BEANS,WHITE,DRY	6-1/4 lbs	3 qts 2 cup	
WATER,COLD	16-3/4 lbs	2 gal	
HAM BROTH (FROM MIX)		3 gal 1 qts	
BACON,RAW	1 lbs		
CARROTS,FRESH,SHREDDED	1 lbs	1 qts 1/8 cup	1-1/4 lbs
ONIONS,FRESH,CHOPPED	2 lbs	1 qts 1-5/8 cup	2-1/4 lbs
POTATOES,FRESH,PEELED,CUBED	5 lbs	3 qts 2-1/2 cup	6-1/8 lbs
PEPPER,BLACK,GROUND	1/8 oz	1/3 tsp	
TOMATOES,CANNED,CRUSHED,INCL LIQUIDS	7-1/4 lbs	1 #10cn	

Method

1 Pick over beans, removing discolored beans and foreign matter. Wash thoroughly in cold water.
2 Cover with cold water; bring to a boil; boil 2 minutes. Turn off heat. Cover; let stand 1 hour.
3 Prepare stock according to recipe. Add to beans; bring to a boil; cover; simmer 2 hours or until beans are tender.
4 Chop raw bacon and brown lightly. Add carrots, onions, pepper and potatoes. Cook 10 minutes, stirring occasionally; add to bean mixture. Crush tomatoes and add. Simmer 25 minutes or until vegetables are tender. CCP: Internal temperature must reach 165 F. or higher for 15 seconds. Hold for service at 140 F. or higher.

SOUPS No.P 008 03

OLD FASHIONED BEAN SOUP

Yield 100   **Portion** 1 Cup

Calories	Carbohydrates	Protein	Fat	Cholesterol	Sodium	Calcium
130 cal	25 g	8 g	0 g	1 mg	535 mg	88 mg

Ingredient	Weight	Measure	Issue
BEANS,WHITE,DRY	6-1/4 lbs	3 qts 2 cup	
WATER,COLD	16-3/4 lbs	2 gal	
HAM BROTH (FROM MIX)		4 gal 1 qts	
CARROTS,FRESH,SHREDDED	1 lbs	1 qts 1/8 cup	1-1/4 lbs
ONIONS,FRESH,CHOPPED	2 lbs	1 qts 1-5/8 cup	2-1/4 lbs
TOMATOES,CANNED,CRUSHED,INCL LIQUIDS	6-1/2 lbs	2 qts 3-3/4 cup	
PEPPER,BLACK,GROUND	1/8 oz	1/3 tsp	
FLOUR,WHEAT,GENERAL PURPOSE	13-1/4 oz	3 cup	
WATER,COLD	2-1/8 lbs	1 qts	

Method

1. Pick over beans, removing discolored beans and foreign matter. Wash thoroughly in cold water.
2. Cover with cold water; bring to a boil; boil 2 minutes. Turn off heat. Cover; let stand 1 hour.
3. Prepare broth according to recipe. Add to beans; bring to a boil; cover; simmer 2 hours or until beans are tender.
4. Add carrots, onions and pepper to bean mixture. Add crushed tomatoes to mixture and simmer for 30 minutes.
5. Blend flour and water to form a smooth paste. Stir into soup; cook 10 minutes. CCP: Internal temperature must reach 165 F. or higher for 15 seconds. Hold for service at 140 F. or higher.

SOUPS No.P 009 00

BEEF WITH VEGETABLES AND BARLEY SOUP (CANNED)

Yield 100 **Portion** 1 Cup

Calories	Carbohydrates	Protein	Fat	Cholesterol	Sodium	Calcium
86 cal	12 g	6 g	2 g	9 mg	1005 mg	2 mg

Ingredient   **Weight**   **Measure**   **Issue**

SOUP,CONDENSED,BEEF W/VEGETABLE AND BARLEY   31-1/4 lbs   3 gal 2-3/4 qts
WATER   23 lbs   2 gal 3 qts

Method

1. Place soup in steam-jacketed kettle or stock pot.
2. Add water to soup. Mix well.
3. Heat to serving temperature. Do not boil. CCP: Internal temperature must reach 165 F. or higher for 15 seconds. Hold for service at 140 F. or higher.

SOUPS No.P 009 01

BEAN WITH BACON SOUP (CANNED)

Yield 100   Portion   1 Cup

Calories	Carbohydrates	Protein	Fat	Cholesterol	Sodium	Calcium
166 cal	26 g	9 g	3 g	4 mg	1046 mg	2 mg

Ingredient	**Weight**	**Measure**	**Issue**
SOUP,CONDENSED,BEAN WITH BACON	31-1/4 lbs	3 gal 1-3/8 qts	
WATER	23 lbs	2 gal 3 qts	

Method

1. Place soup in steam-jacketed kettle or stock pot.
2. Add water to soup. Mix well.
3. Heat, do not boil. CCP: Internal temperature must reach 165 F. or higher for 15 seconds. Hold for service at 140 F. or higher.

SOUPS No.P 009 02

BEEF NOODLE SOUP (CANNED)

Yield 100   Portion 1 Cup

Calories	Carbohydrates	Protein	Fat	Cholesterol	Sodium	Calcium
95 cal	10 g	6 g	4 g	6 mg	1079 mg	19 mg

Ingredient | Weight | Measure | Issue

SOUP,CONDENSED,BEEF NOODLE   31-1/4 lbs   3 gal 2-1/8 qts
WATER   23 lbs   2 gal 3 qts

Method

1 Place soup in steam-jacketed kettle or stock pot.
2 Add water to soup. Mix well.
3 Heat, do not boil. CCP: Internal temperature must reach 165 F. or higher for 15 seconds. Hold for service at 140 F. or higher.

SOUPS No.P 009 03

CHICKEN NOODLE SOUP (CANNED)

Yield 100   Portion   1 Cup

Calories	Carbohydrates	Protein	Fat	Cholesterol	Sodium	Calcium
86 cal	11 g	5 g	3 g	7 mg	1076 mg	18 mg

Ingredient   Weight   Measure   Issue

SOUP,CONDENSED,CHICKEN NOODLE   31-1/4 lbs   3 gal 2-3/8 qts
WATER   23 lbs   2 gal 3 qts

Method

1 Place soup in steam-jacketed kettle or stock pot.
2 Add water to soup. Mix well.
3 Heat, do not boil. CCP: Internal temperature must reach 165 F. or higher for 15 seconds. Hold for service at 140 F. or higher.

SOUPS No.P 009 04

CHICKEN WITH RICE SOUP (CANNED)

Yield 100   **Portion** 1 Cup

Calories	Carbohydrates	Protein	Fat	Cholesterol	Sodium	Calcium
69 cal	8 g	4 g	2 g	7 mg	946 mg	22 mg

Ingredient	**Weight**	**Measure**	**Issue**
SOUP,CONDENSED,CHICKEN WITH RICE	31-1/4 lbs	3 gal 2-3/8 qts	
WATER	23 lbs	2 gal 3 qts	

Method

1. Place soup in steam-jacketed kettle or stock pot.
2. Add water to soup. Mix well.
3. Heat, do not boil. CCP: Internal temperature must reach 165 F. or higher for 15 seconds. Hold for service at 140 F. or higher.

SOUPS No.P 009 05

MANHATTAN CLAM CHOWDER (CANNED)

Yield 100   **Portion** 1 Cup

Calories	Carbohydrates	Protein	Fat	Cholesterol	Sodium	Calcium
79 cal	11 g	4 g	2 g	9 mg	594 mg	42 mg

Ingredient	Weight	Measure	Issue
SOUP,CONDENSED,MANHATTAN CLAM CHOWDER	31-1/4 lbs	3 gal 2-3/4 qts	
WATER	23 lbs	2 gal 3 qts	

Method

1. Place soup in steam-jacketed kettle or stock pot.
2. Add water to soup. Mix well.
3. Heat, do not boil. CCP: Internal temperature must reach 165 F. or higher for 15 seconds. Hold for service at 140 F. or higher.

SOUPS No.P 009 06

MINESTRONE SOUP (CANNED)

Yield 100 **Portion** 1 Cup

Calories	Carbohydrates	Protein	Fat	Cholesterol	Sodium	Calcium
96 cal	13 g	5 g	3 g	1 mg	1058 mg	42 mg

Ingredient   **Weight**   **Measure**   **Issue**

SOUP,CONDENSED,MINESTRONE   31-1/4 lbs   3 gal 2-3/8 qts
WATER   23 lbs   2 gal 3 qts

Method

1 Place soup in steam-jacketed kettle or stock pot.
2 Add water to soup. Mix well.
3 Heat, do not boil. CCP: Internal temperature must reach 165 F. or higher for 15 seconds. Hold for service at 140 F. or higher.

SOUPS No.P 009 07

SPLIT PEA SOUP WITH HAM (CANNED)

Yield 100					Portion 1 Cup

Calories	Carbohydrates	Protein	Fat	Cholesterol	Sodium	Calcium
200 cal	30 g	11 g	5 g	9 mg	1066 mg	25 mg

Ingredient | **Weight** | **Measure** | **Issue**
SOUP,CONDENSED,SPLIT PEA & HAM | 31-1/4 lbs | 3 gal 1-1/8 qts |
WATER | 23 lbs | 2 gal 3 qts |

Method

1 Place soup in steam-jacketed kettle or stock pot.
2 Add water to soup. Mix well.
3 Heat, do not boil. CCP: Internal temperature must reach 165 F. or higher for 15 seconds. Hold for service at 140 F. or higher.

SOUPS No.P 009 08

TOMATO SOUP (CANNED)

Yield 100   Portion   1 Cup

Calories	Carbohydrates	Protein	Fat	Cholesterol	Sodium	Calcium
96 cal	19 g	2 g	2 g	0 mg	788 mg	18 mg

Ingredient

Ingredient	Weight	Measure	Issue
SOUP,CONDENSED,TOMATO	31-1/4 lbs	3 gal 2-1/8 qts	
WATER	23 lbs	2 gal 3 qts	

Method

1. Place soup in steam-jacketed kettle or stock pot.
2. Add water to soup. Mix well.
3. Heat, do not boil. CCP: Internal temperature must reach 165 F. or higher for 15 seconds. Hold for service at 140 F. or higher.

SOUPS No.P 009 09

VEGETABLE SOUP (CANNED)

Yield 100   Portion   1 Cup

Calories	Carbohydrates	Protein	Fat	Cholesterol	Sodium	Calcium
84 cal	14 g	2 g	2 g	0 mg	956 mg	26 mg

Ingredient   **Weight**   **Measure**   **Issue**

SOUP,CONDENSED,VEGETABLE   31-1/4 lbs   3 gal 2-3/8 qts
WATER   23 lbs   2 gal 3 qts

Method

1 Place soup in steam-jacketed kettle or stock pot.
2 Add water to soup. Mix well.
3 Heat, do not boil. CCP: Internal temperature must reach 165 F. or higher for 15 seconds. Hold for service at 140 F. or higher.

SOUPS No.P 009 10

VEGETABLE WITH BEEF SOUP (CANNED)

Yield 100   **Portion** 1 Cup

Calories	Carbohydrates	Protein	Fat	Cholesterol	Sodium	Calcium
89 cal	12 g	6 g	2 g	6 mg	898 mg	21 mg

Ingredient | **Weight** | **Measure** | **Issue**

SOUP,CONDENSED,VEGETABLE WITH BEEF — 31-1/4 lbs — 3 gal 2-1/8 qts
WATER — 23 lbs — 2 gal 3 qts

Method

1 Place soup in steam-jacketed kettle or stock pot.
2 Add water to soup. Mix well.
3 Heat, do not boil. CCP: Internal temperature must reach 165 F. or higher for 15 seconds. Hold for service at 140 F. or higher.

SOUPS No.P 010 00

CHICKEN GUMBO SOUP

Yield 100   Portion 1 Cup

Calories	Carbohydrates	Protein	Fat	Cholesterol	Sodium	Calcium
117 cal	14 g	5 g	5 g	7 mg	1376 mg	49 mg

Ingredient	Weight	Measure	Issue
ONIONS,FRESH,CHOPPED	1 lbs	2-7/8 cup	1-1/8 lbs
MARGARINE	1 lbs	2 cup	
FLOUR,WHEAT,GENERAL PURPOSE	14-1/3 oz	3-1/4 cup	
GARLIC POWDER	1/8 oz	1/8 tsp	
CHICKEN BROTH		4 gal 2 qts	
CHICKEN,COOKED,DICED	1-1/2 lbs		
TOMATOES,CANNED,DICED,INCL LIQUIDS	13-3/4 lbs	1 gal 2 qts	
CELERY,FRESH,CHOPPED	1-2/3 oz	1/4 cup 2-2/3 tbsp	2-1/4 oz
OKRA,FROZEN,CUT	2-1/2 lbs	1 qts 2 cup	
PEPPERS,GREEN,FRESH,CHOPPED	1-1/2 lbs	1 qts 1/2 cup	1-7/8 lbs
RICE,LONG GRAIN	1 lbs	2-3/8 cup	
BAY LEAF,WHOLE,DRIED	1/8 oz	5 each	
PAPRIKA,GROUND	1/8 oz	1/3 tsp	
PEPPER,BLACK,GROUND	1/8 oz	1/3 tsp	
THYME,GROUND	<1/16th oz	1/8 tsp	

Method

1. Saute onions in margarine or butter until tender.
2. Blend flour with onion-fat mixture to form a roux using wire whip; add garlic powder.
3. Prepare broth according to package directions. Add broth to roux, stirring constantly. Bring to a boil; reduce heat.
4. Add chicken, tomatoes, celery, okra, peppers, rice, bay leaves, paprika, pepper, and thyme; mix well.
5. Bring to a boil; reduce heat; simmer 30 minutes. Remove bay leaves. CCP: Internal temperature must reach 165 F. or higher for 15 seconds. Hold for service at 140 F. or higher.

SOUPS No.P 010 01

SHRIMP GUMBO

Yield 100　　　　　　　　　　　　　　　　　**Portion** 1 Cup

Calories	Carbohydrates	Protein	Fat	Cholesterol	Sodium	Calcium
115 cal	14 g	5 g	5 g	22 mg	1397 mg	53 mg

Ingredient	Weight	Measure	Issue
ONIONS,FRESH,CHOPPED	1 lbs	2-7/8 cup	1-1/8 lbs
MARGARINE	1 lbs	2 cup	
FLOUR,WHEAT,GENERAL PURPOSE	14-1/3 oz	3-1/4 cup	
GARLIC POWDER	1/8 oz	1/8 tsp	
CHICKEN BROTH		4 gal 2 qts	
TOMATOES,CANNED,DICED,INCL LIQUIDS	13-3/4 lbs	1 gal 2 qts	
CELERY,FRESH,CHOPPED	12-2/3 oz	3 cup	1-1/8 lbs
OKRA,FROZEN,CUT	2-1/2 lbs	1 qts 2 cup	
PEPPERS,GREEN,FRESH,CHOPPED	1-1/2 lbs	1 qts 1/2 cup	1-7/8 lbs
RICE,LONG GRAIN	1 lbs	2-3/8 cup	
BAY LEAF,WHOLE,DRIED	1/8 oz	5 each	
PAPRIKA,GROUND	1/8 oz	1/3 tsp	
PEPPER,BLACK,GROUND	1/8 oz	1/3 tsp	
THYME,GROUND	<1/16th oz	1/8 tsp	
SHRIMP,RAW,PEELED,DEVEINED,CHOPPED	3 lbs		

Method

1　Saute onions in margarine or butter until tender.
2　Blend flour with onion-fat mixture to form a roux using wire whip; add garlic powder.
3　Prepare broth according to package directions. Add stock to roux, stirring constantly. Bring to a boil; reduce heat.
4　Add tomatoes, celery, okra, peppers, rice, bay leaves, paprika, pepper, and thyme; mix well.
5　Bring to a boil; reduce heat; simmer 27 minutes. Add raw, peeled, deveined shrimp cut into quarters. Boil an additional 2 to 3 minutes. CCP: Internal temperature must reach 165 F. or higher for 15 seconds. Hold for service at 140 F. or higher.

SOUPS No.P 011 00

CORN CHOWDER

Yield 100   Portion   1 Cup

Calories	Carbohydrates	Protein	Fat	Cholesterol	Sodium	Calcium
136 cal	25 g	5 g	3 g	2 mg	761 mg	102 mg

Ingredient	Weight	Measure	Issue
BACON,RAW	8 oz		
CELERY,FRESH,CHOPPED	8 oz	1-7/8 cup	11 oz
ONIONS,FRESH,CHOPPED	1 lbs	2-7/8 cup	1-1/8 lbs
PEPPERS,GREEN,FRESH,CHOPPED	8 oz	1-1/2 cup	9-3/4 oz
OIL,SALAD	1-7/8 oz	1/4 cup 1/3 tbsp	
WATER	16-3/4 lbs	2 gal	
POTATOES,FRESH,PEELED,CUBED	4 lbs	2 qts 3-5/8 cup	5 lbs
SALT	3-3/4 oz	1/4 cup 2-1/3 tbsp	
PEPPER,BLACK,GROUND	1/8 oz	1/3 tsp	
CORN,CANNED,CREAM STYLE	20-1/3 lbs	2 gal 1 qts	
MILK,NONFAT,DRY	1-2/3 lbs	2 qts 3 cup	
WATER,WARM	12-1/2 lbs	1 gal 2 qts	
MARGARINE	8 oz	1 cup	

Method

1. Prepare bacon according to Recipe Nos. L 002 00 or L 002 02. Chop bacon. Set aside for use in Step 3.
2. Saute celery, onions and peppers in salad oil 3 minutes or until tender.
3. Add water, potatoes, salt and pepper to steam-jacketed kettle or stock pot. Add sauteed vegetables. Mix thoroughly. Bring to a boil; reduce heat; simmer 10 minutes or until potatoes are tender.
4. Add corn. Bring to a boil; simmer 5 minutes, stirring occasionally.
5. Reconstitute milk. Add milk and butter or margarine to mixture. Heat slowly to serving temperature. DO NOT BOIL. CCP: Internal temperature must reach 165 F. or higher for 15 seconds. Hold for service at 140 F. or higher.

SOUPS No.P 011 01

CHICKEN CORN CHOWDER

Yield 100	Portion 1 Cup

Calories	Carbohydrates	Protein	Fat	Cholesterol	Sodium	Calcium
179 cal	25 g	6 g	8 g	9 mg	1088 mg	37 mg

Ingredient

Ingredient	Weight	Measure	Issue
SOUP,CONDENSED,CREAM OF CHICKEN	25 lbs	2 gal 3-1/4 qts	
WATER	18-1/4 lbs	2 gal 3/4 qts	
CORN,CANNED,WHOLE KERNEL,DRAINED	20 lbs	3 gal 1-7/8 qts	
PEPPER,BLACK,GROUND	1/8 oz	1/4 tsp	

Method

1 Combine soup and water; mix well.
2 Add canned, whole kernel corn and black pepper. Heat slowly; DO NOT BOIL. CCP: Internal temperature must reach 165 F. or higher for 15 seconds. Hold for service at 140 F. or higher.

SOUPS No.P 012 00

MANHATTAN CLAM CHOWDER

Yield 100   Portion 1 Cup

Calories	Carbohydrates	Protein	Fat	Cholesterol	Sodium	Calcium
80 cal	17 g	3 g	1 g	2 mg	442 mg	51 mg

Ingredient	Weight	Measure	Issue
BACON,RAW	12 oz		
ONIONS,FRESH,CHOPPED	2 lbs	1 qts 1-5/8 cup	2-1/4 lbs
CELERY,FRESH,CHOPPED	2 lbs	1 qts 3-1/2 cup	2-3/4 lbs
CLAMS,CANNED,CHOPPED	12 lbs	1 gal 1-2/3 qts	
TOMATOES,CANNED,CRUSHED,INCL LIQUIDS	19-7/8 lbs	2 gal 1 qts	
CARROTS,FRESH,CHOPPED	1-1/2 lbs	1 qts 1-3/8 cup	1-7/8 lbs
POTATOES,FRESH,PEELED,CUBED	5 lbs	3 qts 2-1/2 cup	6-1/8 lbs
SALT	1 oz	1 tbsp	
PEPPER,BLACK,GROUND	1/4 oz	1 tbsp	
THYME,GROUND	1/8 oz	1/3 tsp	
WORCESTERSHIRE SAUCE	8-1/2 oz	1 cup	
BAY LEAF,WHOLE,DRIED	1/8 oz	2 each	
CATSUP	1 lbs	2 cup	
RESERVED LIQUID	18-3/4 lbs	2 gal 1 qts	
FLOUR,WHEAT,GENERAL PURPOSE	11 oz	2-1/2 cup	
WATER,COLD	2-1/8 lbs	1 qts	

Method

1. Cook bacon until crisp using Recipe No. L 002 00 or L 002 02. Remove bacon; drain; reserve 1/2 cup fat per each 100 servings for use in Step 2. Finely chop bacon. Set aside for use in Step 4.
2. Saute onions and celery in bacon fat about 7 minutes or until tender crisp.
3. Drain clams and reserve clam juice for use in Step 4, clams for use in Step 8.
4. Combine bacon, sauteed vegetables, tomatoes, carrots, potatoes, salt, pepper, thyme, Worcestershire sauce, bay leaves, and catsup with reserved clam juice and water.
5. Bring to a boil; reduce heat; simmer 20 minutes or until vegetables are tender.
6. Blend flour and water to form a smooth paste. Stir into chowder.
7. Bring to a boil; reduce heat; simmer 10 minutes or until thickened.
8. Add clams to chowder; bring to a boil; reduce heat; simmer 10 minutes. Remove bay leaves. CCP: Internal temperature must reach 165 F. or higher for 15 seconds. Hold for service at 140 F. or higher.

SOUPS No.P 013 00

NEW ENGLAND FISH CHOWDER

Yield 100   Portion 1 Cup

Calories	Carbohydrates	Protein	Fat	Cholesterol	Sodium	Calcium
168 cal	15 g	12 g	6 g	39 mg	348 mg	95 mg

Ingredient	Weight	Measure	Issue
BACON,RAW	8 oz		
BACON FAT,RENDERED	2-3/8 oz	1/4 cup 1-2/3 tbsp	
ONIONS,FRESH,CHOPPED	2 lbs	1 qts 1-5/8 cup	2-1/4 lbs
CELERY,FRESH,CHOPPED	1 lbs	3-3/4 cup	1-3/8 lbs
POTATOES,FRESH,PEELED,CUBED	7 lbs	1 gal 1-1/8 qts	8-5/8 lbs
WATER	16-3/4 lbs	2 gal	
BUTTER	1-1/4 lbs	2-1/2 cup	
FLOUR,WHEAT,GENERAL PURPOSE	1-3/8 lbs	1 qts 1 cup	
MILK,NONFAT,DRY	1-1/3 lbs	2 qts 1 cup	
WATER,WARM	23 lbs	2 gal 3 qts	
FISH,FLOUNDER/SOLE FILLET,RAW,2 INCH PIECES	10 lbs		
PEPPER,WHITE,GROUND	1/4 oz	1 tbsp	
THYME,GROUND	1/8 oz	1 tbsp	
PARSLEY,DEHYDRATED,FLAKED	<1/16th oz	1 tbsp	
SALT	1-7/8 oz	3 tbsp	

Method

1 Cook bacon until crisp using Recipe Nos. L 002 00 or L 002 02. Drain; finely chop; set aside for use in Step 6. Reserve appropriate amount of bacon fat for use in Step 2.
2 Saute onions and celery in bacon fat about 7 minutes or until crisp.
3 Add potatoes and water to onion-celery mixture; cook until potatoes are almost tender but still firm, about 10 minutes.
4 Blend butter or margarine and flour to form a roux; set aside for use in Step 6.
5 Reconstitute milk; add to potato mixture. Heat to just below boiling. DO NOT BOIL.
6 Add roux and cooked bacon to milk and potato mixture. Cook until thickened or about 10 minutes.
7 Add fish, pepper, thyme, parsley and salt to mixture. Simmer 10 minutes. CCP: Internal temperature must reach 165 F. or higher for 15 seconds. Hold for service at 140 F. or higher.

SOUPS No.P 013 01

NEW ENGLAND CLAM CHOWDER

Yield 100   Portion   1 Cup

Calories	Carbohydrates	Protein	Fat	Cholesterol	Sodium	Calcium
128 cal	15 g	4 g	6 g	16 mg	333 mg	94 mg

Ingredient	Weight	Measure	Issue
BACON,RAW	8 oz		
BACON FAT,RENDERED	2-3/8 oz	1/4 cup 1-2/3 tbsp	
ONIONS,FRESH,CHOPPED	2 lbs	1 qts 1-5/8 cup	2-1/4 lbs
CELERY,FRESH,CHOPPED	1 lbs	3-3/4 cup	1-3/8 lbs
CLAMS,CANNED,CHOPPED	12 lbs	1 gal 1-2/3 qts	
POTATOES,FRESH,PEELED,CUBED	7 lbs	1 gal 1-1/8 qts	8-5/8 lbs
BUTTER	1-1/4 lbs	2-1/2 cup	
FLOUR,WHEAT,GENERAL PURPOSE	1-3/8 lbs	1 qts 1 cup	
MILK,NONFAT,DRY	1-1/3 lbs	2 qts 1 cup	
WATER,WARM	23 lbs	2 gal 3 qts	
PEPPER,WHITE,GROUND	1/4 oz	1 tbsp	
THYME,GROUND	1/8 oz	1 tbsp	
PARSLEY,DEHYDRATED,FLAKED	<1/16th oz	1 tbsp	
SALT	1 oz	1 tbsp	

Method

1. Cook bacon until crisp using Recipe Nos. L 002 00 or L 002 02. Drain; finely chop; set aside for use in Step 6. Reserve bacon fat for use in Step 2.
2. Saute onions and celery in bacon fat about 7 minutes or until crisp.
3. Add potatoes to onion-celery mixture; cook until potatoes are almost tender but still firm, about 10 minutes. Drain minced clams. Reserve the liquid and combine with water to equal 2 gal per 100 portions. Combine with potato mixture. Reserve drained clams for Step 7.
4. Blend butter or margarine and flour to form a roux; set aside for use in Step 6.
5. Reconstitute milk; add to potato mixture. Heat to just below boiling. DO NOT BOIL.
6. Add roux and cooked bacon to milk and potato mixture. Cook until thickened about 10 minutes.
7. Add clams, pepper, thyme, parsley and salt to mixture. Simmer 10 minutes. CCP: Internal temperature must reach 165 F. or higher for 15 seconds. Hold for service at 140 F. or higher.

SOUPS No.P 014 00

CREAM OF MUSHROOM SOUP

Yield 100　　　　　　　　　　　　　　　　　　**Portion** 1 Cup

Calories	Carbohydrates	Protein	Fat	Cholesterol	Sodium	Calcium
153 cal	14 g	5 g	8 g	22 mg	1316 mg	115 mg

Ingredient　　　　　　　　　　　　　　　　　Weight　　　　　Measure　　　　　Issue

Ingredient	Weight	Measure	Issue
MUSHROOMS,CANNED,SLICED,INCL LIQUIDS	7-1/4 lbs	1 gal 1-1/4 qts	
ONIONS,FRESH,CHOPPED	12-2/3 oz	2-1/4 cup	14-1/8 oz
BUTTER	2 lbs	1 qts	
FLOUR,WHEAT,GENERAL PURPOSE	2-1/4 lbs	2 qts	
PEPPER,BLACK,GROUND	1/8 oz	1/3 tsp	
CHICKEN BROTH		4 gal	
MILK,NONFAT,DRY	1-2/3 lbs	2 qts 3 cup	
WATER,WARM	14-5/8 lbs	1 gal 3 qts	

Method

1. Drain and chop mushrooms. Reserve liquid for use in Step 4.
2. Saute onions and mushrooms in butter or margarine until onions are tender. Remove from fat. Set aside for use in Step 5.
3. Blend fat, flour and pepper to form a roux.
4. Prepare stock according to recipe using both water and reserved mushroom liquid. Gradually blend hot stock mixture into roux stirring constantly until smooth.
5. Add mushroom-onion mixture. Bring to a boil; reduce heat; simmer 15 minutes.
6. Reconstitute milk. Add to soup.
7. Heat to serving temperature. DO NOT BOIL. CCP: Internal temperature must reach 165 F. or higher for 15 seconds. Hold for service at 140 F. or higher.

SOUPS No.P 014 01

CREAM OF BROCCOLI SOUP

Yield 100   Portion   1 Cup

Calories	Carbohydrates	Protein	Fat	Cholesterol	Sodium	Calcium
128 cal	14 g	6 g	6 g	16 mg	1156 mg	117 mg

Ingredient

Ingredient	Weight	Measure	Issue
BROCCOLI,FROZEN,CHOPPED	10 lbs	1 gal 2-2/3 qts	
ONIONS,FRESH,CHOPPED	12-2/3 oz	2-1/4 cup	14-1/8 oz
BUTTER	1-3/8 lbs	2-3/4 cup	
FLOUR,WHEAT,GENERAL PURPOSE	1-7/8 lbs	1 qts 3 cup	
PEPPER,BLACK,GROUND	1/4 oz	1 tbsp	
CHICKEN BROTH		4 gal	
MILK,NONFAT,DRY	1-1/3 lbs	2 qts 1 cup	
WATER,WARM	12-1/2 lbs	1 gal 2 qts	

Method

1. Thaw and chop broccoli. Set aside for use in Step 5.
2. Saute onions in butter or margarine until onions are tender. Do not remove onions from fat.
3. Blend fat with onions, flour and pepper to form a roux.
4. Prepare stock according to package directions. Gradually blend hot stock mixture into roux stirring constantly until smooth.
5. Add broccoli. Bring to a boil; reduce heat; simmer 15 minutes.
6. Reconstitute milk. Add to soup.
7. Heat to serving temperature. DO NOT BOIL. CCP: Internal temperature must reach 165 F. or higher for 15 seconds. Hold for service at 140 F. or higher.

SOUPS No.P 015 00

CREAM OF POTATO SOUP (DEHYDRATED SLICED POTATOES)

Yield 100   **Portion** 1 Cup

Calories	Carbohydrates	Protein	Fat	Cholesterol	Sodium	Calcium
63 cal	10 g	4 g	1 g	2 mg	1162 mg	96 mg

Ingredient	**Weight**	**Measure**	**Issue**
CHICKEN BROTH		4 gal 1 qts	
POTATO,WHITE,DEHYDRATED,SLICED	5 lbs		
ONIONS,FRESH,CHOPPED	3-1/8 lbs	2 qts 1 cup	3-1/2 lbs
PEPPER,BLACK,GROUND	1/8 oz	1/3 tsp	
WATER,WARM	16-3/4 lbs	2 gal	
MILK,NONFAT,DRY	1-1/3 lbs	2 qts 3/4 cup	
PARSLEY,DEHYDRATED,FLAKED	3/8 oz	1/2 cup	

Method

1. Prepare broth according to package directions. Combine broth, potatoes, onions, and pepper in a steam jacketed kettle or stock pot. Bring to a boil. Reduce heat, cover; simmer 1 hour, stirring occasionally. Break up or mash potatoes as necessary.
2. Reconstitute milk; stir milk and parsley into soup. Simmer for 5 minutes.
3. CCP: Internal temperature must reach 165 F. or higher for 15 seconds. Hold for service at 140 F. or higher.

SOUPS No.P 015 01

CREAM OF POTATO SOUP (FRESH WHITE POTATOES)

Yield 100   Portion   1 Cup

Calories	Carbohydrates	Protein	Fat	Cholesterol	Sodium	Calcium
138 cal	28 g	5 g	1 g	2 mg	1034 mg	101 mg

Ingredient	Weight	Measure	Issue
CHICKEN BROTH		3 gal 3 qts	
POTATOES,FRESH,PEELED,CUBED	24-3/4 lbs	4 gal 2 qts	30-5/8 lbs
ONIONS,FRESH,CHOPPED	3-1/8 lbs	2 qts 1 cup	3-1/2 lbs
PEPPER,BLACK,GROUND	1/8 oz	1/3 tsp	
WATER,WARM	16-3/4 lbs	2 gal	
MILK,NONFAT,DRY	1-1/3 lbs	2 qts 3/4 cup	
PARSLEY,DEHYDRATED,FLAKED	3/8 oz	1/2 cup	

Method

1 Prepare broth according to package directions. Combine chicken broth, potatoes, onions, and pepper in a steam jacketed kettle or stock pot. Bring to a boil. Reduce heat; cover; simmer 1 hour or until potatoes are mushy, stirring occasionally.
2 Reconstitute milk; stir milk and parsley into soup. Simmer for 5 minutes.
3 CCP: Internal temperature must reach 165 F. or higher for 15 seconds. Hold for service at 140 F. or higher.

SOUPS No.P 016 00

CREAM OF POTATO SOUP (INSTANT POTATOES)

Yield 100   Portion 1 Cup

Calories	Carbohydrates	Protein	Fat	Cholesterol	Sodium	Calcium
51 cal	8 g	3 g	1 g	2 mg	768 mg	89 mg

Ingredient	Weight	Measure	Issue
CHICKEN BROTH		2 gal 3 qts	
ONIONS,FRESH,CHOPPED	2-1/8 lbs	1 qts 2 cup	2-1/3 lbs
PEPPER,BLACK,GROUND	1/8 oz	1/3 tsp	
WATER	16-3/4 lbs	2 gal	
MILK,NONFAT,DRY	1-1/3 lbs	2 qts 3/4 cup	
PARSLEY,DEHYDRATED,FLAKED	3/8 oz	1/2 cup	
POTATO,WHITE,INSTANT,GRANULES	1 lbs	2 qts 1 cup	

Method

1. Prepare broth according to package directions. Combine chicken broth, onions, and pepper in a steam jacketed kettle or stock pot. Bring to a boil. Reduce heat; cover; simmer for 10 minutes or until onions are tender.
2. Reconstitute milk; stir milk and parsley into soup. Bring to a simmer.
3. Stir potatoes rapidly into soup. Mix until smooth. Simmer for 5 minutes, stirring occasionally.
4. CCP: Internal temperature must reach 165 F. or higher for 15 seconds. Hold for service at 140 F. or higher.

SOUPS No.P 017 00

SPANISH SOUP (DEHYDRATED ONION SOUP)

Yield 100   Portion 1 Cup

Calories	Carbohydrates	Protein	Fat	Cholesterol	Sodium	Calcium
88 cal	10 g	4 g	4 g	11 mg	1087 mg	37 mg

Ingredient	Weight	Measure	Issue
SOUP,DEHYDRATED,ONION	2 lbs	1 qts 3 cup	
WATER,BOILING	37-5/8 lbs	4 gal 2 qts	
SAUSAGE,ITALIAN,HOT	4 lbs		
PEPPERS,GREEN,FRESH,CHOPPED	7-7/8 oz	1-1/2 cup	9-5/8 oz
TOMATOES,CANNED,DICED,INCL LIQUIDS	15 lbs	1 gal 2-1/2 qts	
BAY LEAF,WHOLE,DRIED	1/8 oz	2 each	

Method

1. Stir soup mix into boiling water.
2. Chop sausage. Cook sausage until thoroughly browned. CCP: Internal temperature must reach 145 F. or higher for 15 seconds. Drain well.
3. Mix peppers, tomatoes and bay leaves with browned sausage. Add to soup mixture.
4. Return soup mixture to a boil. Reduce heat. Cover; simmer 20 minutes. Remove bay leaves. CCP: Internal temperature must reach 165 F. or higher for 15 seconds. Hold for service at 140 F. or higher.

SOUPS No.P 017 01

ONION SOUP (DEHYDRATED MIX)

Yield 100   Portion 1 Cup

Calories	Carbohydrates	Protein	Fat	Cholesterol	Sodium	Calcium
27 cal	5 g	1 g	0 g	0 mg	820 mg	18 mg

Ingredient   **Weight**   **Measure**   **Issue**

SOUP,DEHYDRATED,ONION   2 lbs   1 qts 3 cup
WATER,BOILING   52-1/4 lbs   6 gal 1 qts

Method

1 Stir soup mix into boiling water.
2 Simmer 20 minutes. CCP: Internal temperature must reach 165 F. or higher for 15 seconds. Hold for service at 140 F. or higher.

SOUPS No.P 017 02

MEXICAN ONION CORN SOUP (DEHYDRATED MIX)

Yield 100   Portion 1 Cup

Calories	Carbohydrates	Protein	Fat	Cholesterol	Sodium	Calcium
76 cal	16 g	3 g	1 g	0 mg	951 mg	20 mg

Ingredient

Ingredient	Weight	Measure	Issue
SOUP,DEHYDRATED,ONION	2 lbs	1 qts 3 cup	
WATER,BOILING	43-7/8 lbs	5 gal 1 qts	
CORN,CANNED,WHOLE KERNEL,DRAINED	13-1/4 lbs	2 gal 1-1/8 qts	
PIMIENTO,CANNED,DRAINED,CHOPPED	14 oz	2-1/8 cup	
HOT SAUCE	3/8 oz	3/8 tsp	

Method

1. Stir soup mix into boiling water.
2. Drain corn; add to soup mixture.
3. Add canned chopped pimientos and hot sauce. Stir to mix.
4. Stir and simmer 20 minutes. CCP: Internal temperature must reach 165 F. or higher for 15 seconds. Hold for service at 140 F. or higher.

SOUPS No.P 018 00

TOMATO VEGETABLE SOUP (DEHYDRATED)

Yield 100　　　　　　　　　　　　　　　　　　　Portion　1 Cup

Calories	Carbohydrates	Protein	Fat	Cholesterol	Sodium	Calcium
65 cal	12 g	2 g	2 g	1 mg	609 mg	39 mg

Ingredient　　　　　　　　　　　　　　　　　　Weight　　　Measure　　　　Issue

SOUP,DEHYDRATED,TOMATO VEGETABLE W/NOODLES　　4 lbs　　　　3 qts 2 cup
WATER,BOILING　　　　　　　　　　　　　　　　　52-1/4 lbs　　6 gal 1 qts

Method

1. Stir soup mix into boiling water.
2. Return soup mixture to a boil. Cover; simmer 10 minutes or until vegetables are tender, stirring occasionally. CCP: Internal temperature must reach 165 F. or higher for 15 seconds. Hold for service at 140 F. or higher.

SOUPS No.P 018 01

BEEF NOODLE SOUP WITH VEGETABLES (DEHYDRATED)

Yield 100   Portion   1 Cup

Calories	Carbohydrates	Protein	Fat	Cholesterol	Sodium	Calcium
52 cal	8 g	3 g	1 g	2 mg	1342 mg	12 mg

Ingredient	**Weight**	**Measure**	**Issue**
SOUP,DEHYDRATED,BEEF NOODLE W/VEGETABLES	3-1/2 lbs	2 qts 2-3/8 cup	
WATER,BOILING	52-1/4 lbs	6 gal 1 qts	

Method

1 Stir soup mix into boiling water.
2 Return soup mixture to a boil. Cover; simmer 10 minutes or until vegetables are tender, stirring occasionally. CCP: Internal temperature must reach 165 F. or higher for 15 seconds. Hold for service at 140 F. or higher.

SOUPS No.P 018 02

CHICKEN NOODLE SOUP (DEHYDRATED)

Yield 100 **Portion** 1 Cup

Calories	Carbohydrates	Protein	Fat	Cholesterol	Sodium	Calcium
74 cal	10 g	4 g	2 g	3 mg	1815 mg	50 mg

Ingredient   **Weight**   **Measure**   **Issue**

SOUP,DEHYDRATED,CHICKEN NOODLE   4-3/4 lbs   3 qts 2 cup
WATER,BOILING   54-1/3 lbs   6 gal 2 qts

Method

1. Stir soup mix into boiling water.
2. Return soup mixture to a boil. Cover; simmer 7 minutes. Stir occasionally. CCP: Internal temperature must reach 165 F. or higher for 15 seconds. Hold for service at 140 F. or higher.

SOUPS No.P 018 03

CHICKEN NOODLE SOUP WITH VEGETABLES (DEHYDRATED)

Yield 100 **Portion** 1 Cup

Calories	Carbohydrates	Protein	Fat	Cholesterol	Sodium	Calcium
89 cal	14 g	5 g	2 g	3 mg	1826 mg	55 mg

Ingredient

Ingredient	Weight	Measure	Issue
SOUP,DEHYDRATED,CHICKEN NOODLE	4-3/4 lbs	3 qts 2 cup	
WATER,BOILING	54-1/3 lbs	6 gal 2 qts	
VEGETABLES,MIXED,FROZEN	5 lbs	3 qts 1/2 cup	

Method

1. Stir soup mix into boiling water. Add frozen mixed vegetables.
2. Return soup mixture to a boil. Cover; simmer 7 minutes or until vegetables are tender. CCP: Internal temperature must reach 165 F. or higher for 15 seconds. Hold for service at 140 F. or higher.

SOUPS No.P 019 00

PEPPER POT SOUP

Yield 100　　　　　　　　　　　　　　　　　　　Portion　1 Cup

Calories	Carbohydrates	Protein	Fat	Cholesterol	Sodium	Calcium
97 cal	11 g	3 g	5 g	1 mg	1228 mg	38 mg

Ingredient	Weight	Measure	Issue
SHORTENING,VEGETABLE,MELTED	14-1/2 oz	2 cup	
ONIONS,FRESH,CHOPPED	8 oz	1-3/8 cup	8-7/8 oz
PEPPERS,GREEN,FRESH,CHOPPED	1-1/2 lbs	1 qts 1/2 cup	1-7/8 lbs
CELERY,FRESH,SLICED	1-1/2 lbs	1 qts 1-5/8 cup	2 lbs
BEEF BROTH		5 gal	
FLOUR,WHEAT,GENERAL PURPOSE	8-7/8 oz	2 cup	
POTATOES,FRESH,PEELED,CUBED	6 lbs	1 gal 3/8 qts	7-3/8 lbs
PEPPER,BLACK,GROUND	1/8 oz	1/3 tsp	
MILK,NONFAT,DRY	6-5/8 oz	2-3/4 cup	
WATER,WARM	3-1/8 lbs	1 qts 2 cup	
PIMIENTO,CANNED,DRAINED,CHOPPED	6-3/4 oz	1 cup	

Method

1. Saute vegetables in salad oil, shortening or olive oil 10 minutes. Do not brown. Remove vegetables from fat; set aside for use in Step 3 and vegetables aside for use in Step 4.
2. Prepare broth according to package directions.
3. Blend fat and flour together; stir until smooth. Add roux to broth, stirring constantly. Cook until blended.
4. Add sauteed vegetables, potatoes and pepper. Cook about 20 minutes or until vegetables are tender. CCP: Internal temperature must reach 165 F. or higher for 15 seconds.
5. Reconstitute milk.
6. Just before serving, remove soup from heat; slowly add milk, stirring constantly.
7. Add pimientos. CCP: Hold for service at 140 F. or higher.

SOUPS No.P 020 00

CHICKEN VEGETABLE (MULLIGATAWNY) SOUP

Yield 100 Portion 1 Cup

Calories	Carbohydrates	Protein	Fat	Cholesterol	Sodium	Calcium
80 cal	9 g	4 g	3 g	11 mg	1521 mg	37 mg

Ingredient	**Weight**	**Measure**	**Issue**
ONIONS,FRESH,CHOPPED	1 lbs	2-7/8 cup	1-1/8 lbs
PEPPERS,GREEN,FRESH,CHOPPED	1-1/3 lbs	1 qts	1-5/8 lbs
BUTTER	6 oz	3/4 cup	
FLOUR,WHEAT,GENERAL PURPOSE	13-1/4 oz	3 cup	
CHICKEN BROTH		5 gal 2 qts	
CHICKEN,COOKED,DICED	1-1/2 lbs		
TOMATOES,CANNED,CRUSHED,INCL LIQUIDS	6-5/8 lbs	3 qts	
CARROTS,FRESH,CHOPPED	1 lbs	3-1/2 cup	1-1/4 lbs
CELERY,FRESH,CHOPPED	1 lbs	3-3/4 cup	1-3/8 lbs
APPLES,FRESH,PEELED,SLICED	1-1/2 lbs	1 qts 1-1/2 cup	1-7/8 lbs
CURRY POWDER	2/3 oz	3 tbsp	
CLOVES,GROUND	<1/16th oz	1/8 tsp	
PEPPER,BLACK,GROUND	1/8 oz	1/3 tsp	

Method

1. Saute onions and peppers in butter or margarine until tender. Remove from fat; set aside for use in Step 4. Reserve fat for use in Step 2.
2. Blend fat and flour to form a roux.
3. Prepare broth according to package directions. Add broth to roux, stirring constantly. Cook until blended.
4. Add chicken, sauteed onions and peppers, tomatoes, carrots, celery, apples, curry powder, cloves and pepper.
5. Simmer 45 minutes or until vegetables are tender. CCP: Internal temperature must reach 165 F. or higher for 15 seconds. Hold for service at 140 F. or higher.

SOUPS No.P 021 00

ZESTY BEAN SOUP

Yield 100   Portion 1 Cup

Calories	Carbohydrates	Protein	Fat	Cholesterol	Sodium	Calcium
110 cal	20 g	6 g	1 g	0 mg	1211 mg	61 mg

Ingredient	Weight	Measure	Issue
BEANS,KIDNEY,DARK RED,CANNED,INCL LIQUIDS	6-3/4 lbs	3 qts	
BEANS,LIMA,CANNED,INCL LIQUIDS	6-1/2 lbs	3 qts	
BEANS,PINTO,CANNED,INCL LIQUIDS	7-3/8 lbs	3 qts 2 cup	
TOMATOES,CANNED,DICED,INCL LIQUIDS	13-3/4 lbs	1 gal 2 qts	
ONIONS,FRESH,CHOPPED	3 lbs	2 qts 1/2 cup	3-1/3 lbs
CELERY,FRESH,SLICED	2 lbs	1 qts 3-1/2 cup	2-3/4 lbs
BEEF BROTH		3 gal 1 qts	
PEPPER,BLACK,GROUND	3/8 oz	1 tbsp	
PAPRIKA,GROUND	3/8 oz	1 tbsp	
THYME,GROUND	1/4 oz	1 tbsp	
BAY LEAF,WHOLE,DRIED	1/4 oz	6 each	

Method

1 Drain beans.
2 Combine beans, tomatoes, onions, celery, beef broth, pepper, paprika, thyme and bay leaves in stock pot or steam-jacketed kettle; stir well. Bring to a boil; reduce heat. Cover; simmer 45 minutes or until vegetables are tender. Remove bay leaves.
3 CCP: Internal temperature must reach 165 F. or higher for 15 seconds. CCP: Hold for service at 140 F. or higher.

SOUPS No.P 021 01

ZESTY BEAN SOUP (DRY BEANS)

Yield 100   Portion 1 Cup

Calories	Carbohydrates	Protein	Fat	Cholesterol	Sodium	Calcium
127 cal	23 g	8 g	1 g	1 mg	1058 mg	78 mg

Ingredient	Weight	Measure	Issue
BEANS,KIDNEY,DRY	2 lbs	1 qts 7/8 cup	
BEANS,WHITE,DRY	2 lbs	1 qts 1/2 cup	
BEANS,PINTO,DRY	1-7/8 lbs	1 qts 1/2 cup	
WATER	16-3/4 lbs	2 gal	
BEEF BROTH		3 gal 3 qts	
BAY LEAF,WHOLE,DRIED	1/4 oz	6 each	
TOMATOES,CANNED,DICED,INCL LIQUIDS	13-3/4 lbs	1 gal 2 qts	
ONIONS,FRESH,CHOPPED	3 lbs	2 qts 1/2 cup	3-1/3 lbs
CELERY,FRESH,SLICED	2 lbs	1 qts 3-1/2 cup	2-3/4 lbs
PEPPER,BLACK,GROUND	3/8 oz	1 tbsp	
PAPRIKA,GROUND	3/8 oz	1 tbsp	
THYME,GROUND	1/4 oz	1 tbsp	

Method

1. Pick over beans, removing discolored beans and foreign matter. Wash thoroughly in cold water. Cover with cold water in stock pot or steam-jacketed kettle; bring to a boil; boil 2 minutes. Turn off heat. Cover; let stand 1 hour.
2. Prepare broth according to package directions. Add bay leaves. Bring to a boil; cover; simmer 3 hours or until beans are tender.
3. Stir occasionally. Add tomatoes, onions, celery, pepper, paprika and thyme; stir well.
4. Bring to a boil; reduce heat. Cover; simmer 45 minutes or until vegetables are tender. Remove bay leaves. CCP: Internal temperature must reach 165 F. or higher for 15 seconds. CCP: Hold for service at 140 F. or higher.

SOUPS No.P 022 00

CHICKEN MUSHROOM SOUP (CANNED)

Yield 100   **Portion** 1 Cup

Calories	Carbohydrates	Protein	Fat	Cholesterol	Sodium	Calcium
154 cal	13 g	4 g	10 g	7 mg	1080 mg	88 mg

Ingredient

Ingredient	Weight	Measure	Issue
SOUP,CONDENSED,CREAM OF CHICKEN	15-3/4 lbs	1 gal 3-1/8 qts	
SOUP,CONDENSED,CREAM OF MUSHROOM	15-3/4 lbs	1 gal 3-1/8 qts	
MILK,NONFAT,DRY	13-3/4 oz	1 qts 1-3/4 cup	
WATER	23 lbs	2 gal 3 qts	
NUTMEG,GROUND	<1/16th oz	1/8 tsp	

Method

1. Add soups to steam-jacketed kettle or stock pot; mix well.
2. Reconstitute milk; stir into combined soups.
3. Add nutmeg; mix well.
4. CCP: Internal temperature must reach 165 F. or higher for 15 seconds. DO NOT BOIL. CCP: Hold for service at 140 F. or higher.

SOUPS No.P 022 01

DOUBLY GOOD CHICKEN SOUP (CANNED)

Yield 100   **Portion** 1 Cup

Calories	Carbohydrates	Protein	Fat	Cholesterol	Sodium	Calcium
106 cal	10 g	4 g	5 g	10 mg	1047 mg	48 mg

Ingredient	**Weight**	**Measure**	**Issue**
SOUP,CONDENSED,CREAM OF CHICKEN	15-3/4 lbs	1 gal 3-1/8 qts	
SOUP,CONDENSED,CHICKEN WITH RICE	15-3/4 lbs	1 gal 3-1/4 qts	
MILK,NONFAT,DRY	4-3/4 oz	2 cup	
WATER	23 lbs	2 gal 3 qts	
NUTMEG,GROUND	<1/16th oz	1/8 tsp	

Method

1. Add soups to steam-jacketed kettle or stock pot; mix well.
2. Reconstitute milk; stir into combined soups.
3. Add nutmeg; mix well.
4. CCP: Internal temperature must reach 165 F. or higher for 15 seconds. DO NOT BOIL. CCP: Hold for service at 140 F. or higher.

SOUPS No.P 022 02

LOGGING SOUP (CANNED)

Yield 100 Portion 1 Cup

Calories	Carbohydrates	Protein	Fat	Cholesterol	Sodium	Calcium
129 cal	19 g	8 g	3 g	5 mg	980 mg	11 mg

Ingredient

Ingredient	Weight	Measure	Issue
SOUP,CONDENSED,VEGETABLE WITH BEEF	15-3/4 lbs	1 gal 3-1/8 qts	
SOUP,CONDENSED,BEAN WITH BACON	15-3/4 lbs	1 gal 2-3/4 qts	
WATER	23 lbs	2 gal 3 qts	

Method

1. Add soups to steam-jacketed kettle or stock pot; mix well.
2. Add water and mix well.
3. Heat to serving temperature. DO NOT BOIL. CCP: Internal temperature must reach 165 F. or higher for 15 seconds. Hold for service at 140 F. or higher.

SOUPS No.P 022 03

TOMATO NOODLE SOUP (CANNED)

Yield 100   **Portion** 1 Cup

Calories	Carbohydrates	Protein	Fat	Cholesterol	Sodium	Calcium
96 cal	15 g	4 g	3 g	3 mg	941 mg	19 mg

Ingredient	**Weight**	**Measure**	**Issue**
SOUP,CONDENSED,TOMATO	15-3/4 lbs	1 gal 3-1/8 qts	
SOUP,CONDENSED,BEEF NOODLE	15-3/4 lbs	1 gal 3-1/8 qts	
WATER	23 lbs	2 gal 3 qts	

Method

1. Add soups to steam-jacketed kettle or stock pot; mix well.
2. Add water; mix well.
3. Heat to serving temperature. DO NOT BOIL. CCP: Internal temperature must reach 165 F. or higher for 15 seconds. Hold for service at 140 F. or higher.

SOUPS No.P 022 04

VEGETABLE BEEF SUPREME SOUP (CANNED)

Yield 100   **Portion** 1 Cup

Calories	Carbohydrates	Protein	Fat	Cholesterol	Sodium	Calcium
94 cal	15 g	4 g	2 g	3 mg	850 mg	19 mg

Ingredient

Ingredient	Weight	Measure	Issue
SOUP,CONDENSED,VEGETABLE WITH BEEF	15-3/4 lbs	1 gal 3-1/8 qts	
SOUP,CONDENSED,TOMATO	15-3/4 lbs	1 gal 3-1/8 qts	
WATER	23 lbs	2 gal 3 qts	
GINGER,GROUND	1/8 oz	3/8 tsp	

Method

1. Add soups to steam-jacketed kettle or stock pot; mix well.
2. Add water; mix well.
3. Add ginger if desired, mix well.
4. Heat to serving temperature. DO NOT BOIL. CCP: Internal temperature must reach 165 F. or higher for 15 seconds. Hold for service at 140 F. or higher.

SOUPS No.P 023 00

SPLIT PEA SOUP WITH HAM

Yield 100   Portion   1 Cup

Calories	Carbohydrates	Protein	Fat	Cholesterol	Sodium	Calcium
150 cal	24 g	11 g	1 g	5 mg	585 mg	30 mg

Ingredient | **Weight** | **Measure** | **Issue**

Ingredient	Weight	Measure	Issue
PEAS,SPLIT,DRY	7-7/8 lbs	1 gal 1/2 qts	
PORK,HAM,CURED,CHOPPED	2 lbs		
WATER,COLD	25-1/8 lbs	3 gal	
HAM BROTH (FROM MIX)		4 gal	
ONIONS,FRESH,GRATED	2 lbs	1 qts 1-5/8 cup	2-1/4 lbs
CARROTS,FRESH,GRATED	1-1/8 lbs	1 qts 5/8 cup	1-3/8 lbs
BAY LEAF,WHOLE,DRIED	1/8 oz	4 each	
SUGAR,GRANULATED	1-3/4 oz	1/4 cup 1/3 tbsp	
PEPPER,BLACK,GROUND	1/8 oz	1/8 tsp	

Method

1. Pick over peas, removing any foreign matter. Wash thoroughly in cold water.
2. Cover peas with water. Bring to a boil.
3. Prepare broth according to package directions. Add broth, ham, onions, carrots, bay leaves, sugar and pepper to peas.
4. Bring soup mixture to a boil, reduce heat; simmer gently about 2-1/2 hours or until peas are mushy. Remove bay leaves.
5. Whip until mixture is smooth. Add boiling water, if needed, for a thinner consistency. CCP: Internal temperature must reach 165 F. or higher for 15 seconds. Hold for service at 140 F. or higher.

SOUPS No.P 023 01

PUREE MONGOLE

Yield 100   **Portion** 1 Cup

Calories	Carbohydrates	Protein	Fat	Cholesterol	Sodium	Calcium
113 cal	20 g	6 g	2 g	3 mg	609 mg	23 mg

Ingredient   Weight   Measure   Issue

Ingredient	Weight	Measure	Issue
SPLIT PEA SOUP WITH HAM		3 gal 1/2 qts	
SOUP,CONDENSED,TOMATO	12-1/2 lbs	1 gal 1-5/8 qts	
WATER	14-5/8 lbs	1 gal 3 qts	

Method

1. Prepare 1/2 recipe Split Pea Soup, Recipe No. P 023 00.
2. Blend in condensed tomato soup and water. Heat to a simmer. CCP: Internal temperature must reach 165 F. or higher for 15 seconds. Hold for service at 140 F. or higher.

SOUPS No.P 024 00

CREAM OF BROCCOLI SOUP (CANNED)

Yield 100   **Portion** 1 Cup

Calories	Carbohydrates	Protein	Fat	Cholesterol	Sodium	Calcium
235 cal	14 g	7 g	16 g	34 mg	825 mg	170 mg

Ingredient   Weight   Measure   Issue

Ingredient	Weight	Measure	Issue
SOUP,CONDENSED,CREAM OF BROCCOLI,CANNED	31-1/4 lbs	3 gal 1-5/8 qts	
MILK,NONFAT,DRY	1 lbs	1 qts 3 cup	
WATER	23 lbs	2 gal 3 qts	

Method

1. Place soup in steam-jacketed kettle or stock pot.
2. Reconstitute milk. Stir into soup.
3. CCP: Heat to 165 F. or higher for 15 seconds. DO NOT BOIL. Hold for service at 140 F. or higher.

SOUPS No.P 024 01

CREAM OF CHICKEN SOUP (CANNED)

Yield 100   Portion 1 Cup

Calories	Carbohydrates	Protein	Fat	Cholesterol	Sodium	Calcium
149 cal	13 g	6 g	8 g	12 mg	1143 mg	99 mg

Ingredient	Weight	Measure	Issue
SOUP,CONDENSED,CREAM OF CHICKEN	31-1/4 lbs	3 gal 2-1/8 qts	
MILK,NONFAT,DRY	1 lbs	1 qts 3 cup	
WATER	23 lbs	2 gal 3 qts	

Method

1. Place soup in steam-jacketed kettle or stock pot.
2. Reconstitute milk. Stir into soup.
3. CCP: Heat to 165 F. or higher for 15 seconds. DO NOT BOIL. Hold for service at 140 F. or higher.

SOUPS No.P 024 02

CREAM OF MUSHROOM SOUP (CANNED)

Yield 100   Portion 1 Cup

Calories	Carbohydrates	Protein	Fat	Cholesterol	Sodium	Calcium
163 cal	13 g	4 g	11 g	2 mg	1010 mg	98 mg

Ingredient   **Weight**   **Measure**   **Issue**

SOUP,CONDENSED,CREAM OF MUSHROOM   31-1/4 lbs   3 gal 2-1/8 qts
MILK,NONFAT,DRY   1 lbs   1 qts 3 cup
WATER   23 lbs   2 gal 3 qts

Method

1. Place soup in steam-jacketed kettle or stock pot.
2. Reconstitute milk. Stir into soup.
3. CCP: Internal temperature must reach 165 F. or higher for 15 seconds. DO NOT BOIL. Hold for service at 140 F. or higher.

SOUPS No.P 025 00

TEXAS TORTILLA SOUP

Yield 100   Portion 1 Cup

Calories	Carbohydrates	Protein	Fat	Cholesterol	Sodium	Calcium
135 cal	22 g	6 g	4 g	8 mg	1256 mg	129 mg

Ingredient	Weight	Measure	Issue
TORTILLAS,CORN,6 INCH	2-3/4 lbs		
CHICKEN BROTH		3 gal	
TOMATOES,CANNED,DICED,INCL LIQUIDS	27-1/3 lbs	2 gal 3-7/8 qts	
PEPPERS,JALAPENOS,CANNED,CHOPPED	3-5/8 oz	3/4 cup	
ONIONS,FRESH,CHOPPED	5-5/8 oz	1 cup	6-1/4 oz
CUMIN,GROUND	1-2/3 oz	1/2 cup	
CILANTRO,DRY	1-1/8 oz	1 cup	
GARLIC POWDER	5/8 oz	2 tbsp	
CORN,CANNED,WHOLE KERNEL,DRAINED	4-1/3 lbs	3 qts	
BEANS,KIDNEY,DARK RED,CANNED,DRAINED	4-1/2 lbs	2 qts 3-1/2 cup	
CHEESE,CHEDDAR,SHREDDED	1-1/2 lbs	1 qts 2 cup	

Method

1. Cut tortillas into strips 1/2-inch by 3-inches. Spread 13 ounces or 1 quart strips on each sheet pan. Using a convection oven, bake at 350 F. for 6-8 minutes or until crisp and lightly browned on low fan, open vent. Reserve for use in Step 4.
2. Prepare broth according to package directions.
3. Add tomatoes, peppers, onions, cumin, cilantro, and garlic powder to broth in steam-jacketed kettle. Simmer 20 minutes.
4. Drain, rinse and drain corn and beans. Add corn and beans to soup mixture; stir and bring back to a simmer for 10 minutes or until tender; stir occasionally. Drain. CCP: Hold for service at 140 F. or higher.
5. Garnish each portion with 1/4 cup tortilla strips and 1 tablespoon cheese.

SOUPS No.P 026 00

TORTELLINI SOUP

Yield 100   Portion   1 Cup

Calories	Carbohydrates	Protein	Fat	Cholesterol	Sodium	Calcium
140 cal	20 g	7 g	4 g	14 mg	1309 mg	133 mg

Ingredient	Weight	Measure	Issue
COOKING SPRAY,NONSTICK	2 oz	1/4 cup 1/3 tbsp	
ONIONS,FRESH,CHOPPED	1 lbs	2-5/8 cup	1 lbs
GARLIC POWDER	5/8 oz	2 tbsp	
BASIL,DRIED,CRUSHED	5/8 oz	1/4 cup 1/3 tbsp	
PEPPER,BLACK,GROUND	1/8 oz	1/3 tsp	
TOMATOES,CANNED,DICED,INCL LIQUIDS	13-3/4 lbs	1 gal 2 qts	
CHICKEN BROTH		4 gal	
TORTELLINI,FROZEN,CHEESE	6 lbs		
SQUASH,FRESH,SUMMER,SLICED	6-7/8 lbs	1 gal 2-7/8 qts	7-1/4 lbs
CHEESE,PARMESAN,GRATED	7 oz	2 cup	

Method

1. Spray steam-jacketed kettle or stock pot with non-stick cooking spray.
2. Add onions, garlic, basil, and pepper; cover; cook 8 to 10 minutes or until onions are tender.
3. Add tomatoes, bring to a boil. Reduce heat; cover; simmer 15 minutes.
4. Prepare broth according to package directions. Add to tomatoes.
5. Add tortellini; simmer covered 12 to 15 minutes.
6. CCP: Internal temperature must reach 165 F. or higher for 15 seconds. Add squash; simmer covered 3 to 5 minutes or until tender. CCP: Hold for service at 140 F. or higher.
7. Garnish each serving with 1 teaspoon parmesan cheese.

SOUPS No.P 027 00

LENTIL VEGETABLE SOUP

Yield 100   **Portion** 1 Cup

Calories	Carbohydrates	Protein	Fat	Cholesterol	Sodium	Calcium
113 cal	22 g	7 g	0 g	0 mg	432 mg	54 mg

Ingredient	Weight	Measure	Issue
BEANS,LENTIL	4-3/4 lbs	2 qts 3-3/8 cup	
WATER	33-1/2 lbs	4 gal	
TOMATOES,CANNED,DICED,INCL LIQUIDS	13-1/2 lbs	1 gal 1-7/8 qts	
ONIONS,FRESH,CHOPPED	7 lbs	1 gal 1 qts	7-3/4 lbs
CARROTS,FRESH,SLICED	4 lbs	3 qts 2-1/8 cup	4-7/8 lbs
CELERY,FRESH,SLICED	1-1/4 lbs	1 qts 3/4 cup	1-3/4 lbs
SALT	2-1/2 oz	1/4 cup 1/3 tbsp	
GARLIC POWDER	1-1/4 oz	1/4 cup 1/3 tbsp	
PEPPER,BLACK,GROUND	1/2 oz	2 tbsp	
BASIL,DRIED,CRUSHED	5/8 oz	1/4 cup 1/3 tbsp	
OREGANO,CRUSHED	5/8 oz	1/4 cup 1/3 tbsp	
BAY LEAF,WHOLE,DRIED	1/4 oz	8 each	
PARSLEY,FRESH,BUNCH,CHOPPED	4-1/4 oz	2 cup	4-1/2 oz

Method

1. Pick over lentils, removing shriveled lentils and foreign matter. Wash thoroughly in cold water.
2. Place lentils, water, tomatoes, onions, carrots, celery, salt, garlic powder, pepper, basil, oregano, and bay leaves in steam-jacketed kettle or stock pot. Bring to a boil; reduce heat; simmer 2 to 2-1/4 hours stirring occasionally or until lentils are tender. CCP: Internal temperature must reach 165 F. or higher for 15 seconds. Hold for service at 140 F. or higher.
3. Add parsley just before serving.

SOUPS No.P 028 00

CURRIED VEGETABLE SOUP

Yield 100   Portion   1 Cup

Calories	Carbohydrates	Protein	Fat	Cholesterol	Sodium	Calcium
66 cal	14 g	2 g	1 g	0 mg	197 mg	26 mg

Ingredient	Weight	Measure	Issue
VEGETABLE BROTH		4 gal	
WATER	10-1/2 lbs	1 gal 1 qts	
ONIONS,FRESH,CHOPPED	6 lbs	1 gal 1/4 qts	6-2/3 lbs
POTATOES,FRESH,PEELED,CUBED	3 lbs	2 qts 3/4 cup	3-2/3 lbs
CELERY,FRESH,SLICED	1-1/2 lbs	1 qts 1-5/8 cup	2 lbs
CURRY POWDER	1-1/3 oz	1/4 cup 2-1/3 tbsp	
GARLIC POWDER	3/8 oz	1 tbsp	
PARSLEY,DEHYDRATED,FLAKED	3/8 oz	1/2 cup	
PEPPER,BLACK,GROUND	1/8 oz	1/3 tsp	
THYME,GROUND	<1/16th oz	1/8 tsp	
RICE,BROWN,LONG GRAIN,RAW PARBOILED	1-1/4 lbs	3 cup	
PEAS & CARROTS,FROZEN	3-7/8 lbs	3 qts 1/2 cup	
CAULIFLOWER,FROZEN	2 lbs		

Method

1 Prepare vegetable stock according to package directions.
2 Add water, onions, potatoes, celery, curry, garlic powder, parsley, black pepper and thyme to stock. Stir. Bring to a boil; add rice. Stir; reduce heat; cook 30 to 35 minutes or until rice is tender.
3 Add peas, carrots and cauliflower, bring to a boil; simmer 5 minutes. CCP: Internal temperature must reach 165 F. or higher for 15 seconds. CCP: Hold for service at 140 F. or higher.

SOUPS No.P 029 00

TURKEY VEGETABLE SOUP

Yield 100						Portion 1 Cup

Calories	Carbohydrates	Protein	Fat	Cholesterol	Sodium	Calcium
91 cal	10 g	7 g	3 g	13 mg	1475 mg	45 mg

Ingredient	Weight	Measure	Issue
CHICKEN BROTH		5 gal	
POTATOES,FRESH,RED BLISS	4-5/8 lbs	3 qts 2 cup	6-1/8 lbs
CARROTS,FROZEN,SLICED	3 lbs	2 qts 2-5/8 cup	
ONIONS,FRESH,CHOPPED	3 lbs	2 qts 1/2 cup	3-1/3 lbs
CELERY,FRESH,SLICED	2 lbs	1 qts 3-1/2 cup	2-3/4 lbs
PARSLEY,DEHYDRATED,FLAKED	1 oz	1-3/8 cup	
PEPPER,BLACK,GROUND	3/8 oz	1 tbsp	
THYME,GROUND	1/4 oz	1 tbsp	
SAGE,GROUND	1/8 oz	1 tbsp	
TURKEY,BNLS,WHITE AND DARK MEAT,DICED	5 lbs		
PEAS,GREEN,FROZEN	1-7/8 lbs	1 qts 2 cup	

Method

1 Prepare chicken broth according to package directions in steam-jacketed kettle or stock pot.
2 Add potatoes, carrots, onions, celery, parsley, black pepper, thyme, and sage to stock. Stir; bring to a boil; reduce heat; simmer 20 minutes or until the vegetables are tender.
3 Add turkey and peas. Bring to a boil; CCP: Internal temperature must reach 165 F. or higher for 15 seconds. Reduce heat; simmer 5 minutes. CCP: Hold for service at 140 F. or higher.

SOUPS No.P 500 00

ASIAN STIR FRY SOUP

Yield 100   Portion   6 Ounces

Calories	Carbohydrates	Protein	Fat	Cholesterol	Sodium	Calcium
112 cal	10 g	7 g	5 g	16 mg	842 mg	86 mg

Ingredient	Weight	Measure	Issue
OIL, CANOLA	5-1/8 oz	1/2 cup 2-2/3 tbsp	
PORK,SHOULDER,LEAN,RAW,DICED	5 lbs		
SOY SAUCE	13-1/2 oz	1-3/8 cup	
WATERCHESTNUTS,CANNED,SLICED,DRAINED	1-1/2 lbs	1 qts 1 cup	
MUSHROOMS,FRESH,WHOLE,SLICED	1-1/2 lbs	2 qts 2 cup	1-2/3 lbs
PEPPERS,GREEN,FRESH,MEDIUM,SLICED,THIN	1-5/8 lbs	1 qts 1 cup	2 lbs
ONIONS,GREEN,FRESH,CHOPPED	11-3/4 oz	3-3/8 cup	13 oz
HAM BROTH (FROM MIX)		5 gal	
RICE,LONG GRAIN	1 lbs	2-1/2 cup	
SPINACH,FROZEN	10-3/4 lbs	1 gal 2-3/8 qts	

Method

1. Heat oil in steam jacketed kettle. Brown pork in oil, drain off excess oil. CCP: Internal temperature must reach 145 F. or higher for 15 seconds.
2. Add soy sauce, water chestnuts, mushrooms, julienne sliced peppers and green onions, stir fry until vegetables are tender-crisp for 3 to 4 minutes.
3. Prepare broth according to package directions. Add broth and stir well and heat to a boil.
4. Reduce heat and add rice. Cover and simmer about 25 minutes or until rice is tender.
5. Stir in thawed drained spinach. CCP: Internal temperature must reach 165 F. or higher for 15 seconds. CCP: Hold for service at 140 F. or higher.

GUIDELINES FOR POTATO BAR

The potato bar is a popular way to serve baked potatoes along with various toppings. Prepare cold toppings. Keep refrigerated until ready to serve. Prepare baked potatoes and hot toppings. Keep hot. Assemble cold items on potato bar along with hot items. Replenish potato bar as needed.

ITEM	APPROXIMATE PORTION SIZE	100 PORTIONS	
		A.P. WEIGHT AND/ OR RECIPE	E. P.
Potatoes, white, baked	1 Potato (6 1/2 Ounces)	55 lb. Use 1 recipe Baked Potatoes (Recipe No. Q 044 00)	
TOPPINGS Butter or margarine	1 pat	1 lb 2 oz (100 pats)	
Broccoli, partially cooked	2 tbsp	20 lb. Use 1 recipe Broccoli (Recipe No. Q 105 02). Chop partially cooked broccoli in 1-inch pieces	
Tomatoes, fresh,	2 tbsp	6 lb 8 oz	6 lb 6 oz (3 1/4 qt)
Cheese, Cheddar or American, Swiss, Mozzarella or Monterey Jack, shredded	2 tbsp	3 lb 2 oz	3 1/4 qt
Cheese, cottage	1/4 cup (1 No. 16 scoop)	12 lb 8 oz	6 1/4 qt

Q-G. VEGETABLES No. 4

GUIDELINES FOR POTATO BAR (continued)

ITEM	APPROXIMATE PORTION SIZE	100 PORTIONS	
		A.P. WEIGHT AND/ OR RECIPE	E. P.
Chili	1/2 cup (1 Size 2 ladle)	Use 1/2 recipe Chili Con Carne with Beans (Recipe No. L 059 00)	3 1/8 gal
Chives, dehydrated	1/2 tsp	1/3 oz (2 3/4-1/8 oz co)	1 cup
Onions, dry, chopped	2 tsp	3 lb 5 oz	3 lb (2 1/4 qt)
Sour cream	1 tbsp	3 lb 3 oz (6 1/4 cups)	
Yogurt, plain	1 tbsp	3 lb 7 oz (6 1/4 cups)	

NOTE: Ensure there are sufficient serving utensils for baked potatoes and each topping.

GUIDELINES FOR HEATING DEHYDRATED, COMPRESSED VEGETABLES

INGREDIENTS	WEIGHTS	MEASURES	APPROX. AMOUNT OF WATER	METHOD
Beans, green, dehydrated, compressed	2 lb 15 oz	3 – No. 2-1/2 cans	4-1/2 gal	1. Bring water to a boil in steam-jacketed kettle or stock pot. 2. Add 1 tbsp salt.
Peas, dehydrated, sweet (green), compressed	7 lb 2 oz	6 – No. 2-1/2 cans	6 gal	3. Prepare according to following rehydration guidelines for type of vegetable selected (see over).
Vegetables, mixed, dehydrated, compressed	4 lb 8 oz	4-1/2 – No. 2-1/2 cans	6-3/4 gal	4. Drain; reserve 3 cups cooking liquid.
Butter or margarine, melted (optional)	1 lb	2 cups		5. Place vegetables in serving pans. 6. Combine butter or margarine and reserved cooking liquid. Pour an equal amount over vegetables in each pan. Garnish as desired.

REHYDRATION GUIDELINES

TYPE VEGETABLE	APPROXIMATE STANDING TIME	PROCEDURE
Beans, green	10 minutes	Simmer until beans separate, stirring occasionally. Remove from heat. Let stand uncovered until rehydrated.
Peas, green	12 to 15 minutes	Turn off heat; stir; cover. Let stand until rehydrated.
Vegetables, mixed	5 to 7 minutes	Return to boil; stir gently to break apart. Simmer 2 minutes; turn off heat; let stand until rehydrated.

GUIDELINES FOR STEAM COOKING VEGETABLES

Canned Vegetables

Place vegetables not more than 2 inches deep in shallow perforated or solid steamer or steam table pans. Add enough liquid for serving. At 5 lb pressure, heat 3 to 5 minutes; at 15 lb pressure, heat 3 to 4 minutes.

Fresh and Frozen Vegetables

For fresh vegetables, fill steamer pans not more than 4 inches deep. For uniform cooking of frozen vegetables, place vegetables no more than 2 inches deep in shallow pans. Use-steam table pans if available. Follow manufacturer's directions for cooking times or use guidelines below. In some cases, it may be necessary to establish your cooking time; note time on appropriate recipe card.

Vegetable	Directions for Cooking	Approximate Cooking Time (Minutes)	
		5 lb pressure	15 lb pressure (high speed type)
Asparagus, fresh	Place whole stalks in single layers in a solid pan, or place stalks flat 2 inches deep in perforated pan.	6-8	3-5
Asparagus, frozen, cuts and tips	Partially thaw. Arrange in single layers in shallow solid pan or perforated pan.	5-8	3-4
Beans, green or wax, frozen	Break frozen blocks into pieces. Place in shallow perforated pan or-shallow solid pan filled 2/3 full.	20-30	3-4
Beans, lima, frozen	Place loose frozen beans in perforated pan or shallow solid pan.	20-25	3-5

NOTE: See Guidelines For Steam Cookers, Recipe No. A-21.

Vegetable	Directions for Cooking	Approximate Cooking Time (Minutes)	
		5 lb pressure	15 lb pressure (high speed type)
Broccoli, frozen	Place partially thawed in shallow solid pan or perforated pan.	8-10	3-5
Brussels sprouts, frozen	Place partially thawed in shallow solid pan or perforated pan.	5-10	2-1/2 - 3
Cabbage, wedges, fresh (3 oz each)	Place in shallow solid pan or perforated pan.	8-15	5-10
Carrots, fresh, cut in 2 inch lengthwise strips	Place in shallow solid pan filled 1/2 full or perforated pan.	12-15	3-5
Carrots, slices, frozen	Place in shallow solid pan filled 1/2 full or perforated pan.	9-10	2-1/2 - 3
Cauliflower, fresh, flowerets	Place in shallow solid pan or perforated pan.	10-12	8-10
Cauliflower, frozen, flowerets	Partially thaw. Break blocks in pieces. Place in shallow solid pan or perforated pan.	6-8	3 - 3-1/2
Corn, fresh, on-the-cob	Place in perforated pan.	10-12	8-10

GUIDELINES FOR STEAM COOKING VEGETABLES

Vegetable	Directions for Cooking	Approximate Cooking Time (Minutes)	
		5 lb pressure	15 lb pressure (high speed type)
Corn, frozen, on-the-cob	Place in perforated pan.	7-9	4-6
Corn, frozen, whole kernel	Place in shallow solid pan.	9-12	2-3
Okra, frozen	Place in shallow solid pan.	3-5	2-3
Onions, dry, whole	Place in perforated pan.	20-30	10-20
Parsnips, fresh, quartered	Place in perforated pan.	15-20	10-20
Peas, frozen	Place in shallow solid pan.	5-8	1-1/2 - 2-1/2
Peas and carrots, frozen	Place in shallow solid pan.	5-8	1-1/2 - 2-1/2
Potatoes, sweet, fresh, whole, unpared	Place in perforated pan.	25-35	20-30

Vegetable	Directions for Cooking	Approximate Cooking Time (Minutes)	
		5 lb pressure	15 lb pressure (high speed type)
Potatoes, white, fresh, halves or quarters	Place in perforated pan.	30-35	20-30
Rutabagas, fresh, cut in 1/2 inch dices or slices	Place in shallow solid pan or perforated pan.	25-30	15-20
Spinach, frozen	Partially thaw and divide into 3 to 4 blocks. Place in shallow solid pan.	5-10	3-4
Squash, fresh, summer	Place in shallow solid pan.	8-12	5-8
Squash, frozen, summer	Place in shallow solid pan.	8-12	1-1/2 - 2
Squash, fresh, fall and winter, cut in 2 inch pieces	Place in shallow solid pan or perforated pan.	20-25	15-20
Turnips, fresh, white, cut in 1/8 inch slices	Place in shallow solid pan 1/2 full or perforated pan.	15-20	5-6

Q. VEGETABLES No. 0(1)

INDEX

Card No.		Card No.	
Q 001 01	Broccoli Combo	Q 008 00	Harvard Beets
Q 001 02	Bean Combo	Q 008 01	Beets in Orange-Lemon Sauce
Q 001 03	Cauliflower Combo	Q 009 00	Hot Spiced Beets
Q 001 04	Brussels Sprouts Combo	Q 010 00	Broccoli Polonaise
Q 001 05	Green Bean Combo	Q 010 01	Brussels Sprouts Polonaise
Q 001 06	Corn Combo	Q 010 02	Cauliflower Polonaise
Q 002 00	Baked Beans (Canned)	Q 011 00	Sprouts Superba
Q 002 01	Baked Beans (Kidney Beans, Canned)	Q 012 00	Fried Cabbage
Q 002 02	Baked Beans (Pinto Beans, Canned)	Q 012 01	Calico Cabbage
Q 003 00	Boston Baked Beans	Q 012 02	Fried Cabbage with Bacon
Q 003 01	Savory Baked Beans	Q 013 00	Scalloped Sweet Potatoes and Apples
Q 004 00	Italian-Style Baked Beans	Q 014 00	Orange Carrots Amandine
Q 004 01	Italian-Style Baked Beans (Canned Beans)	Q 015 00	Oriental Stir-Fry Cabbage
Q 005 00	Simmered Dry Beans with Bacon	Q 016 00	Carrot and Celery Amandine
Q 005 01	Savory Style Beans	Q 017 00	Lyonnaise Carrots
Q 005 02	Simmered Dry Beans	Q 017 01	Glazed Carrots
Q 006 00	Spanish Style Beans	Q 018 00	Cauliflower Au Gratin
Q 007 00	Lyonnaise Green or Wax Beans	Q 019 00	German Potato Griddle Cakes (Dehy)
Q 007 01	Green Beans Creole	Q 020 00	French Fried Cauliflower
Q 007 02	Green Beans with Mushrooms	Q 020 01	French Fried Okra
Q 007 03	Green Beans Nicoise	Q 021 00	Corn Fritters
Q 007 04	Green Beans Southern Style	Q 021 01	Corn Fritters (Pancake Mix)

Q. VEGETABLES No. 0(1)

Card No. ....   Card No.

Card No.	Name	Card No.	Name
Q 022 00	Ratatouille	Q 033 02	Parsley Buttered Potatoes (Canned)
Q 023 00	Scalloped Cream Style Corn	Q 033 03	Paprika Buttered Potatoes (Canned)
Q 023 01	Scalloped Whole Kernel Corn	Q 034 00	Spanish Onions
Q 024 00	Broccoli Parmesan	Q 035 00	French Fried Onion Rings
Q 024 01	Brussels Sprouts Parmesan	Q 035 01	French Fried Onion Rings (Frozen)
Q 024 02	Cauliflower Parmesan	Q 035 02	Tempura Fried Onion Rings
Q 025 00	Vegetable Stir Fry	Q 036 00	Fried Onions
Q 026 00	Herbed Green Beans	Q 037 00	Smothered Onions (Dehydrated Onions)
Q 027 00	Calico Corn	Q 038 00	Refried Beans with Cheese
Q 027 01	Corn O'Brien	Q 038 01	Refried Beans (Canned Beans)
Q 027 02	Mexican Corn	Q 038 02	Refried Beans with Cheese (Canned Beans)
Q 028 00	Eggplant Parmesan	Q 039 00	Green Beans with Corn (Frozen Beans)
Q 029 00	Southern Style Greens (Fresh Collards)	Q 039 01	Green Beans with Corn (Canned Beans)
Q 029 01	Southern Style Greens (Frozen)	Q 040 00	Turnips and Bacon
Q 029 02	Sweet Sour Greens	Q 041 00	Peas with Mushrooms (Frozen)
Q 029 03	Southern Style Greens (Fresh Kale)	Q 041 01	Peas with Carrots (Frozen)
Q 030 00	Sauteed Mushrooms	Q 041 02	Peas with Celery (Frozen)
Q 030 01	Sauteed Mushrooms and Onions	Q 041 03	Peas with Onions
Q 031 00	Okra and Tomato Gumbo	Q 041 04	Peas with Mushrooms (Canned Peas)
Q 032 00	Southern Fried Okra	Q 042 00	Green Beans Parisienne (Canned)
Q 033 00	Parsley Buttered Potatoes	Q 042 01	Green Beans Parisienne (Frozen Beans)
Q 033 01	Paprika Buttered Potatoes	Q 043 00	Red Cabbage with Sweet and Sour Sauce

INDEX

Card No.		Card No.	
Q 044 00	Baked Potatoes	Q 050 01	Franconia Potatoes
Q 044 01	Quick Baked Potato Halves	Q 050 02	Oven-Glo Potatoes
Q 045 00	French Fried Potatoes	Q 050 03	Oven-Glo Potatoes (Canned)
Q 045 01	French Fried Potatoes (Frozen)	Q 050 04	Oven Browned Potatoes (Canned)
Q 045 02	French Fried Potatoes (Frozen, Oven Method)	Q 051 00	Potatoes Au Gratin
Q 045 03	French Fried Shoestring Potatoes (Frozen)	Q 051 01	Potatoes Au Gratin (Dehydrated, Slices)
Q 045 04	French Fried Shoestring Potatoes (Frozen, Oven)	Q 052 00	Rissole Potatoes
Q 045 05	French Fried Potatoes (Dehydrated Mix)	Q 053 00	Scalloped Potatoes
Q 045 06	Baked Potato Rounds (Precooked)	Q 053 01	Scalloped Potatoes and Onions
Q 046 00	Hashed Brown Potatoes	Q 054 00	Hashed Brown Potatoes (Dehydrated, Sliced)
Q 046 01	Cottage Fried Potatoes	Q 054 01	Lyonnaise Potatoes (Dehydrated)
Q 046 02	Hashed Brown Potatoes (Frozen, Shredded, 3 Oz)	Q 054 02	O'Brien Potatoes (Dehydrated, Sliced)
Q 046 03	Lyonnaise Potatoes	Q 054 03	Hashed Brown Potatoes (Dehydrated, Shredded)
Q 046 04	Hashed Brown Potatoes (Frozen, Shredded, 2.5 Oz)	Q 054 04	Hashed Brown Potatoes (Dehydrated, Diced)
Q 047 00	Home Fried Potatoes	Q 055 00	Scalloped Potatoes And Onions (Dehydrated, Sliced)
Q 048 00	Mashed Potatoes	Q 055 01	Scalloped Potatoes (Dehydrated, Sliced)
Q 048 01	Grilled Potato Patties	Q 055 02	Scalloped Potatoes (Dehydrated, Diced)
Q 049 00	O'Brien Potatoes	Q 056 00	Golden Potato Balls (Instant)
Q 050 00	Oven Browned Potatoes	Q 057 00	Mashed Potatoes (Instant)
		Q 057 01	Grilled Potato Cakes

Q. VEGETABLES No. 0(2)

Card No.		Card No.	
Q 058 00	Stewed Tomatoes	Q 074 00	Squash and Carrot Medley (Fresh)
Q 058 01	Stewed Tomatoes with Croutons	Q 074 01	Squash and Carrot Medley (Frozen)
Q 059 00	German Sauerkraut	Q 075 00	Deviled Oven Fries
Q 060 00	Club Spinach	Q 076 00	South of the Border Broccoli
Q 061 00	Baked Hubbard Squash	Q 076 01	South of the Border Medley
Q 062 00	Creole Summer Squash	Q 077 00	Baked Potato Pancakes (Frozen Shredded Potato)
Q 063 00	Tangy Spinach		
Q 064 00	Louisiana Style Smothered Squash	Q 077 01	Baked Potato Pancakes
Q 064 01	Savory Summer Squash	Q 078 00	Potatoes and Herbs
Q 065 00	Herbed Broccoli	Q 079 00	Hacienda Potatoes
Q 066 00	Baked Sweet Potatoes	Q 080 00	Hacienda Corn and Black Beans
Q 067 00	Candied Sweet Potatoes	Q 081 00	Hacienda Green Beans
Q 067 01	Glazed Sweet Potatoes	Q 082 00	Honey Dijon Vegetables
Q 067 02	Glazed Sweet Potatoes (Syrup)	Q 083 00	Corn and Green Bean Casserole
Q 068 00	Tempura Vegetables	Q 084 00	Garlic-Lemon Roasted Potato Wedges
Q 069 00	Mashed Sweet Potatoes	Q 100 00	Asparagus (Frozen)
Q 069 01	Sweet Potatoes Southern Style	Q 100 01	Asparagus (Canned)
Q 069 02	Marshmallow Sweet Potatoes	Q 100 02	Asparagus (Fresh)
Q 070 00	Garlic Roasted Potato Wedges	Q 101 00	Beans, Green (Frozen)
Q 071 00	Rosemary Roasted Potato Wedges	Q 101 01	Beans, Green (Canned)
Q 072 00	Sesame Glazed Green Beans	Q 101 02	Beans, Wax (Canned)
Q 073 00	Japanese Vegetable Stir Fry	Q 101 03	Beans, French Style Cut (Frozen)

Q. VEGETABLES No. 0(3)

INDEX

Card No.		Card No.	
Q 101 04	Beans, Wax (Frozen)	Q 113 03	Greens, Kale (Fresh)
Q 102 00	Beans, Lima (Frozen)	Q 114 00	Okra (Frozen)
Q 102 01	Beans, Lima (Canned)	Q 114 01	Okra (Canned)
Q 103 01	Beans, White in Tomato Sauce (Canned)	Q 115 01	Onions (Canned)
Q 104 01	Beets (Canned)	Q 115 02	Onions (Fresh)
Q 105 00	Broccoli (Frozen)	Q 116 00	Peas (Frozen)
Q 105 02	Broccoli (Fresh)	Q 116 01	Peas (Canned)
Q 106 00	Brussels Sprouts (Frozen)	Q 117 01	Black-Eyed Peas (Canned)
Q 107 02	Cabbage (Fresh)	Q 118 01	Potatoes, Sweet (Canned)
Q 108 00	Carrot Slices (Frozen)	Q 118 02	Potatoes, Sweet (Fresh)
Q 108 01	Carrot Slices (Canned)	Q 119 01	Potatoes, White (Canned)
Q 108 02	Carrots (1/4 Inch Slices) (Fresh)	Q 119 02	Potatoes, White (Fresh)
Q 108 03	Carrot Strips (Fresh)	Q 120 01	Sauerkraut (Canned)
Q 109 00	Cauliflower (Frozen)	Q 121 00	Spinach (Frozen)
Q 109 02	Cauliflower (Fresh)	Q 121 01	Spinach (Canned)
Q 110 00	Corn, Whole Kernel (Frozen)	Q 121 02	Spinach (Fresh)
Q 110 01	Corn, Whole Kernel (Canned)	Q 122 00	Squash, Summer (Frozen)
Q 111 00	Corn on the Cob (Frozen)	Q 122 02	Squash, Summer (Fresh)
Q 111 02	Corn on the Cob (Fresh)	Q 123 02	Squash, Fall and Winter (Fresh)
Q 112 01	Cream Style Corn (Canned)	Q 124 00	Succotash (Frozen)
Q 113 00	Greens, Collard (Frozen)	Q 125 01	Tomatoes (Canned)
Q 113 02	Greens, Collard (Fresh)	Q 126 00	Mixed Vegetables (Frozen)

Q. VEGETABLES No. 0(3)

Card No. ....

Q 127 00	Peas and Carrots (Frozen)
Q 128 00	Rutabagas (Fresh)
Q 129 00	Turnips (Fresh)
Q 500 00	Garlic Cheese Potatoes
Q 500 01	Garlic Cheese Potatoes (Instant)
Q 502 00	Italian Roasted Potatoes
Q 503 00	Okra Melange
Q 504 00	Roasted Pepper Potatoes
Q 504 01	Roasted Pepper Potatoes (Instant)
Q 800 00	Okra, Frozen, Breaded
Q 801 00	Cajun Oven Fries
Q 803 00	Cauliflower with Cheese Sauce
Q 804 00	Mushrooms, Frozen, Breaded
Q 808 00	Potatoes Au Gratin using Prepared Sauce
Q 809 00	Vegetable Stir Fry using Frozen Vegetables

VEGETABLES No.Q 001 01

BROCCOLI COMBO

Yield 100   **Portion** 3/4 Cup

Calories	Carbohydrates	Protein	Fat	Cholesterol	Sodium	Calcium
107 cal	17 g	4 g	4 g	0 mg	67 mg	39 mg

Ingredient	Weight	Measure	Issue
BROCCOLI,FROZEN,SPEARS	12 lbs	2 gal 3/4 qts	
CORN,FROZEN,WHOLE KERNEL	12 lbs	2 gal 1/4 qts	
CARROTS,FROZEN,SLICED	6 lbs		
MARGARINE,MELTED	1 lbs	2 cup	
RESERVED LIQUID	1-5/8 lbs	3 cup	

Method

1. Cook broccoli for 6 to 8 minutes, corn for 5 to 10 minutes and carrots for 10 to 13 minutes.
2. Drain; reserve liquid for use in Step 3.
3. Combine cooked vegetables; toss lightly; place in steam table pans. Combine melted butter and reserved cooking liquid. Pour an equal quantity over each pan.
4. CCP: Vegetables must be heated to 145 F. or higher for 15 seconds. Hold at 140 F. or higher for service.

VEGETABLES   No.Q 001 02

BEAN COMBO

Yield 100   Portion   3/4 Cup

Calories	Carbohydrates	Protein	Fat	Cholesterol	Sodium	Calcium
113 cal	17 g	4 g	4 g	0 mg	83 mg	53 mg

Ingredient   Weight   Measure   Issue

Ingredient	Weight	Measure
BEANS,GREEN,FROZEN,FRENCH STYLE	15 lbs	3 gal 1-3/4 qts
BEANS,LIMA,FROZEN	7-1/2 lbs	1 gal 1-1/4 qts
CARROTS,FROZEN,SLICED	7-1/2 lbs	1 gal 2-5/8 qts
MARGARINE,MELTED	1 lbs	2 cup
RESERVED LIQUID	1-5/8 lbs	3 cup

Method

1. Cook green beans for 5 to 8 minutes, lima beans for 6 to 12 minutes and carrots for 10 to 13 minutes.
2. Drain; reserve liquid for use in Step 3.
3. Combine cooked vegetables; toss lightly; place in steam table pans. Combine melted butter and reserved cooking liquid. Pour an equal quantity over each pan.
4. CCP: Vegetables must be heated to 145 F. or higher for 15 seconds. Hold at 140 F. or higher for service.

VEGETABLES No.Q 001 03

CAULIFLOWER COMBO

Yield 100   **Portion** 3/4 Cup

Calories	Carbohydrates	Protein	Fat	Cholesterol	Sodium	Calcium
93 cal	13 g	4 g	4 g	0 mg	130 mg	35 mg

Ingredient   **Weight**   **Measure**   **Issue**

Ingredient	Weight	Measure
CAULIFLOWER,FROZEN	7-1/2 lbs	
PEAS & CARROTS,FROZEN	22-1/2 lbs	4 gal 2-1/4 qts
MARGARINE,MELTED	1 lbs	2 cup
RESERVED LIQUID	1-5/8 lbs	3 cup

Method

1. Cook cauliflower 4 to 8 minutes and peas and carrots 6 to 8 minutes.
2. Drain; reserve liquid for use in Step 3.
3. Combine cooked vegetables; toss lightly; place in steam table pans. Combine melted butter and reserved cooking liquid. Pour an equal quantity over each pan.
4. CCP: Vegetables must be heated to 145 F. or higher for 15 seconds. Hold at 140 F. or higher for service.

VEGETABLES   No.Q 001 04

BRUSSELS SPROUTS COMBO

Yield 100   Portion   3/4 Cup

Calories	Carbohydrates	Protein	Fat	Cholesterol	Sodium	Calcium
114 cal	18 g	4 g	4 g	0 mg	73 mg	25 mg

Ingredient	Weight	Measure	Issue
BRUSSELS SPROUTS,FROZEN	12 lbs	2 gal 3/4 qts	
CORN,FROZEN,WHOLE KERNEL	12 lbs	2 gal 1/4 qts	
CARROTS,FROZEN,SLICED	6 lbs	1 gal 1-1/3 qts	
MARGARINE,MELTED	1 lbs	2 cup	
RESERVED LIQUID	1-5/8 lbs	3 cup	

Method

1 Cook brussels sprouts 7 to 9 minutes, corn for 4 to 6 minutes and carrots for 10 to 13 minutes.
2 Drain; reserve liquid for use in Step 3.
3 Combine cooked vegetables; toss lightly; place in steam table pans. Combine melted butter and reserved cooking liquid. Pour an equal quantity over each pan.
4 Vegetables must be heated to 145 F. or higher for 15 seconds. CCP: Hold at 140 F. or higher for service.

VEGETABLES No.Q 001 05

GREEN BEAN COMBO

Yield 100   Portion   3/4 Cup

Calories	Carbohydrates	Protein	Fat	Cholesterol	Sodium	Calcium
77 cal	10 g	2 g	4 g	0 mg	81 mg	54 mg

Ingredient	Weight	Measure	Issue
BEANS,GREEN,FROZEN,FRENCH STYLE	18 lbs	4 gal 1/2 qts	
CARROTS,FROZEN,SLICED	9 lbs	1 gal 4 qts	
CELERY,FRESH,SLICED	3 lbs	2 qts 3-3/8 cup	4-1/8 lbs
MARGARINE,MELTED	1 lbs	2 cup	
RESERVED LIQUID	1-5/8 lbs	3 cup	

Method

1. Cook green beans 5 to 8 minutes and carrots and celery 10 to 13 minutes.
2. Drain; reserve liquid for use in Step 3.
3. Combine cooked vegetables; toss lightly; place in steam table pans. Combine melted butter and reserved cooking liquid. Pour an equal quantity over each pan.
4. CCP: Vegetables must be heated to 145 F. or higher for 15 seconds. Hold at 140 F. or higher for service.

VEGETABLES No.Q 001 06

CORN COMBO

Yield 100   Portion   3/4 Cup

Calories	Carbohydrates	Protein	Fat	Cholesterol	Sodium	Calcium
107 cal	18 g	3 g	4 g	0 mg	66 mg	36 mg

Ingredient | Weight | Measure | Issue

Ingredient	Weight	Measure	Issue
CORN,FROZEN,WHOLE KERNEL	11-1/4 lbs	1 gal 3-3/4 qts	
BEANS,GREEN,FROZEN,CUT	11-1/4 lbs	2 gal 2-1/4 qts	
CARROTS,FROZEN,SLICED	7-1/2 lbs	1 gal 2-5/8 qts	
MARGARINE,MELTED	1 lbs	2 cup	
RESERVED LIQUID	1-5/8 lbs	3 cup	

Method

1. Cook corn for 4 to 6 minutes, beans for 5 to 8 minutes and carrots for 10 to 13 minutes.
2. Drain; reserve liquid for use in Step 3.
3. Combine cooked vegetables; toss lightly; place in steam table pans. Combine melted butter and reserved cooking liquid. Pour an equal quantity over each pan.
4. CCP: Heat to 145 F. or higher for 15 seconds. Hold at 140 F. or higher for service.

VEGETABLES No.Q 002 00

BAKED BEANS (CANNED)

Yield 100   **Portion** 1/2 Cup

Calories	Carbohydrates	Protein	Fat	Cholesterol	Sodium	Calcium
172 cal	32 g	8 g	3 g	11 mg	672 mg	79 mg

Ingredient

Ingredient	Weight	Measure	Issue
BACON,RAW	1 lbs		
ONIONS,FRESH,CHOPPED	1-7/8 lbs	1 qts 1-1/4 cup	2 lbs
BEANS,BAKED,W/PORK,CANNED	30-1/8 lbs	3 gal 1-1/2 qts	
CATSUP	1-1/4 lbs	2-1/4 cup	
SUGAR,BROWN,PACKED	9 oz	1-3/4 cup	
MUSTARD,PREPARED	6-5/8 oz	3/4 cup	

Method

1. Cook bacon according to Recipe Nos. L 002 00 or L 002 02. Drain. Finely chop.
2. Combine onions, beans, catsup, sugar, mustard and bacon. Mix well.
3. Pour 7-3/4 quarts bean mixture into each steam table pan.
4. Using a convection oven, bake at 325 F. for 1-1/2 hours on high fan, open vent. CCP: Heat to 145 F. or higher for 15 seconds. Hold at 140 F. or higher for service.

VEGETABLES No.Q 002 01

BAKED BEANS (KIDNEY BEANS, CANNED)

Yield 100   Portion   1/2 Cup

Calories	Carbohydrates	Protein	Fat	Cholesterol	Sodium	Calcium
139 cal	26 g	7 g	1 g	1 mg	543 mg	43 mg

Ingredient	Weight	Measure	Issue
BACON,RAW	1 lbs		
BEANS,KIDNEY,DARK RED,CANNED,DRAINED	27-1/2 lbs	4 gal 1-5/8 qts	
ONIONS,FRESH,CHOPPED	1-7/8 lbs	1 qts 1-1/4 cup	2 lbs
CATSUP	1-1/4 lbs	2-3/8 cup	
SUGAR,BROWN,PACKED	1-1/4 lbs	3-3/4 cup	
MUSTARD,PREPARED	6-5/8 oz	3/4 cup	

Method

1. Cook bacon according to Recipe Nos. L 002 00 or L 002 02. Drain. Finely chop.
2. Combine onions, beans, catsup, sugar, mustard and bacon. Mix well.
3. Pour 7-3/4 quarts bean mixture into each steam table pan.
4. Bake in a convection oven at 325 F. for 1-1/2 hours on high fan, open vent. CCP: Heat to 145 F. or higher for 15 seconds. Hold at 140 F. or higher for service.

VEGETABLES No.Q 002 02

BAKED BEANS (PINTO BEANS, CANNED)

Yield 100   Portion   1/2 Cup

Calories	Carbohydrates	Protein	Fat	Cholesterol	Sodium	Calcium
147 cal	27 g	7 g	2 g	1 mg	483 mg	64 mg

Ingredient	Weight	Measure	Issue
BACON,RAW	1 lbs		
ONIONS,FRESH,CHOPPED	1-7/8 lbs	1 qts 1-1/4 cup	2 lbs
BEANS,PINTO,CANNED,DRAINED	28 lbs	3 gal 1-1/4 qts	
CATSUP	1-1/4 lbs	2-3/8 cup	
SUGAR,BROWN,PACKED	1-1/4 lbs	3-3/4 cup	
MUSTARD,PREPARED	6-5/8 oz	3/4 cup	

Method

1. Cook bacon according to Recipe Nos. L 002 00 or L 002 02. Drain. Finely chop.
2. Combine onions, beans, catsup, sugar, mustard and bacon. Mix well.
3. Pour 7-3/4 quarts bean mixture into each steam table pan.
4. Using a convection oven, bake at 325 F. for 1-1/2 hours on high fan, open vent. CCP: Heat to 145 F. or higher for 15 seconds. Hold for service at 140 F. or higher.

VEGETABLES No.Q 003 00

BOSTON BAKED BEANS

Yield 100   Portion   1/2 Cup

Calories	Carbohydrates	Protein	Fat	Cholesterol	Sodium	Calcium
179 cal	32 g	10 g	2 g	1 mg	204 mg	83 mg

Ingredient	Weight	Measure	Issue
BEANS,KIDNEY,DRY	8-7/8 lbs	1 gal 1-1/2 qts	
WATER,COLD	46 lbs	5 gal 2 qts	
BACON,RAW	1 lbs		
SALT	1-1/2 oz	2-1/3 tbsp	
MUSTARD,DRY	2-1/2 oz	1/4 cup 2-2/3 tbsp	
SUGAR,BROWN,PACKED	10-7/8 oz	2-1/8 cup	
VINEGAR,DISTILLED	2-1/8 oz	1/4 cup 1/3 tbsp	
MOLASSES	1-1/2 lbs	2 cup	
COOKING SPRAY,NONSTICK	2 oz	1/4 cup 1/3 tbsp	

Method

1. Pick over beans, removing discolored beans and foreign matter. Wash beans thoroughly. Cover; let soak 1 hour.
2. Cover with water. Bring beans to a boil; add more water if necessary to keep beans covered. Turn down heat, simmer 1-1/2 hours or until tender, but not mushy. Drain beans. Reserve liquid and beans for use in Step 4.
3. Cook bacon by arranging slices in rows down the length of 18x26 sheet pan, with fat edges slightly overlapping lean edges. Using a convection oven, bake 25 minutes at 325 F. on high fan, open vent. Drain excess fat. Bake an additional 5 to 10 minutes or until bacon is slightly crisp. DO NOT OVERCOOK. Drain thoroughly. Finely chop.
4. Take reserved bean liquid and add water to equal 1 gallon and combine with salt, mustard, brown sugar, vinegar, molasses and chopped bacon. Add to beans; mix well.
5. Lightly spray pans with non-stick cooking spray. Pour 20 pounds or 7-1/2 quarts bean mixture into each lightly sprayed pan; cover. Using a convection oven, bake at 325 F., 1 hour to 1 hour 15 minutes, or until sauce is just below surface of beans, on high fan, closed vent. Uncover; stir; bake additional 15 minutes or until set, on low fan. CCP: Heat to 145 F. or higher for 15 seconds. Hold at 140 F. or higher for service.

VEGETABLES No.Q 003 01

SAVORY BAKED BEANS

Yield 100 **Portion** 1/2 Cup

Calories	Carbohydrates	Protein	Fat	Cholesterol	Sodium	Calcium
195 cal	36 g	10 g	2 g	1 mg	374 mg	85 mg

Ingredient	Weight	Measure	Issue
BEANS,KIDNEY,DRY	8-7/8 lbs	1 gal 1-1/2 qts	
WATER,COLD	46 lbs	5 gal 2 qts	
BACON,RAW	1 lbs		
SALT	1-1/2 oz	2-1/3 tbsp	
MUSTARD,DRY	2-1/2 oz	1/4 cup 2-2/3 tbsp	
CATSUP	3-1/8 lbs	1 qts 2 cup	
ONIONS,FRESH,CHOPPED	11-1/4 oz	2 cup	12-1/2 oz
SUGAR,BROWN,PACKED	10-7/8 oz	2-1/8 cup	
VINEGAR,DISTILLED	2-1/8 oz	1/4 cup 1/3 tbsp	
MOLASSES	1-1/2 lbs	2 cup	
COOKING SPRAY,NONSTICK	2 oz	1/4 cup 1/3 tbsp	

Method

1. Pick over beans, removing discolored beans and foreign matter. Wash beans thoroughly. Cover; let soak 1 hour.
2. Cover with water. Bring beans to a boil; add more water if necessary to keep beans covered. Turn down heat, simmer 1-1/2 hours or until tender, but not mushy. Drain beans. Reserve liquid and beans for use in Step 4.
3. Cook bacon by arranging slices in rows down the length of 18x26 sheet pan, with fat edges slightly overlapping lean edges. Using a convection oven, bake 25 minutes at 325 F. on high fan, open vent. Drain excess fat. Bake an additional 5 to 10 minutes or until bacon is slightly crisp. DO NOT OVERCOOK. Drain thoroughly. Finely chop.
4. Take reserved bean liquid and add water to equal 2-1/2 quarts per 100 portions and combine with salt, mustard, catsup, onions, brown sugar, vinegar, molasses, and chopped bacon. Add to beans; mix well.
5. Lightly spray each steam table pan with non-stick cooking spray. Pour 20-1/8 pounds or 7-1/2 quarts bean mixture into each lightly sprayed steam table pan; cover. Using a convection oven, bake at 325 F., 1 hour to 1 hour 15 minutes stir; bake additional 15 minutes or until set on low fan. CCP: Heat to 145 F. or higher for 15 seconds. Hold at 140 F. or higher for service.

VEGETABLES No.Q 004 00

ITALIAN-STYLE BAKED BEANS

Yield 100   Portion   1/2 Cup

Calories	Carbohydrates	Protein	Fat	Cholesterol	Sodium	Calcium
133 cal	23 g	8 g	2 g	1 mg	424 mg	79 mg

Ingredient	Weight	Measure	Issue
BEANS,KIDNEY,DRY	6-1/8 lbs	3 qts 3 cup	
WATER,COLD	31-1/3 lbs	3 gal 3 qts	
ONIONS,FRESH,CHOPPED	1-1/3 lbs	3-3/4 cup	1-1/2 lbs
CELERY,FRESH,CHOPPED	1-1/4 lbs	1 qts 3/4 cup	1-3/4 lbs
OIL,OLIVE	2-7/8 oz	1/4 cup 2-1/3 tbsp	
PARSLEY,FRESH,BUNCH,CHOPPED	2-1/8 oz	1 cup	2-1/4 oz
THYME,GROUND	<1/16th oz	1/8 tsp	
OREGANO,CRUSHED	1/8 oz	1 tbsp	
SALT	1-1/2 oz	2-1/3 tbsp	
PEPPER,BLACK,GROUND	1/8 oz	1/4 tsp	
GARLIC POWDER	1/8 oz	1/4 tsp	
BASIL,DRIED,CRUSHED	1/8 oz	1/3 tsp	
SUGAR,GRANULATED	1/2 oz	1 tbsp	
TOMATO PASTE,CANNED	6 lbs	2 qts 2-1/2 cup	
CHEESE,PARMESAN,GRATED	5-1/4 oz	1-1/2 cup	

Method

1 Pick over beans, removing discolored beans and foreign matter. Wash beans thoroughly. Cover; let soak 1 hour.
2 Cover with water; bring beans to a boil; add more water to cover beans if necessary. Simmer 1-1/2 hours or until beans are just tender but not mushy. Drain beans; reserve liquid for use in Step 4, and beans for use in Step 5.
3 Saute onions and celery in olive oil or shortening 10 minutes or until tender.
4 Take reserved bean liquid and add water to equal 1 gallon per 100 portions and combine with parsley, thyme, oregano, salt, pepper, sugar, garlic, basil, tomato paste to onion mixture; bring to a boil; reduce heat; simmer 10 minutes.
5 Place 1 gallon cooked beans in each steam table pan; add 3-3/4 quarts sauce; mix carefully. Sprinkle cheese over beans.
6 Using a convection oven, bake in 325 F. oven for 45 minutes on low fan, open vent. CCP: Heat to 145 F. or higher for 15 seconds. Hold at 140 F. or higher for service.

VEGETABLES No.Q 004 01

ITALIAN-STYLE BAKED BEANS (CANNED BEANS)

Yield 100   **Portion** 1/2 Cup

Calories	Carbohydrates	Protein	Fat	Cholesterol	Sodium	Calcium
117 cal	20 g	7 g	2 g	1 mg	741 mg	62 mg

Ingredient	Weight	Measure	Issue
BEANS,KIDNEY,DARK RED,CANNED,DRAINED	20-3/4 lbs	3 gal 1-1/4 qts	
ONIONS,FRESH,CHOPPED	1-1/4 lbs	3-1/2 cup	1-3/8 lbs
CELERY,FRESH,CHOPPED	1-1/2 lbs	1 qts 1-5/8 cup	2 lbs
OIL,SALAD	2-7/8 oz	1/4 cup 2-1/3 tbsp	
PARSLEY,FRESH,BUNCH,CHOPPED	2-1/8 oz	1 cup	2-1/4 oz
THYME,GROUND	<1/16th oz	1/8 tsp	
OREGANO,CRUSHED	1/8 oz	1 tbsp	
SALT	1-1/2 oz	2-1/3 tbsp	
PEPPER,BLACK,GROUND	1/8 oz	1/4 tsp	
GARLIC POWDER	1/8 oz	1/4 tsp	
BASIL,DRIED,CRUSHED	1/8 oz	1/3 tsp	
SUGAR,GRANULATED	1/2 oz	1 tbsp	
TOMATO PASTE,CANNED	6 lbs	2 qts 2-1/2 cup	
CHEESE,PARMESAN,GRATED	5-1/4 oz	1-1/2 cup	

Method

1. Drain beans; reserve liquid for use in Step 3, and beans for use in Step 4.
2. Saute onions and celery in salad oil or shortening 10 minutes or until tender.
3. Take reserved bean liquid and add water to equal 1 gallon per 100 portions and combine with parsley, thyme, oregano, salt, pepper, sugar, garlic, basil, tomato paste, and onion mixture; bring to a boil; reduce heat; simmer 10 minutes.
4. Place 1 gallon cooked beans in each steam table pan; add 3-3/4 quarts sauce; mix carefully. Sprinkle cheese over beans.
5. Using a convection oven, bake at 325 F. for 45 minutes on low fan, open vent. CCP: Heat to 145 F. or higher for 15 seconds. Hold at 140 F. or higher for service.

VEGETABLES No.Q 005 00

SIMMERED DRY BEANS WITH BACON

Yield 100   Portion 2/3 Cup

Calories	Carbohydrates	Protein	Fat	Cholesterol	Sodium	Calcium
137 cal	22 g	10 g	2 g	2 mg	195 mg	57 mg

Ingredient | Weight | Measure | Issue

Ingredient	Weight	Measure	Issue
BEANS,KIDNEY,DRY	8-1/8 lbs	1 gal 1 qts	
WATER,COLD	41-3/4 lbs	5 gal	
BACON,RAW	2 lbs		
SALT	1-1/4 oz	2 tbsp	
PEPPER,BLACK,GROUND	1/4 oz	1 tbsp	

Method

1. Pick over beans, removing discolored beans and foreign matter. Wash beans thoroughly. Cover; let soak 1 hour.
2. Cover with water; bring to a boil in steam-jacketed kettle; boil 2 minutes.
3. Add bacon, salt and pepper to beans.
4. Turn down heat; add more water if necessary to cover beans; cover. Simmer 1-1/2 hours or until beans are just tender. CCP: Heat to 145 F. or higher for 15 seconds. Hold at 140 F. or higher for service.

VEGETABLES No.Q 005 01

SAVORY STYLE BEANS

Yield 100   **Portion** 2/3 Cup

Calories	Carbohydrates	Protein	Fat	Cholesterol	Sodium	Calcium
126 cal	23 g	9 g	0 g	0 mg	21 mg	61 mg

Ingredient   Weight   Measure   Issue

Ingredient	Weight	Measure	Issue
BEANS,KIDNEY,DRY	8-1/8 lbs	1 gal 1 qts	
WATER,COLD	41-3/4 lbs	5 gal	
ONIONS,FRESH,CHOPPED	1-1/3 lbs	3-3/4 cup	1-1/2 lbs
CELERY,FRESH,CHOPPED	1-1/2 lbs	1 qts 1-5/8 cup	2 lbs
GARLIC POWDER	1/4 oz	1/3 tsp	
CUMIN,GROUND	1/8 oz	1/3 tsp	
PEPPER,BLACK,GROUND	1/8 oz	1/3 tsp	

Method

1 Pick over beans, removing discolored beans and foreign matter. Wash beans thoroughly. Cover; let soak 1 hour.
2 Cover with water; bring to a boil in steam-jacketed kettle; boil 2 minutes.
3 Add onions, celery, garlic powder, cumin, and black pepper.
4 Reduce heat; add more water if necessary to cover beans; cover. Simmer 2 hours or until beans are just tender. CCP: Heat to 145 F. or higher for 15 seconds. Hold at 140 F. or higher for service.

VEGETABLES No.Q 005 02

SIMMERED DRY BEANS

Yield 100   Portion   2/3 Cup

Calories	Carbohydrates	Protein	Fat	Cholesterol	Sodium	Calcium
123 cal	22 g	9 g	0 g	0 mg	154 mg	57 mg

Ingredient	Weight	Measure	Issue
BEANS,KIDNEY,DRY	8-1/8 lbs	1 gal 1 qts	
WATER,COLD	41-3/4 lbs	5 gal	
SALT	1-1/4 oz	2 tbsp	
PEPPER,BLACK,GROUND	1/4 oz	1 tbsp	

Method

1 Pick over beans, removing discolored beans and foreign matter. Wash beans thoroughly. Cover; let soak 1 hour.
2 Cover with water; bring to a boil in steam-jacketed kettle; boil 2 minutes.
3 Add salt and pepper to beans.
4 Reduce heat, add more water if necessary to cover beans; cover. Simmer 1-1/2 hours or until beans are just tender. CCP: Heat to 145 F. or higher for 15 seconds. Hold at 140 F. or higher for service.

VEGETABLES No.Q 006 00

SPANISH STYLE BEANS

Yield 100   Portion   1/2 Cup

Calories	Carbohydrates	Protein	Fat	Cholesterol	Sodium	Calcium
161 cal	32 g	9 g	1 g	0 mg	468 mg	63 mg

Ingredient
Ingredient	Weight	Measure	Issue
BEANS,PINTO,DRY	8-1/2 lbs	1 gal 1 qts	
WATER,COLD	41-3/4 lbs	5 gal	
SALT	3-3/4 oz	1/4 cup 2-1/3 tbsp	
ONIONS,FRESH,CHOPPED	1 lbs	3 cup	1-1/8 lbs
TOMATOES,CANNED,CRUSHED,INCL LIQUIDS	6-5/8 lbs	3 qts	
SUGAR,GRANULATED	1 lbs	2-1/4 cup	
CLOVES,GROUND	<1/16th oz	1/8 tsp	
PEPPER,BLACK,GROUND	1/8 oz	1/8 tsp	
MUSTARD,DRY	3/4 oz	2 tbsp	

Method

1 Pick over beans, removing discolored beans and foreign matter. Wash beans thoroughly. Cover; let soak 1 hour.
2 Cover with water; add salt. Bring to a boil in steam-jacketed kettle; boil 2 minutes.
3 Add onions, tomatoes, sugar, mustard, cloves, and pepper. Reduce heat, add more water to cover beans. Simmer 1 hour or until beans are just tender. CCP: Heat to 145 F. or higher for 15 seconds. Hold at 140 F. or higher for service.

VEGETABLES No.Q 007 00

LYONNAISE GREEN OR WAX BEANS

Yield 100   **Portion** 1/2 Cup

Calories	Carbohydrates	Protein	Fat	Cholesterol	Sodium	Calcium
54 cal	7 g	2 g	3 g	7 mg	102 mg	35 mg

Ingredient	**Weight**	**Measure**	**Issue**
ONIONS,FRESH,SLICED	3 lbs	3 qts	3-3/8 lbs
BUTTER	12 oz	1-1/2 cup	
BEANS,GREEN,FROZEN,WHOLE	16 lbs	3 gal 2-5/8 qts	
SALT	5/8 oz	1 tbsp	
WATER,BOILING	12-1/2 lbs	1 gal 2 qts	
PEPPER,BLACK,GROUND	<1/16th oz	1/8 tsp	

Method

1. Saute onions in butter or margarine until tender. Set aside for use in Step 4.
2. Add beans to boiling, salted water. Bring to a boil; cover; simmer 5 to 8 minutes, or until beans are just tender. Drain; reserve 1 quart liquid.
3. Combine onions, beans, bean liquid, and pepper. Mix lightly. Serve. CCP: Heat to 145 F. or higher for 15 seconds. Hold at 140 F. or higher for service.

VEGETABLES No.Q 007 01

GREEN BEANS CREOLE

Yield 100   **Portion** 1/2 Cup

Calories	Carbohydrates	Protein	Fat	Cholesterol	Sodium	Calcium
54 cal	11 g	2 g	1 g	0 mg	221 mg	51 mg

Ingredient	Weight	Measure	Issue
BEANS,GREEN,FROZEN,WHOLE	16 lbs	3 gal 2-5/8 qts	
SALT	5/8 oz	1 tbsp	
WATER	12-1/2 lbs	1 gal 2 qts	
CREOLE SAUCE		1 gal 2 qts	

Method

1. Add beans to salted water.
2. Bring to a boil; cover; simmer 5 to 8 minutes or until beans are tender. Drain; reserve 1 quart liquid. CCP: Heat to 145 F. or higher for 15 seconds. Hold for service at 140 F. or higher.
3. Add Creole Sauce, Recipe No. O 005 00 to drained beans.

VEGETABLES No.Q 007 02

GREEN BEANS WITH MUSHROOMS

Yield 100   Portion   1/2 Cup

Calories	Carbohydrates	Protein	Fat	Cholesterol	Sodium	Calcium
52 cal	6 g	2 g	3 g	7 mg	170 mg	34 mg

Ingredient	Weight	Measure	Issue
MUSHROOMS,CANNED,SLICED,DRAINED	3-1/2 lbs	2 qts 2-3/8 cup	
BUTTER	12 oz	1-1/2 cup	
BEANS,GREEN,FROZEN,WHOLE	16 lbs	3 gal 2-5/8 qts	
SALT	5/8 oz	1 tbsp	
WATER,BOILING	12-1/2 lbs	1 gal 2 qts	
PEPPER,BLACK,GROUND	<1/16th oz	1/8 tsp	

Method

1. Saute mushrooms in butter.
2. Add beans to salted water. Bring to a boil; cover; simmer 5 to 8 minutes, or until beans are just tender. Drain; reserve 1 quart liquid.
3. Combine mushrooms, beans, bean liquid and pepper. Mix lightly; serve. CCP: Heat to 145 F. or higher for 15 seconds. Hold for service at 140 F. or higher.

VEGETABLES No.Q 007 03

GREEN BEANS NICOISE

Yield 100   **Portion** 1/2 Cup

Calories	Carbohydrates	Protein	Fat	Cholesterol	Sodium	Calcium
59 cal	8 g	2 g	3 g	7 mg	144 mg	44 mg

Ingredient	**Weight**	**Measure**	**Issue**
GARLIC POWDER	<1/16th oz	1/8 tsp	
ONIONS,FRESH,SLICED	3 lbs	3 qts	3-3/8 lbs
BUTTER	12 oz	1-1/2 cup	
BEANS,GREEN,FROZEN,WHOLE	16 lbs	3 gal 2-5/8 qts	
SALT	5/8 oz	1 tbsp	
WATER,BOILING	12-1/2 lbs	1 gal 2 qts	
PEPPER,BLACK,GROUND	<1/16th oz	1/8 tsp	
TOMATOES,CANNED,WHOLE,PEELED,DRAINED	6-1/4 lbs	2 qts 3-3/4 cup	

Method

1. Saute onions and garlic powder in butter or margarine until tender.
2. Add beans to salted water. Bring to a boil; cover; simmer 5 to 8 minutes or until beans are tender. Drain; reserve 1 quart liquid.
3. Drain canned tomatoes. Crush tomatoes. Combine onions, garlic, beans, bean liquid, and pepper. Mix lightly. CCP: Heat to 145 F. or higher for 15 seconds. Hold at 140 F. or higher for service.

VEGETABLES No.Q 007 04

GREEN BEANS SOUTHERN STYLE

Yield 100   Portion   1/2 Cup

Calories	Carbohydrates	Protein	Fat	Cholesterol	Sodium	Calcium
41 cal	6 g	2 g	2 g	2 mg	24 mg	32 mg

Ingredient | Weight | Measure | Issue

Ingredient	Weight	Measure	Issue
BACON,RAW	1 lbs		
BACON FAT,RENDERED	3-5/8 oz	1/2 cup	
BEANS,GREEN,FROZEN,WHOLE	16 lbs	3 gal 2-5/8 qts	
WATER,BOILING	12-1/2 lbs	1 gal 2 qts	
PEPPER,BLACK,GROUND	<1/16th oz	1/8 tsp	

Method

1. Cook bacon until crisp; drain; crumble bacon; reserve bacon fat.
2. Add bacon fat to beans and water. Bring to a boil; cover; simmer 5 to 8 minutes or until beans are tender. Drain; reserve 1 quart liquid.
3. Add reserved bean liquid, crumbled bacon and black pepper to beans.
4. Mix lightly. CCP: Heat to 145 F. or higher for 15 seconds. Hold at 140 F. or higher for service.

VEGETABLES No.Q 008 00

HARVARD BEETS

Yield 100 **Portion** 3/4 Cup

Calories	Carbohydrates	Protein	Fat	Cholesterol	Sodium	Calcium
100 cal	20 g	1 g	2 g	0 mg	538 mg	25 mg

Ingredient	Weight	Measure	Issue
BEETS,CANNED,SLICED,INCL LIQUIDS	39 lbs	4 gal 2 qts	
CLOVES,GROUND	1/3 oz	1 tbsp	
CORNSTARCH	6-3/4 oz	1-1/2 cup	
WATER,COLD	1-5/8 lbs	3 cup	
SUGAR,GRANULATED	1-1/2 lbs	3-3/8 cup	
SALT	5/8 oz	1 tbsp	
VINEGAR,DISTILLED	1-1/8 lbs	2-1/4 cup	
MARGARINE	8 oz	1 cup	

Method

1 Drain beets; reserve liquid for use in Step 2 and beets for use in Step 6.
2 Take reserved liquid and add water to equal 4 quarts per 100 portions. Add cloves to liquid; bring to a boil.
3 Dissolve cornstarch in cold water; add to boiling liquid. Cook 5 minutes; stirring constantly until thick and clear.
4 Add sugar, salt, vinegar, and margarine or butter to thickened mixture, stir until blended.
5 Add drained beets to sauce. CCP: Heat to 145 F. or higher for 15 seconds. Hold at 140 F. or higher for service.

VEGETABLES No.Q 008 01

BEETS IN ORANGE-LEMON SAUCE

Yield 100   Portion   3/4 Cup

Calories	Carbohydrates	Protein	Fat	Cholesterol	Sodium	Calcium
103 cal	21 g	2 g	2 g	0 mg	539 mg	26 mg

Ingredient | Weight | Measure | Issue

Ingredient	Weight	Measure	Issue
BEETS,CANNED,SLICED,INCL LIQUIDS	39 lbs	4 gal 2 qts	
CLOVES,GROUND	1/3 oz	1 tbsp	
CORNSTARCH	6-3/4 oz	1-1/2 cup	
WATER,COLD	1-5/8 lbs	3 cup	
SUGAR,GRANULATED	1-1/2 lbs	3-3/8 cup	
SALT	5/8 oz	1 tbsp	
JUICE,LEMON	6-1/2 oz	3/4 cup	
LEMON RIND,GRATED	5/8 oz	3 tbsp	
JUICE,ORANGE	1-2/3 lbs	3 cup	
MARGARINE	8 oz	1 cup	

Method

1. Drain beets; reserve liquid for use in Step 2 and beets for use in Step 6.
2. Take reserved liquid and add water to equal 4 quarts per 100 portions and add cloves; bring to a boil.
3. Dissolve cornstarch in cold water; add to boiling liquid. Cook 5 minutes; stirring constantly until thick and clear.
4. Add sugar, salt, lemon and orange juices, lemon rind, and margarine or butter to thickened mixture, stir until blended.
5. Add drained beets to sauce. CCP: Heat to 145 F. or higher for 15 seconds. Hold at 140 F. or higher for service.

VEGETABLES No.Q 009 00

HOT SPICED BEETS

Yield 100 **Portion** 3/4 Cup

Calories	Carbohydrates	Protein	Fat	Cholesterol	Sodium	Calcium
129 cal	28 g	2 g	2 g	0 mg	542 mg	36 mg

Ingredient	**Weight**	**Measure**	**Issue**
BEETS,CANNED,SLICED,INCL LIQUIDS	39 lbs	4 gal 2 qts	
VINEGAR,DISTILLED	6-1/4 lbs	3 qts	
CINNAMON,GROUND	1/3 oz	1 tbsp	
CLOVES,GROUND	2/3 oz	3 tbsp	
SALT	5/8 oz	1 tbsp	
PEPPER,BLACK,GROUND	1/4 oz	1 tbsp	
SUGAR,GRANULATED	1-1/3 lbs	3 cup	
SUGAR,BROWN,PACKED	2 lbs	1 qts 2-3/8 cup	
MARGARINE	8 oz	1 cup	

Method

1 Drain beets; reserve liquid for use in Step 2 and beets for use in Step 4.
2 Take reserved beet liquid and add water to equal 4-1/2 quarts per 100 portions and add to vinegar, cinnamon, cloves, salt, pepper and sugars; mix well.
3 Bring to a boil; reduce heat; simmer 10 minutes.
4 Add beets and margarine or butter. CCP: Heat to 145 F. or higher for 15 seconds. Hold at 140 F. or higher for service.

VEGETABLES No.Q 010 00

BROCCOLI POLONAISE

Yield 100   Portion   3 Ounces

Calories	Carbohydrates	Protein	Fat	Cholesterol	Sodium	Calcium
60 cal	7 g	4 g	3 g	24 mg	188 mg	55 mg

Ingredient

Ingredient	Weight	Measure	Issue
BROCCOLI,FROZEN,SPEARS	20 lbs	3 gal 2-1/2 qts	
SALT	1 oz	1 tbsp	
WATER,BOILING	16-3/4 lbs	2 gal	
BREADCRUMBS,DRY,GROUND,FINE	1 lbs	1 qts	
BUTTER,MELTED	8 oz	1 cup	
EGG,HARD COOKED,CHOPPED	1 lbs	9 Eggs	

Method

1. Add frozen broccoli to boiling, salted water; return to a boil; cook UNCOVERED 3 minutes. Cover; reduce heat; cook 7 to 9 minutes or until just tender. Drain; place an equal quantity in each pan.
2. Brown crumbs in butter or margarine. Sprinkle 1 cup crumbs over broccoli in each pan.
3. Garnish with hard cooked eggs. CCP: Heat to 145 F. or higher for 15 seconds. Hold at 140 F. or higher for service.

VEGETABLES No.Q 010 01

BRUSSELS SPROUTS POLONAISE

Yield 100 Portion 1/2 Cup

Calories	Carbohydrates	Protein	Fat	Cholesterol	Sodium	Calcium
73 cal	10 g	4 g	3 g	24 mg	187 mg	31 mg

Ingredient	Weight	Measure	Issue
BRUSSELS SPROUTS,FROZEN	20 lbs	3 gal 2-5/8 qts	
WATER,BOILING	16-3/4 lbs	2 gal	
SALT	1 oz	1 tbsp	
BREADCRUMBS,DRY,GROUND,FINE	1 lbs	1 qts	
BUTTER,MELTED	8 oz	1 cup	
EGG,HARD COOKED,CHOPPED	1 lbs	9 Eggs	

Method

1. Add frozen brussels sprouts to boiling, salted water; return to boil; cook UNCOVERED for 7 to 9 minutes. Cover; reduce heat; cook 3 minutes or until tender. Drain. Place an equal quantity in each pan.
2. Brown crumbs in butter or margarine. Sprinkle 1 cup crumbs over brussels sprouts in each pan.
3. Garnish with hard cooked eggs.CCP: Heat to 145 F. or higher for 15 seconds. Hold at 140 F. or higher for service.

VEGETABLES No.Q 010 02

CAULIFLOWER POLONAISE

Yield 100   **Portion** 1/2 Cup

Calories	Carbohydrates	Protein	Fat	Cholesterol	Sodium	Calcium
52 cal	6 g	2 g	3 g	24 mg	182 mg	24 mg

Ingredient	Weight	Measure	Issue
CAULIFLOWER,FROZEN	20 lbs		
WATER,BOILING	16-3/4 lbs	2 gal	
SALT	1 oz	1 tbsp	
BREADCRUMBS,DRY,GROUND,FINE	1 lbs	1 qts	
BUTTER,MELTED	8 oz	1 cup	
EGG,HARD COOKED,CHOPPED	1 lbs	9 Eggs	

Method

1 Add frozen cauliflower to boiling, salted water; return to boil; cover; reduce heat, allow cauliflower to simmer 4 minutes or until tender. Drain. Place an equal quantity in each pan.
2 Brown crumbs in butter or margarine. Sprinkle 1 cup crumbs over cauliflower in each pan.
3 Garnish with hard cooked eggs. CCP: Heat to 145 F. or higher for 15 seconds. Hold at 140 F. or higher for service.

VEGETABLES No.Q 011 00

SPROUTS SUPERBA

Yield 100   Portion   1/2 Cup

Calories	Carbohydrates	Protein	Fat	Cholesterol	Sodium	Calcium
71 cal	10 g	4 g	3 g	0 mg	304 mg	34 mg

Ingredient	Weight	Measure	Issue
BRUSSELS SPROUTS,FROZEN	17 lbs	3 gal 1/2 qts	
SALT	1/2 oz	3/8 tsp	
WATER,BOILING	14-5/8 lbs	1 gal 3 qts	
CELERY,FRESH,CHOPPED	3 lbs	2 qts 3-3/8 cup	4-1/8 lbs
MARGARINE	2 oz	1/4 cup 1/3 tbsp	
SOUP,CONDENSED,CREAM OF MUSHROOM	6-5/8 lbs	3 qts	
WATER	2-1/8 lbs	1 qts	
PIMIENTO,CANNED,DRAINED,CHOPPED	12-2/3 oz	1-7/8 cup	
GARLIC POWDER	5/8 oz	2 tbsp	
PEPPER,WHITE,GROUND	1/8 oz	1/3 tsp	

Method

1. Add brussels sprouts to boiling salted water; return to a boil; cook 8 to 10 minutes.
2. Drain; set aside for use in Step 5.
3. Saute celery in margarine or butter 5 minutes or until tender.
4. Combine soup and water; mix well. Add celery, pimientos, garlic powder and white pepper. Simmer 10 minutes.
5. Add brussels sprouts to soup mixture, mix lightly. Simmer 5 minutes or until hot. Serve. CCP: Heat to 145 F. or higher for 15 seconds. Hold at 140 F. or higher for service.

VEGETABLES No.Q 012 00

FRIED CABBAGE

Yield 100   **Portion** 1/2 Cup

Calories	Carbohydrates	Protein	Fat	Cholesterol	Sodium	Calcium
47 cal	5 g	1 g	3 g	7 mg	184 mg	44 mg

Ingredient

Ingredient	Weight	Measure	Issue
CABBAGE,GREEN,FRESH,SHREDDED	20 lbs	8 gal 3/8 qts	25 lbs
BUTTER	12 oz	1-1/2 cup	
SALT	1-1/4 oz	2 tbsp	
PEPPER,BLACK,GROUND	1/2 oz	2 tbsp	

Method

1. Divide cabbage into equal batches weighing 10 pounds.
2. Fry each batch in butter, margarine or salad oil on 325 F. griddle for 10 minutes or until tender, stirring frequently to avoid scorching. CCP: Heat to 145 F. or higher for 15 seconds.
3. Add salt and pepper to each batch. CCP: Hold at 140 F. or higher for service.

VEGETABLES No.Q 012 01

CALICO CABBAGE

Yield 100   Portion   1/2 Cup

Calories	Carbohydrates	Protein	Fat	Cholesterol	Sodium	Calcium
53 cal	7 g	1 g	3 g	7 mg	187 mg	46 mg

Ingredient	Weight	Measure	Issue
CABBAGE,GREEN,FRESH,SHREDDED	20 lbs	8 gal 3/8 qts	25 lbs
CARROTS,FROZEN,SLICED	8 oz	1-3/4 cup	
CELERY,FRESH,SLICED	8 oz	1-7/8 cup	11 oz
ONIONS,FRESH,CHOPPED	1 lbs	2-7/8 cup	1-1/8 lbs
BUTTER	12 oz	1-1/2 cup	
SUGAR,GRANULATED	3-1/2 oz	1/2 cup	
SALT	1-1/4 oz	2 tbsp	
PEPPER,BLACK,GROUND	1/4 oz	1 tbsp	

Method

1. Add carrots, fresh celery rings and chopped dry onions to cabbage. Divide cabbage into 2 batches.
2. Fry each batch in butter, margarine or salad oil on 325 F. griddle for 10 minutes or until tender, stirring frequently to avoid scorching.
3. Add salt, pepper and sugar to each batch. CCP: Heat to 145 F. or higher for 15 seconds. Hold at 140 F. or higher for service.

VEGETABLES No.Q 012 02

FRIED CABBAGE WITH BACON

Yield 100 **Portion** 1/2 Cup

Calories	Carbohydrates	Protein	Fat	Cholesterol	Sodium	Calcium
61 cal	5 g	2 g	4 g	10 mg	155 mg	44 mg

Ingredient	Weight	Measure	Issue
BACON,RAW	2 lbs		
CABBAGE,GREEN,FRESH,SHREDDED	20 lbs	8 gal 3/8 qts	25 lbs
BUTTER	12 oz	1-1/2 cup	
SALT	5/8 oz	1 tbsp	
PEPPER,BLACK,GROUND	1/4 oz	1 tbsp	

Method

1. Cook bacon until crisp; drain; crumble bacon.
2. Divide cabbage into two batches. Fry each batch in butter, margarine or salad oil on 325 F. griddle for 10 minutes or until tender, stirring frequently to avoid scorching; add bacon.
3. Add salt and pepper to each batch. CCP: Heat to 145 F. or higher for 15 seconds. Hold at 140 F. or higher for service.

VEGETABLES No.Q 013 00

SCALLOPED SWEET POTATOES AND APPLES

Yield 100   Portion   1/2 Cup

Calories	Carbohydrates	Protein	Fat	Cholesterol	Sodium	Calcium
166 cal	35 g	1 g	3 g	0 mg	168 mg	31 mg

Ingredient   Weight   Measure   Issue

SWEET POTATOES,CANNED,W/SYRUP   24-1/8 lbs   3 gal
APPLES,CANNED,SLICED,DRAINED   6 lbs   3 qts
CINNAMON,GROUND   2 oz   1/2 cup 1/3 tbsp
SUGAR,BROWN,PACKED   1-1/2 lbs   1 qts 3/4 cup
SHORTENING,VEGETABLE,MELTED   9 oz   1-1/4 cup
SALT   1 oz   1 tbsp
WATER   2-1/8 lbs   1 qts

Method

1 Arrange 3 quarts drained sweet potatoes and 3 cups apples in alternate layers in each pan.
2 Combine brown sugar, cinnamon, shortening or salad oil, salt and water in steam-jacketed kettle or stock pot. Cook at low heat, stirring constantly until sugar is dissolved. Pour an equal quantity over potatoes in each pan.
3 Using a convection oven, bake at 300 F. for 30 minutes on low fan, open vent, or until apples and potatoes are thoroughly heated. CCP: Heat to 145 F. or higher for 15 seconds. Hold at 140 F. or higher for service.

VEGETABLES No.Q 014 00

ORANGE CARROTS AMANDINE

Yield 100   Portion   1/2 Cup

Calories	Carbohydrates	Protein	Fat	Cholesterol	Sodium	Calcium
76 cal	9 g	2 g	4 g	0 mg	119 mg	40 mg

Ingredient | Weight | Measure | Issue

Ingredient	Weight	Measure
CARROTS,FROZEN,SLICED	16 lbs	3 gal 2-1/8 qts
SALT	3/8 oz	1/3 tsp
WATER,BOILING	16-3/4 lbs	2 gal
MARGARINE,MELTED	10 oz	1-1/4 cup
SUGAR,BROWN,PACKED	5-1/8 oz	1 cup
ORANGE PEEL,FRESH,GRATED	10-1/8 oz	3 cup
JUICE,ORANGE	2-7/8 oz	1/4 cup 1-2/3 tbsp
ALMONDS,SLIVERED	11-3/8 oz	3 cup

Method

1. Cook carrots 10 to 13 minutes. Add carrots to salted boiling water. Return to a boil; reduce heat; simmer 15 minutes or until tender. Drain.
2. Add brown sugar, orange rind, orange juice, and almonds to melted butter or margarine. Blend well.
3. Add glaze to carrots; mix until carrots are well coated. CCP: Heat to 145 F. or higher for 15 seconds. Hold at 140 F. or higher for service.

VEGETABLES No.Q 015 00

ORIENTAL STIR-FRY CABBAGE

Yield 100 **Portion** 3/4 Cup

Calories	Carbohydrates	Protein	Fat	Cholesterol	Sodium	Calcium
56 cal	12 g	3 g	0 g	0 mg	384 mg	61 mg

Ingredient	Weight	Measure	Issue
SOY SAUCE	1-3/8 lbs	2-1/4 cup	
SUGAR,BROWN,PACKED	5-1/8 oz	1 cup	
GARLIC POWDER	1-3/4 oz	1/4 cup 2-1/3 tbsp	
GINGER,GROUND	3/4 oz	1/4 cup 1/3 tbsp	
PEPPER,BLACK,GROUND	3/8 oz	1 tbsp	
WATER	1-1/8 lbs	2-1/4 cup	
CORNSTARCH	7/8 oz	3 tbsp	
CABBAGE,GREEN,FRESH,SHREDDED	24 lbs	9 gal 2-7/8 qts	30 lbs
PEPPERS,RED,FRESH,SLICED	5 lbs	1 gal 2-1/8 qts	6-1/8 lbs
ONIONS,FRESH,SLICED	5 lbs	1 gal 7/8 qts	5-1/2 lbs
COOKING SPRAY,NONSTICK	1 oz	2 tbsp	

Method

1 Combine soy sauce, brown sugar, garlic powder, ginger and pepper; mix thoroughly. Bring to a boil; reduce heat to simmer.
2 Blend cornstarch with water until dissolved; add to soy sauce mixture stirring constantly; simmer 2 minutes or until lightly thickened and clear. Remove from heat.
3 Preheat tilt-fry pan. Spray lightly with non-stick spray. Stir and cook vegetables in 25 portion batches as follows: Cabbage and onions, 5 minutes; add red peppers for 1 minute. Do not overcook!
4 Remove to serving pans. Pour 1-1/4 cups sauce over each 25 portion batch of cabbage. Mix thoroughly to distribute the sauce. CCP: Heat to 145 F. or higher for 15 seconds. Hold for service at 140 F. or higher.

VEGETABLES No.Q 016 00

CARROT AND CELERY AMANDINE

Yield 100   Portion   1/2 Cup

Calories	Carbohydrates	Protein	Fat	Cholesterol	Sodium	Calcium
37 cal	2 g	1 g	3 g	0 mg	213 mg	26 mg

Ingredient	Weight	Measure	Issue
CARROTS,FROZEN,SLICED	10-3/4 oz		
CELERY,FRESH,SLICED	7-3/4 lbs	1 gal 3-1/3 qts	10-5/8 lbs
WATER,BOILING	28-1/4 lbs	3 gal 1-1/2 qts	
SALT	1-1/2 oz	2-1/3 tbsp	
ALMONDS,SLIVERED	11-3/8 oz	3 cup	
JUICE,LEMON	6-1/2 oz	3/4 cup	
MARGARINE,MELTED	5-1/3 oz	1/2 cup 2-2/3 tbsp	

Method

1 Cook carrots and celery in boiling salted water 10 to 13 minutes.
2 Drain; reserve carrots and celery for use in Step 4.
3 Spread almonds on pans in a thin layer. Using a convection oven, bake at 300 F. for 15 minutes on high fan, open vent stirring occasionally until almonds are lightly browned. Remove from oven.
4 Add almonds, lemon juice, and margarine to carrot and celery. Toss or stir lightly. Mix thoroughly. CCP: Heat to 145 F. or higher for 15 seconds. Hold at 140 F. or higher for service.

VEGETABLES No.Q 017 00

LYONNAISE CARROTS

Yield 100   **Portion** 1/2 Cup

Calories	Carbohydrates	Protein	Fat	Cholesterol	Sodium	Calcium
58 cal	10 g	1 g	2 g	5 mg	186 mg	33 mg

Ingredient

Ingredient	Weight	Measure	Issue
CARROTS,FROZEN,SLICED	18 lbs	3 gal 4 qts	
WATER,BOILING	18-3/4 lbs	2 gal 1 qts	
SALT	5/8 oz	1 tbsp	
BUTTER	8 oz	1 cup	
PEPPER,BLACK,GROUND	1/8 oz	1/8 tsp	
ONIONS,FRESH,SLICED	4 lbs	3 qts 3-3/4 cup	4-1/2 lbs
SUGAR,GRANULATED	2-1/3 oz	1/4 cup 1-2/3 tbsp	
SALT	3/8 oz	1/3 tsp	
PARSLEY,FRESH,BUNCH,CHOPPED	1 oz	1/4 cup	1 oz

Method

1. Add carrots to boiling salted water. Bring to a boil; cool 10 minutes.
2. Drain; reserve carrots for use in Step 6. Add pepper and onion to melted butter in steam-jacketed kettle or tilting frying pan. Saute until tender, about 10 minutes.
3. Add sugar, salt and reserved carrots to sauteed onions; mix lightly; cook 5 minutes tossing occasionally.
4. Garnish with parsley before serving. CCP: Heat to 145 F. or higher for 15 seconds. Hold at 140 F. or higher for service

VEGETABLES No.Q 017 01

GLAZED CARROTS

Yield 100   **Portion** 1/2 Cup

Calories	Carbohydrates	Protein	Fat	Cholesterol	Sodium	Calcium
73 cal	14 g	1 g	2 g	5 mg	166 mg	25 mg

Ingredient   **Weight**   **Measure**   **Issue**

CARROTS,FROZEN,SLICED   18 lbs
WATER,BOILING   18-3/4 lbs   2 gal 1 qts
SALT   5/8 oz   1 tbsp
BUTTER   8 oz   1 cup
GINGER,GROUND   5/8 oz   3 tbsp
SUGAR,GRANULATED   1-1/4 lbs   2-3/4 cup
SALT   3/8 oz   1/3 tsp

Method

1. Cook carrots 10 to 13 minutes.
2. Drain; reserve carrots for use in Step 5.
3. Melt butter in a steam-jacketed kettle or tilting frying pan; add ginger and stir until well blended.
4. Add sugar and stir. Mixture will resemble a thick roux.
5. Toss carrots in sauce until well coated; cook 5 minutes, tossing occasionally. CCP: Heat to 145 F. or higher for 15 seconds. Hold at 140 F. or higher for service.

VEGETABLES No.Q 018 00

CAULIFLOWER AU GRATIN

Yield 100　　　　　　　　　　　　　　　　　**Portion**　1/2 Cup

Calories	Carbohydrates	Protein	Fat	Cholesterol	Sodium	Calcium
125 cal	9 g	5 g	8 g	23 mg	226 mg	105 mg

Ingredient	Weight	Measure	Issue
CAULIFLOWER,FROZEN	20 lbs		
SALT	5/8 oz	1 tbsp	
WATER,BOILING	25-1/8 lbs	3 gal	
MILK,NONFAT,DRY	8-3/4 oz	3-5/8 cup	
WATER,WARM	9-3/8 lbs	1 gal 1/2 qts	
BUTTER,MELTED	1 lbs	2 cup	
FLOUR,WHEAT,GENERAL PURPOSE	11 oz	2-1/2 cup	
CHEESE,CHEDDAR,SHREDDED	1-1/2 lbs	1 qts 2 cup	
PEPPER,WHITE,GROUND	<1/16th oz	1/8 tsp	
BREADCRUMBS,DRY,GROUND,FINE	1 lbs	1 qts	
BUTTER,MELTED	8 oz	1 cup	

Method

1. Add cauliflower to salted boiling water. Bring to a boil; cover. Simmer 4 to 8 minutes or until just tender.
2. Drain; place about 3-3/4 quarts cauliflower in each steam table pan. Set aside for use in Step 8.
3. Reconstitute milk; heat to just below boiling. DO NOT BOIL.
4. Blend butter and flour together; stir until smooth.
5. Add flour mixture to milk, stirring constantly. Simmer 5 minutes or until thickened.
6. Add cheese and pepper; stir until blended.
7. Pour 1-1/2 quarts sauce over cauliflower in each pan.
8. Mix crumbs and butter or margarine. Sprinkle 1 cup evenly over cauliflower in each pan.
9. Using a convection oven, bake at 325 F. for 10 minutes or until crumbs are browned. CCP: Heat to 145 F. or higher for 15 seconds. Hold for service at 140 F. or higher.

VEGETABLES No.Q 019 00

GERMAN POTATO GRIDDLE CAKES (DEHY)

Yield 100 **Portion** 2 Cakes

Calories	Carbohydrates	Protein	Fat	Cholesterol	Sodium	Calcium
114 cal	12 g	3 g	6 g	46 mg	244 mg	49 mg

Ingredient	**Weight**	**Measure**	**Issue**
ONIONS,FRESH,CHOPPED	11-1/4 oz	2 cup	12-1/2 oz
WATER,BOILING	29-1/4 lbs	3 gal 2 qts	
POTATO,WHITE,DEHYDRATED,SLICED	4 lbs		
MILK,NONFAT,DRY	6 oz	2-1/2 cup	
WATER,WARM	6-1/4 lbs	3 qts	
EGGS,WHOLE,FROZEN	2 lbs	3-3/4 cup	
FLOUR,WHEAT,GENERAL PURPOSE	1-3/4 lbs	1 qts 2-1/2 cup	
SALT	1-7/8 oz	3 tbsp	
PEPPER,BLACK,GROUND	1/8 oz	1/3 tsp	
NUTMEG,GROUND	<1/16th oz	1/8 tsp	
THYME,GROUND	<1/16th oz	<1/16th tsp	
SHORTENING,VEGETABLE,MELTED	7-1/4 oz	1 cup	
SOUR CREAM	3 lbs	1 qts 2 cup	

Method

1. Add potatoes and onions to boiling water. Bring to a boil; simmer 15 minutes or until soft but not mushy. DO NOT OVERCOOK. Drain immediately or mixture will be too moist.
2. Beat potato and onion mixture in mixer bowl at medium speed 2 minutes.
3. Reconstitute milk; add eggs. Add to potato mixture; blend at low speed 1 minute.
4. Add flour, salt, pepper, nutmeg, thyme and melted shortening or salad oil to mixture; blend at low speed 2 minutes.
5. Drop 1/4 cup, or one No.16 scoop batter onto lightly greased 375 F. griddle. Cook until well browned, about 2-1/2 to 3 minutes on each side.
6. Serve with 1 tablespoon sour cream. CCP: Hold for service at 140 F. or higher.

VEGETABLES No.Q 020 00

FRENCH FRIED CAULIFLOWER

Yield 100　　　　　　　　　　　　　　　　　　**Portion**　3-1/2 Ounces

Calories	Carbohydrates	Protein	Fat	Cholesterol	Sodium	Calcium
159 cal	19 g	6 g	7 g	27 mg	382 mg	86 mg

Ingredient　　　　　　　　　　　　　　　Weight　　　　Measure　　　　　Issue

Ingredient	Weight	Measure	Issue
MILK,NONFAT,DRY	2-3/8 oz	1 cup	
WATER,WARM	2-1/3 lbs	1 qts 1/2 cup	
EGGS,WHOLE,FROZEN	1-1/4 lbs	2-1/4 cup	
CAULIFLOWER,FROZEN	20 lbs		
FLOUR,WHEAT,GENERAL PURPOSE	4-3/8 lbs	1 gal	
SALT	2-1/2 oz	1/4 cup 1/3 tbsp	
PEPPER,BLACK,GROUND	1/4 oz	1 tbsp	
CHEESE,PARMESAN,GRATED	14-1/8 oz	1 qts	

Method

1　Reconstitute milk; add eggs. Mix well.
2　Cut large cauliflower pieces in half. Dip in milk and egg mixture; drain well.
3　Combine flour, salt, pepper and cheese. Dredge cauliflower in flour mixture; shake off excess.
4　Fry in 375 F. deep fat fryer for 3 minutes or until golden brown. Drain on absorbent paper. Serve immediately. CCP: Hold at 140 F. or higher for service.

VEGETABLES No.Q 020 01

FRENCH FRIED OKRA

Yield 100   Portion 3/4 Cup

Calories	Carbohydrates	Protein	Fat	Cholesterol	Sodium	Calcium
196 cal	21 g	5 g	11 g	3 mg	356 mg	125 mg

Ingredient	Weight	Measure	Issue
OKRA,FROZEN,CUT	18 lbs	2 gal 3 qts	
FLOUR,WHEAT,GENERAL PURPOSE	4-3/8 lbs	1 gal	
SALT	2-1/2 oz	1/4 cup 1/3 tbsp	
PEPPER,BLACK,GROUND	1/4 oz	1 tbsp	
CHEESE,PARMESAN,GRATED	14-1/8 oz	1 qts	

Method

1. Partially thaw okra. Break large pieces apart.
2. Combine flour, salt, pepper and cheese. Dredge okra in flour mixture; shake off excess.
3. Fry in 375 F. deep fat fryer for 2 minutes or until golden brown. Drain on absorbent paper. Serve immediately. CCP: Hold at 140 F. or higher for service.

VEGETABLES No.Q 021 00

CORN FRITTERS

Yield 100   **Portion** 2 Fritters

Calories	Carbohydrates	Protein	Fat	Cholesterol	Sodium	Calcium
208 cal	30 g	5 g	8 g	44 mg	565 mg	148 mg

Ingredient   Weight   Measure   Issue

FLOUR,WHEAT,GENERAL PURPOSE   6-5/8 lbs   1 gal 2 qts
SALT   1-7/8 oz   3 tbsp
BAKING POWDER   7-3/4 oz   1 cup
SUGAR,GRANULATED   3-1/2 oz   1/2 cup
MILK,NONFAT,DRY   1-3/4 oz   3/4 cup
WATER,WARM   2 lbs   3-3/4 cup
EGGS,WHOLE,FROZEN   2 lbs   3-3/4 cup
CORN,CANNED,CREAM STYLE   6-3/4 lbs   3 qts
BUTTER,MELTED   8 oz   1 cup

Method

1 Sift together flour, salt, baking powder, sugar and milk into mixer bowl.
2 Combine water, eggs, corn and butter or margarine; mix well.
3 Add corn mixture to dry ingredients; mix until well blended. Batter will not be smooth.
4 Drop 2 tablespoons batter into 350 F. deep fat.
5 Fry 5 minutes or until golden brown.
6 Drain on absorbent paper. CCP: Hold for service at 140 F. or higher.

VEGETABLES No.Q 021 01

CORN FRITTERS (PANCAKE MIX)

Yield 100   **Portion** 2 Fritters

Calories	Carbohydrates	Protein	Fat	Cholesterol	Sodium	Calcium
177 cal	27 g	4 g	6 g	6 mg	460 mg	75 mg

Ingredient	**Weight**	**Measure**	**Issue**
CORN,CANNED,CREAM STYLE	6-3/4 lbs	2 qts 4 cup	
PANCAKE MIX	6-3/4 lbs	1 gal 1-7/8 qts	
WATER	3-1/8 lbs	1 qts 2 cup	

Method

1 Combine canned cream style corn, canned pancake mix and water. Mix well.
2 Drop 2 tablespoons batter into 350 F. deep fat.
3 Fry 5 minutes or until golden brown.
4 Drain on absorbent paper. CCP: Hold for service at 140 F. or higher.

VEGETABLES No.Q 022 00

RATATOUILLE

Yield 100   **Portion** 1/2 Cup

Calories	Carbohydrates	Protein	Fat	Cholesterol	Sodium	Calcium
45 cal	10 g	2 g	0 g	0 mg	407 mg	34 mg

Ingredient	Weight	Measure	Issue
GARLIC POWDER	1/4 oz	1/3 tsp	
TOMATOES,CANNED,CRUSHED,INCL LIQUIDS	13-1/4 lbs	1 gal 2 qts	
SUGAR,GRANULATED	3-1/2 oz	1/2 cup	
SALT	3 oz	1/4 cup 1 tbsp	
BASIL,DRIED,CRUSHED	1/4 oz	1 tbsp	
THYME,GROUND	1/8 oz	1 tbsp	
PEPPER,BLACK,GROUND	1/8 oz	1/3 tsp	
BAY LEAF,WHOLE,DRIED	1/8 oz	3 each	
EGGPLANT,FRESH,CUBES	9-3/8 lbs	3 gal 1 qts	11-5/8 lbs
SQUASH,ZUCCHINI,FRESH,CHOPPED	7-1/8 lbs	1 gal 2-1/2 qts	7-1/2 lbs
PEPPERS,GREEN,FRESH,CHOPPED	2-1/2 lbs	1 qts 3-1/2 cup	3 lbs
ONIONS,FRESH,CHOPPED	1-5/8 lbs	1 qts 1/2 cup	1-3/4 lbs

Method

1 Combine tomatoes, sugar, salt, basil, thyme, garlic, pepper and bay leaves in a stock pot or steam-jacketed kettle. Stir well.
2 Add eggplant, squash, sweet peppers and onions. Bring to a boil stirring constantly. Cover and simmer 45 minutes or until eggplant is tender. Stir occasionally. Remove bay leaves.

VEGETABLES No.Q 023 00

SCALLOPED CREAM STYLE CORN

Yield 100   Portion   1/2 Cup

Calories	Carbohydrates	Protein	Fat	Cholesterol	Sodium	Calcium
148 cal	26 g	3 g	5 g	9 mg	447 mg	23 mg

Ingredient | Weight | Measure | Issue

Ingredient	Weight	Measure	Issue
BUTTER,MELTED	14 oz	1-3/4 cup	
CRACKERS,SODA,SALTED,CRUMBLED	1-3/4 lbs		
PEPPER,BLACK,GROUND	1/8 oz	1/4 tsp	
COOKING SPRAY,NONSTICK	2 oz	1/4 cup 1/3 tbsp	
CORN,CANNED,CREAM STYLE	23-2/3 lbs	2 gal 2-1/2 qts	
MILK,NONFAT,DRY	2-2/3 oz	1-1/8 cup	
WATER,WARM	3 lbs	1 qts 1-3/4 cup	

Method

1. Combine butter or margarine, cracker crumbs, and pepper. Reserve 3 cups buttered crumbs for use in Step 4.
2. Pour 2-3/4 quarts corn into each lightly sprayed steam table pan. Stir in 2-1/2 cups buttered crumbs in each pan. Mix until just combined.
3. Reconstitute milk; pour 1-1/2 cups milk evenly over top of mixture in each pan. Mix until just combined.
4. Sprinkle 3/4 cup reserved buttered crumbs over top of corn mixture.
5. Using a convection oven, bake in 300 F. oven for 30 minutes or until lightly browned. CCP: Heat to 145 F. or higher for 15 seconds. Hold for service at 140 F. or higher.

VEGETABLES No.Q 023 01

SCALLOPED WHOLE KERNEL CORN

Yield 100   **Portion** 1/2 Cup

Calories	Carbohydrates	Protein	Fat	Cholesterol	Sodium	Calcium
125 cal	21 g	3 g	4 g	9 mg	340 mg	22 mg

Ingredient   Weight   Measure   Issue

Ingredient	Weight	Measure	Issue
BUTTER,MELTED	14 oz	1-3/4 cup	
CRACKERS,SODA,SALTED,CRUMBLED	1-1/3 lbs	100 each	
PEPPER,BLACK,GROUND	1/8 oz	1/4 tsp	
CORN,CANNED,WHOLE KERNEL,INCL LIQUIDS	23-1/4 lbs	2 gal 2-1/4 qts	
MILK,NONFAT,DRY	2-2/3 oz	1-1/8 cup	
WATER,WARM	3 lbs	1 qts 1-3/4 cup	

Method

1. Combine butter or margarine, cracker crumbs, and pepper. Reserve 3 cups buttered crumbs for use in Step 4.
2. Drain corn; reserve liquid. Pour drained corn into lightly greased pans. Stir in 2-1/2 cups buttered crumbs in each steam table pan. Mix until just combined.
3. Reconstitute milk; mix liquid with milk; pour 3 cups milk and drained liquid mixture evenly over top of mixture in each pan. Mix until just combined.
4. Sprinkle 3/4 cup reserved buttered crumbs over top of corn mixture.
5. Bake 30 minutes or until lightly browned in 300 F. convection oven. CCP: Heat to 145 F. or higher for 15 seconds. Hold for service at 140 F. or higher.

VEGETABLES No.Q 024 00

BROCCOLI PARMESAN

Yield 100　　　　　　　　　　　　　　　　　　**Portion** 2 Stalks

Calories	Carbohydrates	Protein	Fat	Cholesterol	Sodium	Calcium
77 cal	10 g	7 g	2 g	5 mg	222 mg	167 mg

Ingredient	Weight	Measure	Issue
COOKING SPRAY,NONSTICK	1/8 oz	1/8 tsp	
ONIONS,FRESH,CHOPPED	1-3/8 lbs	1 qts	1-5/8 lbs
MILK,NONFAT,DRY	7-1/4 oz	3 cup	
WATER	5-3/4 lbs	2 qts 3 cup	
FLOUR,WHEAT,GENERAL PURPOSE	8-7/8 oz	2 cup	
WATER	2-1/8 lbs	1 qts	
CHEESE,PARMESAN,GRATED	1-1/3 lbs	1 qts 2 cup	
BROCCOLI,FROZEN,SPEARS	24 lbs	4 gal 1-1/2 qts	
WATER,BOILING	16-3/4 lbs	2 gal	
SALT	5/8 oz	1 tbsp	

Method

1. Spray steam-jacketed kettle or stock pot with cooking spray. Add onions; stir well; cover; cook 5 to 7 minutes or until tender.
2. Reconstitute milk; add to onions in steam-jacketed kettle or stock pot. Heat to just below boiling. Do not boil.
3. Blend flour with water using wire whip to form slurry; stir until smooth.
4. Add slurry to milk mixture gradually, stirring constantly. Simmer 8 to 10 minutes or until thickened.
5. Add cheese; bring to a simmer, stirring until smooth. Do not boil.
6. Prepare broccoli. Drain; place about 50 spears or 5 pounds broccoli in each steam table pan.
7. Pour about 4-3/4 cups sauce over broccoli in each steam table pan. Using a convection oven, bake at 325 F. for 10 minutes on high fan, open vent. CCP: Heat to 145 F. or higher for 15 seconds. Hold for service at 140 F. or higher. Each Portion: 2 stalks with 3 tablespoons of sauce.

VEGETABLES No.Q 024 01

BRUSSELS SPROUTS PARMESAN

Yield 100   Portion   3/4 Cup

Calories	Carbohydrates	Protein	Fat	Cholesterol	Sodium	Calcium
92 cal	13 g	8 g	2 g	5 mg	221 mg	138 mg

Ingredient

Ingredient	Weight	Measure	Issue
COOKING SPRAY,NONSTICK	1/8 oz	1/8 tsp	
ONIONS,FRESH,CHOPPED	1-3/8 lbs	1 qts	1-5/8 lbs
MILK,NONFAT,DRY	7-1/4 oz	3 cup	
WATER	5-3/4 lbs	2 qts 3 cup	
FLOUR,WHEAT,GENERAL PURPOSE	8-7/8 oz	2 cup	
WATER	2-1/8 lbs	1 qts	
CHEESE,PARMESAN,GRATED	1-1/3 lbs	1 qts 2 cup	
BRUSSELS SPROUTS,FROZEN	24 lbs	4 gal 1-5/8 qts	
WATER,BOILING	16-3/4 lbs	2 gal	
SALT	5/8 oz	1 tbsp	

Method

1. Spray steam-jacketed kettle or stock pot with cooking spray. Add onions; stir well; cover; cook 5 to 7 minutes or until tender.
2. Reconstitute milk; add to onions in steam-jacketed kettle or stock pot. Heat to just below boiling. Do not boil.
3. Blend flour with water using wire whip to form slurry; stir until smooth.
4. Add slurry to milk mixture gradually, stirring constantly. Simmer 8 to 10 minutes or until thickened.
5. Add cheese; bring to a simmer, stirring until smooth. Do not boil.
6. Prepare brussels sprouts. Drain; place about 5-3/4 pounds brussels sprouts in each steam table pan.
7. Pour about 4-3/4 cups sauce over brussels sprouts in each steam table pan. Using a convection oven, bake at 325 F. for 10 minutes on high fan, open vent. CCP: Heat to 145 F. or higher for 15 seconds. Hold at 140 F. or higher for service.

VEGETABLES No.Q 024 02

CAULIFLOWER PARMESAN

Yield 100   Portion   3/4 Cup

Calories	Carbohydrates	Protein	Fat	Cholesterol	Sodium	Calcium
67 cal	8 g	5 g	2 g	5 mg	216 mg	130 mg

Ingredient	Weight	Measure	Issue
COOKING SPRAY,NONSTICK	1/8 oz	1/8 tsp	
ONIONS,FRESH,CHOPPED	1-3/8 lbs	1 qts	1-5/8 lbs
MILK,NONFAT,DRY	7-1/4 oz	3 cup	
WATER	5-3/4 lbs	2 qts 3 cup	
FLOUR,WHEAT,GENERAL PURPOSE	8-7/8 oz	2 cup	
WATER	2-1/8 lbs	1 qts	
CHEESE,PARMESAN,GRATED	1-1/3 lbs	1 qts 2 cup	
CAULIFLOWER,FROZEN	24 lbs		
WATER,BOILING	16-3/4 lbs	2 gal	
SALT	5/8 oz	1 tbsp	

Method

1 Spray steam-jacketed kettle or stock pot with cooking spray. Add onions; stir well; cover; cook 5 to 7 minutes or until tender.
2 Reconstitute milk; add to onions in steam-jacketed kettle or stock pot. Heat to just below boiling. Do not boil.
3 Blend flour with water using wire whip to form slurry; stir until smooth.
4 Add slurry to milk mixture gradually, stirring constantly. Simmer 8 to 10 minutes or until thickened.
5 Add cheese; bring to a simmer, stirring until smooth. Do not boil.
6 Prepare cauliflower. Drain; place about 5-1/2 pounds cauliflower in each steam table pan.
7 Pour about 4-3/4 cups sauce over cauliflower in each pan. Using a convection oven, bake at 325 F. for 20 minutes on high fan, open vent. CCP: Heat to 145 F. or higher for 15 seconds. Hold for service at 140 F. or higher.

VEGETABLES No.Q 025 00

VEGETABLE STIR FRY

Yield 100   **Portion** 1/2 Cup

Calories	Carbohydrates	Protein	Fat	Cholesterol	Sodium	Calcium
55 cal	6 g	1 g	4 g	0 mg	108 mg	27 mg

Ingredient	**Weight**	**Measure**	**Issue**
CARROTS,FRESH,SLICED	3-3/4 lbs	3 qts 1-1/4 cup	4-5/8 lbs
CELERY,FRESH,SLICED	4-1/2 lbs	1 gal 1/4 qts	6-1/8 lbs
CABBAGE,GREEN,FRESH,CHOPPED	4-1/2 lbs	1 gal 3-1/4 qts	5-5/8 lbs
PEPPERS,GREEN,FRESH,MEDIUM,SLICED,THIN	2-1/4 lbs	1 qts 2-7/8 cup	2-3/4 lbs
ONIONS,FRESH	1-1/2 lbs	1 qts 1/4 cup	1-2/3 lbs
MUSHROOMS,CANNED,DRAINED	11 oz	2 cup	
ONIONS,FRESH,CHOPPED	1-1/2 lbs	1 qts 1/4 cup	1-2/3 lbs
CHICKEN BROTH		3 cup	
PEPPER,BLACK,GROUND	<1/16th oz	1/8 tsp	
CORNSTARCH	7/8 oz	3 tbsp	
WATER	3-1/8 oz	1/4 cup 2-1/3 tbsp	
SOY SAUCE	1 oz	1 tbsp	
OIL,SALAD	11-1/2 oz	1-1/2 cup	

Method

1. Wash and trim vegetables. Set aside for use in Step 5.
2. Prepare chicken broth according to recipe. Add pepper. Set aside for use in Step 4.
3. Blend cornstarch with water and soy sauce to make a smooth paste.
4. Slowly add paste to broth stirring constantly. Simmer 2 minutes or until lightly thickened and clear, stirring constantly. Remove from heat.
5. Saute vegetables salad oil as follows: Carrots, 3 minutes; add celery and green peppers, 2 minutes; add remaining vegetables, 4 minutes.
6. Pour sauce over vegetables 15 minutes before serving. CCP: Heat to 145 F. or higher for 15 seconds. Hold at 140 F. or higher for service.

VEGETABLES No.Q 026 00

HERBED GREEN BEANS

Yield 100   **Portion** 3/4 Cup

Calories	Carbohydrates	Protein	Fat	Cholesterol	Sodium	Calcium
58 cal	9 g	2 g	2 g	0 mg	484 mg	60 mg

Ingredient

Ingredient	Weight	Measure	Issue
ONIONS,FRESH,CHOPPED	6 lbs	1 gal 1/4 qts	6-2/3 lbs
CELERY,FRESH,CHOPPED	3 lbs	2 qts 3-3/8 cup	4-1/8 lbs
MARGARINE	9 oz	1-1/8 cup	
GARLIC POWDER	1/2 oz	1 tbsp	
BASIL,DRIED,CRUSHED	3/4 oz	1/4 cup 1-1/3 tbsp	
ROSEMARY,GROUND	1/2 oz	1/4 cup 2/3 tbsp	
BEANS,GREEN,CANNED	38-1/8 lbs	4 gal 2 qts	

Method

1. Saute onions and celery in butter or margarine until tender.
2. Add garlic powder, basil and rosemary to sauteed vegetables; mix well.
3. Drain green beans, reserving liquid. Prepare canned green beans. Add beans and reserved liquid to onion-herb mixture. CCP: Heat to 145 F. or higher for 15 seconds. Hold for service at 140 F. or higher.

VEGETABLES No.Q 027 00

CALICO CORN

Yield 100   **Portion** 3/4 Cup

Calories	Carbohydrates	Protein	Fat	Cholesterol	Sodium	Calcium
114 cal	24 g	4 g	2 g	1 mg	301 mg	7 mg

Ingredient   Weight   Measure   Issue

Ingredient	Weight	Measure	Issue
BACON,RAW	1 lbs		
CORN,CANNED,WHOLE KERNEL,DRAINED	28-7/8 lbs	5 gal	
PEPPER,BLACK,GROUND	1/8 oz	3/8 tsp	
PIMIENTO,CANNED,DRAINED,CHOPPED	7-5/8 oz	1-1/8 cup	

Method

1. Cook bacon until crisp. See Recipe No. L 002 00 or L 002 02. Drain. Set bacon aside for use in Step 2.
2. Drain corn; mix with pepper and pimientos. Crumble bacon. Add to corn mixture. Mix well.
3. Heat at medium heat until hot, stirring constantly. CCP: Heat to 145 F. or higher for 15 seconds. Hold at 140 F. or higher for service.

VEGETABLES No.Q 027 01

CORN O'BRIEN

Yield 100   **Portion** 3/4 Cup

Calories	Carbohydrates	Protein	Fat	Cholesterol	Sodium	Calcium
136 cal	26 g	4 g	4 g	1 mg	302 mg	10 mg

Ingredient	Weight	Measure	Issue
BACON,RAW	1 lbs		
PEPPERS,GREEN,FRESH,CHOPPED	3 lbs	2 qts 1 cup	3-5/8 lbs
ONIONS,FRESH,CHOPPED	2-3/8 lbs	1 qts 2-3/4 cup	2-2/3 lbs
OIL,SALAD	5-3/4 oz	3/4 cup	
CORN,CANNED,WHOLE KERNEL,DRAINED	28-7/8 lbs	5 gal	
PEPPER,BLACK,GROUND	1/8 oz	3/8 tsp	
PIMIENTO,CANNED,DRAINED,CHOPPED	7-5/8 oz	1-1/8 cup	

Method

1. Cook bacon until crisp. See Recipe No. L 002 00 or L 002 02. Drain. Set bacon aside for use in Step 3.
2. Saute chopped onions and sweet green peppers in oil or shortening.
3. Drain corn; mix with pepper and pimientos, and sauteed onions and peppers. Add crumbled bacon.
4. Heat at medium heat until hot, stirring constantly. CCP: Heat to 145 F. or higher for 15 seconds. Hold at 140 F. or higher for service.

VEGETABLES No.Q 027 02

MEXICAN CORN

Yield 100 **Portion** 3/4 Cup

Calories	Carbohydrates	Protein	Fat	Cholesterol	Sodium	Calcium
117 cal	25 g	4 g	2 g	2 mg	288 mg	8 mg

Ingredient

Ingredient	Weight	Measure	Issue
PEPPERS,GREEN,FRESH,CHOPPED	3 lbs	2 qts 1 cup	3-5/8 lbs
BUTTER	3 oz	1/4 cup 2-1/3 tbsp	
CORN,CANNED,WHOLE KERNEL,DRAINED	28-7/8 lbs	5 gal	
PEPPER,BLACK,GROUND	1/8 oz	3/8 tsp	
PIMIENTO,CANNED,DRAINED,CHOPPED	7-5/8 oz	1-1/8 cup	

Method

1 Saute chopped sweet peppers in butter or margarine until tender.
2 Drain corn; mix with pepper and pimientos, and then with sauteed peppers.
3 Heat at medium heat until hot, stirring constantly. CCP: Heat to 145 F. or higher for 15 seconds. Hold at 140 F. or higher for service.

VEGETABLES No.Q 028 00

EGGPLANT PARMESAN

Yield 100   **Portion** 6-1/2 Ounces

Calories	Carbohydrates	Protein	Fat	Cholesterol	Sodium	Calcium
201 cal	34 g	9 g	5 g	31 mg	1209 mg	167 mg

Ingredient	**Weight**	**Measure**	**Issue**
TOMATOES,CANNED,DICED,INCL LIQUIDS	26-1/2 lbs	2 gal 3-1/2 qts	
TOMATO PASTE,CANNED	9-1/4 lbs	1 gal	
WATER	8-1/3 lbs	1 gal	
ONIONS,FRESH,CHOPPED	3-3/4 lbs	2 qts 1 cup	4-1/4 lbs
SUGAR,GRANULATED	7 oz	1 cup	
SALT	2-1/2 oz	1/4 cup 1/3 tbsp	
GARLIC POWDER	1 oz	3-1/3 tbsp	
BASIL,SWEET,WHOLE,CRUSHED	5/8 oz	1/4 cup 1/3 tbsp	
THYME,GROUND	1/3 oz	2 tbsp	
OREGANO,CRUSHED	5/8 oz	1/4 cup 1/3 tbsp	
PEPPER,RED,GROUND	1/8 oz	1/4 tsp	
BAY LEAF,WHOLE,DRIED	3/8 oz	12 lf	
EGGPLANT,FRESH,UNPEELED,SLICED	18-1/2 lbs	6 gal 1-5/8 qts	19-1/8 lbs
SALT	1-7/8 oz	3 tbsp	
FLOUR,WHEAT,GENERAL PURPOSE	1-3/8 lbs	1 qts 1 cup	
MILK,NONFAT,DRY	1-1/3 oz	1/2 cup 1 tbsp	
WATER,WARM	1-1/2 lbs	2-3/4 cup	
EGGS,WHOLE,FROZEN	1 lbs	1-7/8 cup	
BREADCRUMBS,DRY,GROUND,FINE	1-7/8 lbs	2 qts	
CHEESE,PARMESAN,GRATED	3-1/2 oz	1 cup	
CHEESE,MOZZARELLA,SHREDDED	3 lbs	3 qts	

Method

1. Combine tomatoes, tomato paste, water, onions, sugar, salt, garlic powder, basil, thyme, oregano, red pepper and bay leaves; mix well. Bring to a boil; reduce heat; simmer 1 hour or until thickened, stirring occasionally. Remove bay leaves.
2. Sprinkle eggplant with salt. Let stand 30 minutes; drain.
3. Dredge eggplant in flour; shake off excess.
4. Reconstitute milk; combine with eggs.
5. Dip eggplant in milk and egg mixture; drain well.
6. Dredge eggplant in crumbs; shake off excess.
7. Fry 3 minutes in 350 F. deep fat fryer or until golden brown.
8. Place 1 layer eggplant in table pans. Pour 3 cups sauce evenly over eggplant in each steam table pan.
9. Add second layer of eggplant. Cover with remaining sauce, 3 cups per pan.
10. Sprinkle parmesan cheese evenly over sauce in each pan.
11. Sprinkle shredded mozzarella cheese evenly over sauce in each pan.
12. Using a convection oven, bake at 325 F. for 20 minutes or until cheese is melted. CCP: Heat to 145 F. or higher for 15 seconds. Hold at 140 F. or higher for service.

VEGETABLES No.Q 029 00

SOUTHERN STYLE GREENS (FRESH COLLARDS)

Yield 100　　　　　　　　　　　　　　　　　　　　**Portion**　1/2 Cup

Calories	Carbohydrates	Protein	Fat	Cholesterol	Sodium	Calcium
98 cal	5 g	7 g	6 g	15 mg	290 mg	118 mg

Ingredient	**Weight**	**Measure**	**Issue**
PORK,HOCKS,(CURED & SMOKED),FROZEN	10 lbs		
WATER,BOILING	33-1/2 lbs	4 gal	
ONIONS,FRESH,CHOPPED	1-5/8 lbs	1 qts 1/2 cup	1-3/4 lbs
PEPPER,BLACK,GROUND	1/4 oz	1 tbsp	
WATER	33-1/2 lbs	4 gal	
GREENS,COLLARD,FRESH	20 lbs	2 gal 3-7/8 qts	27 lbs

Method

1. Add water to steam-jacketed kettle or stock pot. Add pork hocks and onions to water. Cover; simmer 2-1/2 hours or until tender. Remove; trim meat and fat from bones. Cut meat into small pieces. Add meat and bones to stock.
2. Add greens, pepper and water to stock. Bring to a boil; stir immediately.
3. Simmer 1 hour, uncovered or until greens are tender, stirring occasionally. CCP: Heat to 145 F. or higher for 15 seconds. Remove bones; serve greens with cooking liquid. CCP: Hold at 140 F. or higher for service. NOTES: In Step 1, 2 pounds raw bacon may be used for pork hocks per 100 portions.

VEGETABLES No.Q 029 01

SOUTHERN STYLE GREENS (FROZEN)

Yield 100   Portion 1/2 Cup

Calories	Carbohydrates	Protein	Fat	Cholesterol	Sodium	Calcium
107 cal	7 g	8 g	6 g	15 mg	542 mg	201 mg

Ingredient	Weight	Measure	Issue
PORK,HOCKS,(CURED & SMOKED),FROZEN	10 lbs		
WATER,BOILING	33-1/2 lbs	4 gal	
ONIONS,FRESH,CHOPPED	1-5/8 lbs	1 qts 1/2 cup	1-3/4 lbs
PEPPER,BLACK,GROUND	1/4 oz	1 tbsp	
WATER	37-5/8 lbs	4 gal 2 qts	
GREENS,COLLARD,FROZEN	20 lbs	3 gal 1-3/8 qts	

Method

1. Add water to steam-jacketed kettle or stock pot. Add pork hocks and onions to water. Cover; simmer 2-1/2 hours or until tender. Remove; trim meat and fat from bones. Cut meat into small pieces. Add meat and bones to stock.
2. Add greens, pepper and water to stock. Bring to a boil; stir immediately.
3. Break through frozen greens several times to hasten cooking. Simmer 25 minutes, uncovered or until greens are tender, stirring occasionally. CCP: Heat to 145 F. or higher for 15 seconds. Remove bones; serve greens with cooking liquid. CCP: Hold at 140 F. or higher for service.

Notes

1. In Step 1, 2 pounds of bacon may be used for pork hocks per 100 portions.

VEGETABLES No.Q 029 02

SWEET SOUR GREENS

Yield 100   **Portion** 1/2 Cup

Calories	Carbohydrates	Protein	Fat	Cholesterol	Sodium	Calcium
137 cal	14 g	7 g	6 g	16 mg	295 mg	121 mg

Ingredient	**Weight**	**Measure**	**Issue**
PORK,HOCKS,(CURED & SMOKED),FROZEN	10 lbs		
WATER,BOILING	33-1/2 lbs	4 gal	
ONIONS,FRESH,CHOPPED	1-5/8 lbs	1 qts 1/2 cup	1-3/4 lbs
PEPPER,BLACK,GROUND	1/4 oz	1 tbsp	
WATER	33-1/2 lbs	4 gal	
GREENS,COLLARD,FRESH	20 lbs	2 gal 3-7/8 qts	27 lbs
ONIONS,FRESH,CHOPPED	1-3/8 lbs	1 qts	1-5/8 lbs
BUTTER	2 oz	1/4 cup 1/3 tbsp	
SUGAR,GRANULATED	1-3/4 lbs	1 qts	
VINEGAR,DISTILLED	3-1/8 lbs	1 qts 2 cup	

Method

1. Add water to steam-jacketed kettle or stock pot. Add pork hocks and onions to water. Cover; simmer 2-1/2 hours or until tender. Remove; trim meat and fat from bones. Cut meat into small pieces. Add meat and bones to stock.
2. Add greens, pepper and water to stock. Bring to a boil; stir immediately.
3. Simmer 1 hour, uncovered or until greens are tender, stirring occasionally.
4. Saute chopped onions in butter or margarine until tender; add granulated sugar and vinegar; stir to mix well. Cook 3 minutes. Add to cooked greens. CCP: Heat to 145 F. or higher for 15 seconds.
5. Remove bones; serve greens with cooking liquid. CCP: Hold at 140 F. or higher for service.

Notes

1. In Step 1, 2 pounds of bacon may be used for pork hocks per 100 portions.

VEGETABLES No.Q 029 03

SOUTHERN STYLE GREENS (FRESH KALE)

Yield 100   **Portion** 1/2 Cup

Calories	Carbohydrates	Protein	Fat	Cholesterol	Sodium	Calcium
101 cal	6 g	7 g	6 g	15 mg	306 mg	85 mg

Ingredient   Weight   Measure   Issue

Ingredient	Weight	Measure	Issue
PORK,HOCKS,(CURED & SMOKED),FROZEN	10 lbs		
ONIONS,FRESH,CHOPPED	1-5/8 lbs	1 qts 1/2 cup	1-3/4 lbs
WATER,BOILING	33-1/2 lbs	4 gal	
KALE,FRESH,CHOPPED	12 lbs	5 gal 1/3 qts	16-7/8 lbs
PEPPER,BLACK,GROUND	1/4 oz	1 tbsp	
WATER	41-3/4 lbs	5 gal	

Method

1. Add water to steam-jacketed kettle or stock pot. Add pork hocks and onions to water. Cover; simmer 2-1/2 hours or until tender. Remove; trim meat and fat from bones. Cut meat into small pieces. Add meat and bones to stock.
2. Add kale, pepper and water to stock. Bring to a boil; stir immediately.
3. Simmer 20 minutes, uncovered or until greens are tender, stirring occasionally. CCP: Heat to 145 F. or higher for 15 seconds.
4. Remove bones; serve greens with cooking liquid. CCP: Hold at 140 F. or higher for service.

Notes

1. In Step 1, 2 pounds of bacon may be used for pork hocks per 100 servings.

VEGETABLES No.Q 030 00

SAUTEED MUSHROOMS

Yield 100   Portion   2 Tablespoons

Calories	Carbohydrates	Protein	Fat	Cholesterol	Sodium	Calcium
21 cal	1 g	0 g	2 g	5 mg	98 mg	3 mg

Ingredient

Ingredient	Weight	Measure	Issue
MUSHROOMS,CANNED,DRAINED	4-1/8 lbs	3 qts	
BUTTER	8 oz	1 cup	

Method

1. Drain mushrooms.
2. Saute mushrooms lightly in butter or margarine. CCP: Heat to 145 F. or higher for 15 seconds. Hold at 140 F. or higher for service.

VEGETABLES No.Q 030 01

SAUTEED MUSHROOMS AND ONIONS

Yield 100   Portion 2 Ounces

Calories	Carbohydrates	Protein	Fat	Cholesterol	Sodium	Calcium
67 cal	4 g	1 g	6 g	15 mg	137 mg	11 mg

Ingredient   Weight   Measure   Issue

MUSHROOMS,CANNED,DRAINED   4-1/8 lbs   3 qts
ONIONS,FRESH,SLICED   8-1/8 lbs   2 gal   9 lbs
BUTTER   1-1/2 lbs   3 cup

Method

1 Drain mushrooms.
2 Saute onions in butter until tender; add mushrooms. Heat thoroughly. CCP: Heat to 145 F. or higher for 15 seconds. Hold at 140 F. or higher for service.

VEGETABLES No.Q 031 00

OKRA AND TOMATO GUMBO

Yield 100   **Portion**   1/2 Cup

Calories	Carbohydrates	Protein	Fat	Cholesterol	Sodium	Calcium
100 cal	14 g	3 g	4 g	9 mg	391 mg	71 mg

Ingredient | Weight | Measure | Issue

Ingredient	Weight	Measure	Issue
ONIONS,FRESH,CHOPPED	2-1/8 lbs	1 qts 2 cup	2-1/3 lbs
BACON,RAW	1 lbs		
OKRA,FROZEN,CUT	10 lbs	1 gal 2-1/8 qts	
FLOUR,WHEAT,GENERAL PURPOSE	4-3/8 oz	1 cup	
SUGAR,GRANULATED	1-3/4 oz	1/4 cup 1/3 tbsp	
SALT	1-7/8 oz	3 tbsp	
CHILI POWDER,DARK,GROUND	1 oz	1/4 cup 1/3 tbsp	
PEPPER,BLACK,GROUND	1/8 oz	1/8 tsp	
TOMATOES,CANNED,CRUSHED,INCL LIQUIDS	13-1/4 lbs	1 gal 2 qts	
WATER,BOILING	3-1/8 lbs	1 qts 2 cup	
BREAD,WHITE,STALE,SLICED	2 lbs	1 gal 2-1/2 qts	
BUTTER,MELTED	12 oz	1-1/2 cup	
GARLIC CLOVES,FRESH,MINCED	1/8 oz	1/4 tsp	

Method

1 Saute onions and bacon until onions are tender and bacon is crisp.
2 Add okra to onions and bacon. Cook 5 minutes, stirring frequently.
3 Add flour, sugar, salt, chili powder, and pepper; stir until blended.
4 Add tomatoes and water; mix well.
5 Bring to a boil. Reduce heat; simmer 15 minutes or until okra is tender. CCP: Heat to 145 F. or higher for 15 seconds. Hold at 140 F. or higher for service.
6 Prepare Garlic Croutons. Trim crusts from bread; cut bread into 1/2-inch cubes. Place bread cubes on sheet pans. Brown lightly in 325 F. oven, about 20 to 25 minutes or in 375 F. convection oven for about 6 minutes on high fan, open vent. Melt butter or margarine; blend in minced garlic. Pour mixture evenly over lightly browned croutons in steam table pans; toss lightly.

Notes

1 In Step 1, 2 lbs bread will yield about 1 gallon lightly browned croutons.

VEGETABLES No.Q 032 00

SOUTHERN FRIED OKRA

Yield 100   **Portion** 1/3 Cup

Calories	Carbohydrates	Protein	Fat	Cholesterol	Sodium	Calcium
184 cal	15 g	2 g	13 g	0 mg	212 mg	57 mg

Ingredient	Weight	Measure	Issue
OKRA,FROZEN,CUT	15 lbs	2 gal 1-1/4 qts	
CORN MEAL	1-7/8 lbs	1 qts 2 cup	
FLOUR,WHEAT,GENERAL PURPOSE	1-1/4 lbs	1 qts 1/2 cup	
SALT	1-7/8 oz	3 tbsp	
PEPPER,BLACK,GROUND	1/8 oz	1/3 tsp	
SHORTENING	1-3/4 lbs	1 qts	

Method

1. Thaw okra. Mix cornmeal, flour, salt and pepper. Dredge okra in mixture.
2. Fry on well greased 375 F. griddle 10 minutes or until golden brown. CCP: Hold at 140 F. or higher for service.

VEGETABLES No.Q 033 00

PARSLEY BUTTERED POTATOES

Yield 100   Portion   4 Pieces

Calories	Carbohydrates	Protein	Fat	Cholesterol	Sodium	Calcium
170 cal	32 g	3 g	4 g	10 mg	609 mg	19 mg

Ingredient	Weight	Measure	Issue
POTATOES,FRESH,PEELED,CUBED	35 lbs	6 gal 1-1/2 qts	43-1/4 lbs
WATER	33-1/2 lbs	4 gal	
SALT	5-1/8 oz	1/2 cup	
BUTTER,MELTED	1 lbs	2 cup	
RESERVED LIQUID	1 lbs	2 cup	
PARSLEY,FRESH,BUNCH,CHOPPED	4-1/4 oz	2 cup	4-1/2 oz

Method

1. Cover potatoes with salted water; bring to a boil; reduce heat. Cover; simmer 20 to 25 minutes or until tender.
2. Drain; reserve 2 cups of liquid for use in Step 4.
3. Place an equal quantity of potatoes in steam table pans.
4. Combine butter or margarine and reserved liquid; pour 1 cup over potatoes in each pan. CCP: Heat to 145 F. or higher for 15 seconds.
5. Sprinkle 1/2 cup parsley over potatoes in each pan. CCP: Hold at 140 F. or higher for service.

VEGETABLES No.Q 033 01

PAPRIKA BUTTERED POTATOES

Yield 100 **Portion** 4 Pieces

Calories	Carbohydrates	Protein	Fat	Cholesterol	Sodium	Calcium
170 cal	32 g	3 g	4 g	10 mg	608 mg	18 mg

Ingredient	Weight	Measure	Issue
POTATOES,FRESH,PEELED,CUBED	35 lbs	6 gal 1-1/2 qts	43-1/4 lbs
WATER	33-1/2 lbs	4 gal	
SALT	5-1/8 oz	1/2 cup	
BUTTER,MELTED	1 lbs	2 cup	
RESERVED LIQUID	1 lbs	2 cup	
PAPRIKA,GROUND	1 oz	1/4 cup 1/3 tbsp	

Method

1. Cover potatoes with salted water; bring to a boil; reduce heat. Cover; simmer 20 to 25 minutes or until tender.
2. Drain; reserve 2 cups of liquid for use in Step 4.
3. Place an equal quantity of potatoes in steam table pans.
4. Combine butter or margarine and reserved liquid; pour 1 cup over potatoes in each pan. CCP: Heat to 145 F. or higher for 15 seconds.
5. Sprinkle 1 tablespoon paprika over potatoes in each pan. CCP: Hold at 140 F. or higher for service.

VEGETABLES No.Q 033 02

PARSLEY BUTTERED POTATOES (CANNED)

Yield 100 **Portion** 4 Pieces

Calories	Carbohydrates	Protein	Fat	Cholesterol	Sodium	Calcium
95 cal	12 g	2 g	5 g	13 mg	302 mg	48 mg

Ingredient | **Weight** | **Measure** | **Issue**

POTATOES, CANNED, WHOLE — 34 lbs — 2 gal 1-5/8 qts
RESERVED LIQUID — 1-3/8 lbs — 2-5/8 cup
BUTTER, MELTED — 1-1/3 lbs — 2-5/8 cup
PARSLEY, DEHYDRATED, FLAKED — 3/8 oz — 1/2 cup

Method

1. Drain potatoes; reserve 2-2/3 cups liquid for use in Step 3.
2. Place 1-1/3 gal potatoes in each pan.
3. Combine margarine or butter and reserved liquid; pour 1-/3 cup over potatoes in each pan.
4. Sprinkle 2 tbsp parsley over potatoes in each pan.
5. Using a convection oven, bake at 350 F. 25-30 minutes or until browned on high fan, open vent. CCP: Temperature must reach 145 F. or higher for 15 seconds. Hold at 140 F. or higher for service.

VEGETABLES No.Q 033 03

PAPRIKA BUTTERED POTATOES (CANNED)

Yield 100 **Portion** 4 Pieces

Calories	Carbohydrates	Protein	Fat	Cholesterol	Sodium	Calcium
95 cal	12 g	2 g	5 g	13 mg	301 mg	47 mg

Ingredient	Weight	Measure	Issue
POTATOES, CANNED, WHOLE	34 lbs	2 gal 1-5/8 qts	
RESERVED LIQUID	1-3/8 lbs	2-5/8 cup	
BUTTER, MELTED	1-1/3 lbs	2-5/8 cup	
PAPRIKA, GROUND	1 oz	1/4 cup 1/3 tbsp	

Method

1 Drain potatoes; reserve 2-2/3 cups liquid for use in Step 3.
2 Place 1-1/3 gal potatoes in each pan.
3 Combine margarine or butter and reserved liquid; pour 1-1/3 cup over potatoes in each pan.
4 Sprinkle 4 tbsp paprika over potatoes in each pan.
5 Using a convection oven, bake at 350 F. 25-30 minutes or until browned on high fan, open vent. CCP: Temperature must reach 145 F. or higher for 15 seconds. Hold at 140 F. or higher for service.

VEGETABLES No.Q 034 00

SPANISH ONIONS

Yield 100   Portion   1/2 Cup

Calories	Carbohydrates	Protein	Fat	Cholesterol	Sodium	Calcium
97 cal	13 g	2 g	5 g	0 mg	446 mg	44 mg

Ingredient	Weight	Measure	Issue
ONIONS,FRESH,QUARTERED	15 lbs	3 gal 2-7/8 qts	16-2/3 lbs
WATER,BOILING	25-1/8 lbs	3 gal	
SALT	3-1/8 oz	1/4 cup 1-1/3 tbsp	
TOMATOES,CANNED,CRUSHED,INCL LIQUIDS	13-1/4 lbs	1 gal 2 qts	
PEPPERS,GREEN,FRESH,CHOPPED	4 lbs	3 qts 1/8 cup	4-7/8 lbs
CELERY,FRESH,CHOPPED	3 lbs	2 qts 3-3/8 cup	4-1/8 lbs
SUGAR,GRANULATED	1-3/4 oz	1/4 cup 1/3 tbsp	
PEPPER,BLACK,GROUND	1/8 oz	1/3 tsp	
OIL,SALAD	1 lbs	2 cup	
FLOUR,WHEAT,GENERAL PURPOSE	4-3/8 oz	1 cup	

Method

1 Cook onions in salted water 15 minutes or until tender; drain.
2 Spread 1-1/4 gallon onions in each steam table pan.
3 Combine tomatoes, peppers, celery, sugar, and pepper. Heat to boiling; simmer until vegetables are tender.
4 Blend salad oil and flour together; stir until smooth; add to tomatoes, stirring constantly. Cook 10 minutes or until slightly thickened.
5 Pour 3 quarts tomato mixture over onions in each pan. Bake in 350 F. oven for 15 minutes. CCP: Heat to 145 F. or higher for 15 seconds. Hold at 140 F. or higher for service.

VEGETABLES No.Q 035 00

FRENCH FRIED ONION RINGS

Yield 100   **Portion** 2-1/2 Ounces

Calories	Carbohydrates	Protein	Fat	Cholesterol	Sodium	Calcium
274 cal	40 g	7 g	10 g	1 mg	656 mg	75 mg

Ingredient

Ingredient	Weight	Measure	Issue
ONIONS,FRESH,SLICED	20 lbs	4 gal 3-3/4 qts	22-1/4 lbs
WATER,COLD	16-3/4 lbs	2 gal	
FLOUR,WHEAT,GENERAL PURPOSE	8-7/8 lbs	2 gal	
SALT	5-3/4 oz	1/2 cup 1 tbsp	
PEPPER,BLACK,GROUND	1/8 oz	1/3 tsp	
MILK,NONFAT,DRY	13-3/4 oz	1 qts 1-3/4 cup	
WATER,WARM	7-7/8 lbs	3 qts 3 cup	

Method

1 Separate onion slices into rings. Cover with cold water. Let stand 10 to 15 minutes. Drain.
2 Dredge onion rings in mixture of flour, salt and pepper; shake off excess. Reserve remaining seasoned flour for use in Step 4.
3 Reconstitute milk; dip floured onion rings into milk. Drain well.
4 Dredge onion rings in seasoned flour until well coated; shake off excess.
5 Fry 2 minutes in 350 F. deep fat or until golden brown.
6 Drain well in basket or on absorbent paper. CCP: Hold for service at 140 F. or higher.

VEGETABLES No.Q 035 01

FRENCH FRIED ONION RINGS (FROZEN)

Yield 100   Portion 3 Ounces

Calories	Carbohydrates	Protein	Fat	Cholesterol	Sodium	Calcium
333 cal	35 g	4 g	20 g	0 mg	279 mg	52 mg

Ingredient	Weight	Measure	Issue
ONION RINGS,RAW,BREADED,FROZEN	25 lbs		

Method

1. Fry according to directions on package.
2. Drain well in basket or an absorbent paper. CCP: Hold at 140 F. or higher for service.

VEGETABLES No.Q 035 02

TEMPURA FRIED ONION RINGS

Yield 100   Portion   2-1/2 Ounces

Calories	Carbohydrates	Protein	Fat	Cholesterol	Sodium	Calcium
190 cal	29 g	5 g	6 g	45 mg	547 mg	91 mg

Ingredient	Weight	Measure	Issue
ONIONS,FRESH,SLICED	20 lbs	4 gal 3-3/4 qts	22-1/4 lbs
WATER,COLD	16-3/4 lbs	2 gal	
TEMPURA BATTER		2 gal	

Method

1. Separate onions slices into rings. Cover with cold water. Let stand 10 to 15 minutes. Drain.
2. Prepare Tempura Batter, Recipe No. D 038 00. Dip individual onion rings into batter.
3. Drop onion rings gently into 350 F. deep fat; fry about 1-1/2 minutes or until golden brown.
4. Drain well in basket or on absorbent paper. CCP: Hold at 140 F. or higher for service.

VEGETABLES No.Q 036 00

FRIED ONIONS

Yield 100　　　　　　　　　　　　　　　　　　　**Portion**　1/4 Cup

Calories	Carbohydrates	Protein	Fat	Cholesterol	Sodium	Calcium
101 cal	10 g	1 g	7 g	0 mg	3 mg	23 mg

Ingredient	Weight	Measure	Issue
OIL,SALAD	1-1/2 lbs	3 cup	
ONIONS,FRESH,SLICED	25 lbs	6 gal 5/8 qts	27-3/4 lbs

Method

1. Heat 1-1/2 cups salad oil in each steam table pan.
2. Place 12 pounds 8 ounces onions in each pan. Cook 40 minutes in 400 F. oven or until tender and lightly brown, stirring occasionally to prevent burning. CCP: Heat to 145 F. or higher for 15 seconds. Hold at 140 F. or higher for service.

VEGETABLES No.Q 037 00

SMOTHERED ONIONS (DEHYDRATED ONIONS)

Yield 100 Portion 1/2 Cup

Calories	Carbohydrates	Protein	Fat	Cholesterol	Sodium	Calcium
117 cal	19 g	2 g	4 g	0 mg	358 mg	61 mg

Ingredient	Weight	Measure	Issue
ONIONS,DEHYDRATED,CHOPPED	5 lbs	2 gal 2 qts	
WATER,WARM	33-1/2 lbs	4 gal	
OIL,SALAD	1 lbs	2 cup	
SALT	3-1/8 oz	1/4 cup 1-1/3 tbsp	
PEPPER,BLACK,GROUND	1/8 oz	1/8 tsp	

Method

1 Rehydrate onions in water 1 hour; drain well.
2 Blend salad oil, salt and pepper with onions in steam-jacketed kettle or stock pot.
3 Cover; bring to a boil. Reduce heat; simmer 20 minutes or until tender and slightly browned, stirring occasionally.
4 Drain well. CCP: Heat to 145 F. or higher for 15 seconds. Hold at 140 F. or higher for service.

VEGETABLES No.Q 038 00

REFRIED BEANS WITH CHEESE

Yield 100   **Portion** 1/2 Cup

Calories	Carbohydrates	Protein	Fat	Cholesterol	Sodium	Calcium
145 cal	15 g	8 g	6 g	14 mg	382 mg	142 mg

Ingredient   Weight   Measure   Issue

Ingredient	Weight	Measure	Issue
BEANS,PINTO,CANNED,INCL LIQUIDS	21 lbs	2 gal 1-7/8 qts	
CHILI POWDER,DARK,GROUND	3-1/8 oz	3/4 cup	
GARLIC POWDER	1/3 oz	1 tbsp	
CHEESE,CHEDDAR,SHREDDED	2 lbs	2 qts	
ONIONS,GREEN,FRESH,GRATED	7 oz	2 cup	7-7/8 oz
HOT SAUCE	1 oz	2 tbsp	
RESERVED LIQUID	5-1/4 lbs	2 qts 2 cup	
COOKING SPRAY,NONSTICK	2 oz	1/4 cup 1/3 tbsp	
CHEESE,CHEDDAR,SHREDDED	1 lbs	1 qts	

Method

1. Drain beans. Reserve beans for use in Step 2; stock for use in Step 3.
2. Place beans in mixer bowl; beat at low speed until mashed.
3. Add chili powder, garlic powder, 1 quart cheese, onions, hot sauce and 1-1/2 quarts bean stock per 100 servings. Whip at medium speed, adding more liquid to obtain consistency of mashed potatoes.
4. Spread an equal quantity of bean mixture in each sprayed steam table pan. Bake in 350 F. oven for 30 minutes.
5. Sprinkle an equal quantity of remaining 1 quart cheese over bean mixture in each pan. CCP: Heat to 145 F. or higher for 15 seconds. Hold for service at 140 F. or higher.

VEGETABLES No.Q 038 01

REFRIED BEANS (CANNED BEANS)

Yield 100 **Portion** 1/2 Cup

Calories	Carbohydrates	Protein	Fat	Cholesterol	Sodium	Calcium
107 cal	17 g	6 g	2 g	9 mg	325 mg	38 mg

Ingredient	**Weight**	**Measure**	**Issue**
BEANS,REFRIED	24 lbs	2 gal 2-7/8 qts	
COOKING SPRAY,NONSTICK	2 oz	1/4 cup 1/3 tbsp	

Method

1 Use canned refried beans.
2 Lightly spray each steam table pan with non-stick cooking spray. Spread an equal quantity of bean mixture in each sprayed pan. Bake in 350 F. oven for 30 minutes. CCP: Internal temperature must reach 145 F. or higher for 15 seconds. Hold at 140 F. or higher for service.
3 If desired, mashed bean mixture may be fried on greased 350 F. griddle.

VEGETABLES No.Q 038 02

REFRIED BEANS WITH CHEESE (CANNED BEANS)

Yield 100   Portion 1/2 Cup

Calories	Carbohydrates	Protein	Fat	Cholesterol	Sodium	Calcium
164 cal	18 g	10 g	6 g	23 mg	426 mg	139 mg

Ingredient	Weight	Measure	Issue
BEANS,REFRIED	24 lbs	2 gal 2-7/8 qts	
CHILI POWDER,DARK,GROUND	3-1/8 oz	3/4 cup	
HOT SAUCE	1 oz	2 tbsp	
GARLIC POWDER	1/3 oz	1 tbsp	
CHEESE,CHEDDAR,SHREDDED	1 lbs	1 qts	
COOKING SPRAY,NONSTICK	2 oz	1/4 cup 1/3 tbsp	
CHEESE,CHEDDAR,SHREDDED	2 lbs	2 qts	

Method

1. Add chili powder, garlic, cheese and hot sauce to canned beans. Mix well.
2. Lightly spray each steam table pan with non-stick cooking spray. Spread an equal quantity of bean mixture in each sprayed pan. Bake in 350 F. oven for 30 minutes. CCP: Internal temperature must reach 145 F. or higher for 15 seconds.
3. Sprinkle an equal quantity of cheese over bean mixture in each pan. CCP: Hold at 140 F. or higher for service.

VEGETABLES No.Q 039 00

GREEN BEANS WITH CORN (FROZEN BEANS)

Yield 100   Portion   1/2 Cup

Calories	Carbohydrates	Protein	Fat	Cholesterol	Sodium	Calcium
63 cal	13 g	2 g	1 g	1 mg	220 mg	26 mg

Ingredient	Weight	Measure	Issue
BACON,RAW	12 oz		
BACON FAT,RENDERED	1-3/4 oz	1/4 cup 1/3 tbsp	
ONIONS,FRESH,SLICED	1 lbs	1 qts	1-1/8 lbs
BEANS,GREEN,FROZEN,CUT	12 lbs	2 gal 3 qts	
SALT	5/8 oz	1 tbsp	
WATER	6-1/4 lbs	3 qts	
PEPPER,RED,GROUND	<1/16th oz	1/8 tsp	
CORN,CANNED,CREAM STYLE	10-1/8 lbs	1 gal 1/2 qts	

Method

1 Cook bacon until partially done. Drain fat; set aside 1/4 cup of bacon fat for use in Step 2; set aside bacon for use in Step 4.
2 Saute onions in bacon fat until tender.
3 Cook green beans 5 minutes. Drain beans and reserve liquid.
4 Combine beans, bacon, onions, red pepper and corn. Combine reserved liquid and water to equal 2-1/2 qts per 100 portions. Add bean and vegetable mixture to liquid; cover and continue cooking 10 minutes. CCP: Internal temperature must reach 145 F. or higher for 15 seconds.
5 Serve with cooking liquid. CCP: Hold at 140 F. or higher for service.

VEGETABLES No.Q 039 01

GREEN BEANS WITH CORN (CANNED BEANS)

Yield 100   **Portion** 1/2 Cup

Calories	Carbohydrates	Protein	Fat	Cholesterol	Sodium	Calcium
58 cal	12 g	2 g	1 g	1 mg	368 mg	23 mg

Ingredient

Ingredient	Weight	Measure	Issue
BACON,RAW	12 oz		
BACON FAT,RENDERED	1-3/4 oz	1/4 cup 1/3 tbsp	
ONIONS,FRESH,SLICED	1 lbs	1 qts	1-1/8 lbs
BEANS,GREEN,CANNED	18-3/4 lbs	2 gal 7/8 qts	
PEPPER,RED,GROUND	<1/16th oz	1/8 tsp	
RESERVED LIQUID	5-1/4 lbs	2 qts 2 cup	
CORN,CANNED,CREAM STYLE	10-1/8 lbs	1 gal 1/2 qts	

Method

1. Cook bacon until partially done. Drain bacon; set aside 1/4 cup bacon fat. Set aside bacon for use in Step 4.
2. Saute onions in bacon fat until tender.
3. Drain beans. Reserve 2-1/2 quarts of liquid for use in Step 4.
4. Combine beans, bacon, onions, red pepper, reserved liquid and corn. Cover; continue cooking 15 minutes. CCP: Heat to 145 F. or higher for 15 seconds.
5. Serve with cooking liquid. CCP: Hold for service at 140 F. or higher.

VEGETABLES No.Q 040 00

TURNIPS AND BACON

Yield 100 **Portion** 1/2 Cup

Calories	Carbohydrates	Protein	Fat	Cholesterol	Sodium	Calcium
30 cal	5 g	1 g	1 g	1 mg	217 mg	26 mg

Ingredient | Weight | Measure | Issue

Ingredient	Weight	Measure	Issue
BACON,RAW	1 lbs		
WATER,BOILING	12-1/2 lbs	1 gal 2 qts	
SALT	1-1/4 oz	2 tbsp	
PEPPER,BLACK,GROUND	1/8 oz	1/4 tsp	
TURNIPS,WHITE,FRESH,CUBES	18-1/3 lbs	4 gal	22-2/3 lbs

Method

1. Add bacon to water; simmer 30 minutes.
2. Add salt, pepper, and turnips to bacon and water.
3. Cover; bring to a boil. Remove cover; simmer 15 to 20 minutes or until just tender. CCP: Internal temperature must reach 145 F. or higher for 15 seconds. Hold at 140 F. or higher for service.

VEGETABLES No.Q 041 00

PEAS WITH MUSHROOMS (FROZEN)

Yield 100   Portion   3/4 Cup

Calories	Carbohydrates	Protein	Fat	Cholesterol	Sodium	Calcium
126 cal	21 g	7 g	2 g	0 mg	216 mg	38 mg

Ingredient	Weight	Measure	Issue
PEAS,GREEN,FROZEN	27 lbs	5 gal 1-1/4 qts	
SALT	5/8 oz	1 tbsp	
WATER,BOILING	16-3/4 lbs	2 gal	
MUSHROOMS,CANNED,DRAINED	6-1/4 lbs	1 gal 1/2 qts	
MARGARINE	8 oz	1 cup	

Method

1. Add peas to boiling salted water.
2. Bring to a boil; cover; cook gently 6 to 8 minutes or until tender. Drain.
3. Saute mushrooms in margarine or butter.
4. Combine hot peas and mushrooms; mix gently. CCP: Internal temperature must reach 145 F. or higher for 15 seconds. Hold at 140 F. or higher for service.

VEGETABLES No.Q 041 01

PEAS WITH CARROTS (FROZEN)

Yield 100   Portion 3/4 Cup

Calories	Carbohydrates	Protein	Fat	Cholesterol	Sodium	Calcium
133 cal	23 g	6 g	2 g	0 mg	215 mg	56 mg

Ingredient	Weight	Measure	Issue
PEAS,GREEN,FROZEN	22-1/3 lbs	4 gal 1-5/8 qts	
SALT	5/8 oz	1 tbsp	
WATER,BOILING	16-3/4 lbs	2 gal	
CARROTS,FROZEN,SLICED	18 lbs	3 gal 4 qts	
SALT	5/8 oz	1 tbsp	
WATER,BOILING	6-1/4 lbs	3 qts	
MARGARINE	8 oz	1 cup	

Method

1. Add frozen peas to boiling salted water.
2. Bring to a boil; cover; cook gently 6 to 8 minutes or until tender. Drain.
3. Place carrots and salt in boiling water; cook 10 to 13 minutes or until tender; drain.
4. Combine hot peas and carrots with melted butter or margarine; mix gently. CCP: Internal temperature must reach 145 F. or higher for 15 seconds. Hold at 140 F. or higher for service.

VEGETABLES No.Q 041 02

PEAS WITH CELERY (FROZEN)

Yield 100   Portion 3/4 Cup

Calories	Carbohydrates	Protein	Fat	Cholesterol	Sodium	Calcium
111 cal	18 g	6 g	2 g	0 mg	148 mg	53 mg

Ingredient   Weight   Measure   Issue

PEAS,GREEN,FROZEN   22-1/2 lbs   4 gal 1-3/4 qts
SALT   5/8 oz   1 tbsp
WATER,BOILING   16-3/4 lbs   2 gal
CELERY,FRESH,CHOPPED   12-3/4 lbs   3 gal <1/16th qts   17-1/2 lbs
WATER,BOILING   6-1/4 lbs   3 qts
MARGARINE   8 oz   1 cup

Method

1. Add frozen peas to boiling salted water.
2. Bring to a boil; cover; cook gently 6 to 8 minutes or until tender. Drain.
3. Place celery in boiling water. Cook 10 to 15 minutes or until tender; drain.
4. Combine hot peas and celery with melted butter or margarine; mix gently. CCP: Internal temperature must reach 145 F. or higher for 15 seconds. Hold at 140 F. or higher for service.

VEGETABLES No.Q 041 03

PEAS WITH ONIONS

Yield 100   **Portion** 3/4 Cup

Calories	Carbohydrates	Protein	Fat	Cholesterol	Sodium	Calcium
130 cal	22 g	7 g	2 g	0 mg	96 mg	40 mg

Ingredient	Weight	Measure	Issue
PEAS,GREEN,FROZEN	27 lbs	5 gal 1-1/4 qts	
SALT	5/8 oz	1 tbsp	
WATER,BOILING		2 gal	
ONIONS,FRESH,CHOPPED	6-1/3 lbs	1 gal 1/2 qts	7 lbs
MARGARINE	8 oz	1 cup	

Method

1. Add peas to salted boiling water.
2. Bring to a boil; cover; cook gently 6 to 8 minutes or until tender. Drain.
3. Saute onions in butter or margarine until tender.
4. Combine hot peas and sauteed onions; mix gently. CCP: Internal temperature must reach 145 F. or higher for 15 seconds. Hold at 140 F. or higher for service.

VEGETABLES No.Q 041 04

PEAS WITH MUSHROOMS (CANNED PEAS)

Yield 100 **Portion** 3/4 Cup

Calories	Carbohydrates	Protein	Fat	Cholesterol	Sodium	Calcium
47 cal	6 g	2 g	2 g	0 mg	252 mg	12 mg

Ingredient	**Weight**	**Measure**	**Issue**
PEAS,GREEN,CANNED,INCL LIQUIDS	9-7/8 lbs	1 gal 1/2 qts	
MUSHROOMS,CANNED,DRAINED	6-1/4 lbs	1 gal 1/2 qts	
MARGARINE	8 oz	1 cup	

Method

1. Drain peas.
2. Saute mushrooms in butter or margarine.
3. Heat peas; drain and combine with mushrooms; mix gently. CCP: Internal temperature must reach 145 F. or higher for 15 seconds. Hold for service at 140 F. or higher.

VEGETABLES No.Q 042 00

GREEN BEANS PARISIENNE (CANNED)

Yield 100   Portion 1/2 Cup

Calories	Carbohydrates	Protein	Fat	Cholesterol	Sodium	Calcium
64 cal	6 g	2 g	4 g	6 mg	384 mg	54 mg

Ingredient	Weight	Measure	Issue
ONIONS,FRESH,SLICED	1-1/8 lbs	1 qts 1/2 cup	1-1/4 lbs
BUTTER	2 oz	1/4 cup 1/3 tbsp	
SOUP,CONDENSED,CREAM OF MUSHROOM	4-3/4 lbs	2 qts 5/8 cup	
WATER	1-1/3 lbs	2-1/2 cup	
WORCESTERSHIRE SAUCE	1/2 oz	1 tbsp	
BEANS,GREEN,CANNED,DRAINED	14-1/4 lbs	3 gal	
BREADCRUMBS,DRY,GROUND,FINE	7-5/8 oz	2 cup	
BUTTER,MELTED	4 oz	1/2 cup	
CHEESE,PARMESAN,GRATED	7 oz	2 cup	

Method

1 Saute onions in butter or margarine until tender.
2 Blend soup, water, and Worcestershire sauce into onion mixture.
3 Drain beans; add beans to soup mixture; mix lightly.
4 Place 6-1/4 quarts mixture in each steam table pan.
5 Combine bread crumbs and melted butter or margarine.
6 Sprinkle 1 cup over mixture in each pan.
7 Sprinkle 1 cup cheese over bread crumbs in each pan.
8 Using a convection oven, bake in 350 F. oven for 15 minutes on high fan, open vent or until sauce is bubbling and cheese is melted. CCP: Internal temperature must reach 145 F. or higher for 15 seconds. Hold for service at 140 F. or higher.

VEGETABLES No.Q 042 01

GREEN BEANS PARISIENNE (FROZEN BEANS)

Yield 100   Portion   1/2 Cup

Calories	Carbohydrates	Protein	Fat	Cholesterol	Sodium	Calcium
75 cal	9 g	3 g	4 g	6 mg	216 mg	68 mg

Ingredient   Weight   Measure   Issue

Ingredient	Weight	Measure	Issue
ONIONS,FRESH,SLICED	1-1/8 lbs	1 qts 1/2 cup	1-1/4 lbs
BUTTER	2 oz	1/4 cup 1/3 tbsp	
SOUP,CONDENSED,CREAM OF MUSHROOM	4-3/4 lbs	2 qts 5/8 cup	
WATER	1-1/3 lbs	2-1/2 cup	
WORCESTERSHIRE SAUCE	1/2 oz	1 tbsp	
BEANS,GREEN,FROZEN,CUT	16 lbs	3 gal 2-5/8 qts	
BREADCRUMBS,DRY,GROUND,FINE	7-5/8 oz	2 cup	
BUTTER,MELTED	4 oz	1/2 cup	
CHEESE,PARMESAN,GRATED	7 oz	2 cup	

Method

1. Saute onions in butter or margarine until tender.
2. Blend soup, water and Worcestershire sauce into onion mixture.
3. Use frozen green beans.
4. Place about 6-1/4 quarts in each steam table pan.
5. Combine bread crumbs and melted butter or margarine.
6. Sprinkle 1 cup over mixture in each pan.
7. Sprinkle 1 cup cheese over breadcrumbs in each pan.
8. Using a convection oven, bake in 350 F. oven for 15 minutes on high fan, open vent or until sauce is bubbly and cheese is melted.
   CCP: Internal temperature must reach 145 F. or higher for 15 seconds. Hold at 140 F. or higher for service.

VEGETABLES No.Q 043 00

RED CABBAGE WITH SWEET AND SOUR SAUCE

Yield 100   Portion   1/2 Cup

Calories	Carbohydrates	Protein	Fat	Cholesterol	Sodium	Calcium
81 cal	10 g	1 g	5 g	12 mg	337 mg	48 mg

Ingredient	Weight	Measure	Issue
BUTTER,MELTED	1-1/4 lbs	2-1/2 cup	
CABBAGE,RED,FRESH,CHOPPED	18 lbs	5 gal 2-1/2 qts	22-1/2 lbs
APPLES,FRESH,MEDIUM,UNPEELED,DICED	2 lbs	1 qts 3-1/4 cup	2-1/3 lbs
VINEGAR,DISTILLED	1-5/8 lbs	3 cup	
SUGAR,BROWN,PACKED	10-7/8 oz	2-1/8 cup	
SALT	2-1/2 oz	1/4 cup 1/3 tbsp	
CLOVES,GROUND	7/8 oz	1/4 cup 1/3 tbsp	
BAY LEAF,WHOLE,DRIED	1/4 oz	7 each	

Method

1. Place 1-1/4 cups butter or margarine in each roasting pan.
2. Add 9 pounds or 11-1/4 quarts cabbage and 5-1/2 cups apples to each pan. Mix thoroughly.
3. Cook at low heat 30 minutes, stirring frequently to avoid scorching.
4. Combine vinegar, brown sugar, salt, cloves and bay leaves.
5. Pour vinegar mixture evenly over hot cabbage and apples in each pan.
6. Simmer 2 to 3 minutes to blend seasonings. Remove bay leaves. CCP: Internal temperature must reach 145 F. or higher for 15 seconds. Hold at 140 F. or higher for service.

VEGETABLES No.Q 044 00

BAKED POTATOES

Yield 100　　　　　　　　　　　　　　　　　　　Portion 1 Each

Calories	Carbohydrates	Protein	Fat	Cholesterol	Sodium	Calcium
146 cal	34 g	3 g	0 g	0 mg	9 mg	14 mg

Ingredient　　　　　　　　　　　　　　　Weight　　　Measure　　　Issue

POTATOES,WHITE,FRESH　　　　　　　　37-1/2 lbs　　100 each

Method

1. Scrub potatoes well; remove any blemishes. Place on sheet pans. Prick skin with fork to allow steam to escape.
2. Using a convection oven, bake at 400 F. for 35 minutes on high fan, closed vent or until done. Potatoes are done when 208 F. to 211 F. internal temperature is reached. When done, a fork will easily pierce a potato. CCP: Hold at 140 F. or higher for service.

VEGETABLES No.Q 044 01

QUICK BAKED POTATO HALVES

Yield 100 **Portion** 2 Halves

Calories	Carbohydrates	Protein	Fat	Cholesterol	Sodium	Calcium
151 cal	34 g	3 g	1 g	0 mg	9 mg	14 mg

Ingredient

Ingredient	Weight	Measure	Issue
POTATOES,WHITE,FRESH	37-1/2 lbs	100 each	
COOKING SPRAY,NONSTICK	2 oz	1/4 cup 1/3 tbsp	

Method

1. Scrub potatoes well; remove any blemishes.
2. Cut potatoes in half lengthwise. Dry cut sides on paper towels.
3. Lightly spray sheet pans with non-stick cooking spray. Place cut sides down, in rows 5x6, on sprayed sheet pans.
4. Using a convection oven, bake 30 minutes at 400 F. or until done or cut sides are evenly browned on high fan, closed vent. CCP: Heat to 145 F. or higher for 15 seconds. Hold at 140 F. or higher for service.

VEGETABLES   No.Q 045 00

FRENCH FRIED POTATOES

Yield 100   **Portion**   3-1/2 Ounces

Calories	Carbohydrates	Protein	Fat	Cholesterol	Sodium	Calcium
265 cal	34 g	3 g	14 g	0 mg	11 mg	15 mg

Ingredient   **Weight**   **Measure**   **Issue**

POTATOES,WHITE,FRESH,PEELED,FRENCH-FRY CUT   37 lbs   6 gal 2-7/8 qts   45-2/3 lbs
WATER,COLD   16-3/4 lbs   2 gal

Method

1. Hold peeled potatoes in cold water until needed to prevent discoloration.
2. Drain; dry well.
3. Fill fryer basket about 2/3 full; fry about 7 minutes in 365 F. deep fat or until golden brown.
4. Drain well in basket or on absorbent paper. CCP: Hold at 140 F. or higher for service.

VEGETABLES No.Q 045 01

FRENCH FRIED POTATOES (FROZEN)

Yield 100   Portion   3-1/2 Ounces

Calories	Carbohydrates	Protein	Fat	Cholesterol	Sodium	Calcium
288 cal	39 g	4 g	14 g	0 mg	37 mg	10 mg

Ingredient   **Weight**   **Measure**   **Issue**

POTATO,WHITE,FROZEN,FRENCH FRIED   35 lbs

Method

1. Use frozen French fried potatoes.
2. Fill fryer basket about 2/3 full; fry about 4 minutes at 375 F. or until golden brown.
3. Drain well in basket or on absorbent paper. Do not cover fries. CCP: Hold at 140 F. or higher for service.

VEGETABLES No.Q 045 02

FRENCH FRIED POTATOES (FROZEN, OVEN METHOD)

Yield 100   Portion   3-1/2 Ounces

Calories	Carbohydrates	Protein	Fat	Cholesterol	Sodium	Calcium
252 cal	39 g	4 g	10 g	0 mg	37 mg	10 mg

Ingredient   Weight   Measure   Issue

POTATO,WHITE,FROZEN,FRENCH FRIED   35 lbs
COOKING SPRAY,NONSTICK   2 oz   1/4 cup 1/3 tbsp

Method

1 Use frozen French fried potatoes.
2 Lightly spray sheet pans with non-stick cooking spray.
3 Place about 3 pounds 14 ounces potatoes on each sheet pan.
4 Using a convection oven, bake at 450 F. 20 to 25 minutes on high fan, open vent. CCP: Hold at 140 F. or higher for service.

VEGETABLES   No.Q 045 03

FRENCH FRIED SHOESTRING POTATOES (FROZEN)

Yield 100   Portion   3-1/2 Ounces

Calories	Carbohydrates	Protein	Fat	Cholesterol	Sodium	Calcium
252 cal	33 g	3 g	13 g	0 mg	31 mg	8 mg

Ingredient   **Weight**   **Measure**   **Issue**

POTATO,WHITE,FROZEN,SHOESTRING   30 lbs

Method

1 Fry about 3 minutes at 365 F. or until golden brown.
2 Drain well in basket or on absorbent paper. CCP: Hold at 140 F. or higher for service.

VEGETABLES No.Q 045 04

FRENCH FRIED SHOESTRING POTATOES (FROZEN, OVEN)

Yield 100 **Portion** 3-1/2 Ounces

Calories	Carbohydrates	Protein	Fat	Cholesterol	Sodium	Calcium
217 cal	33 g	3 g	8 g	0 mg	31 mg	8 mg

Ingredient | Weight | Measure | Issue

Ingredient	Weight	Measure
POTATO,WHITE,FROZEN,SHOESTRING	30 lbs	
COOKING SPRAY,NONSTICK	2 oz	1/4 cup 1/3 tbsp

Method

2 Lightly spray sheet pans with non-stick cooking spray.
3 Place about 2 pounds 8 ounces potatoes on each sheet pan.
4 Using a convection oven, bake in 400 F. for 7 to 10 minutes on high fan, open vent until golden brown. CCP: Hold at 140 F. or higher for service.

VEGETABLES No.Q 045 05

FRENCH FRIED POTATOES (DEHYDRATED MIX)

Yield 100   Portion 3-1/2 Ounces

Calories	Carbohydrates	Protein	Fat	Cholesterol	Sodium	Calcium
209 cal	30 g	3 g	9 g	0 mg	39 mg	9 mg

Ingredient	Weight	Measure	Issue
POTATO,WHITE,INSTANT,GRANULES	8 lbs	4 gal 2-7/8 qts	

Method

1 Use dehydrated potato mix. Rehydrate, dispense and fry mix according to manufacturer's directions. CCP: Hold at 140 F. or higher for service.

VEGETABLES No.Q 045 06

BAKED POTATO ROUNDS (PRECOOKED)

Yield 100   **Portion** 3/4 Cup

Calories	Carbohydrates	Protein	Fat	Cholesterol	Sodium	Calcium
88 cal	20 g	3 g	0 g	0 mg	28 mg	9 mg

Ingredient   Weight   Measure   Issue

POTATO,ROUND,FROZEN   25 lbs   3 gal 3-5/8 qts

Method

1. Place 5 pounds potatoes on each sheet pan.
2. Using a convection oven, bake at 450 F. for 8 minutes on high fan, open vent or until golden brown. CCP: Hold at 140 F. or higher for service.

VEGETABLES No.Q 046 00

HASHED BROWN POTATOES

Yield 100   Portion 2/3 Cup

Calories	Carbohydrates	Protein	Fat	Cholesterol	Sodium	Calcium
175 cal	28 g	2 g	6 g	0 mg	242 mg	13 mg

Ingredient	Weight	Measure	Issue
POTATOES,FRESH,PEELED,CUBED	31 lbs	5 gal 2-1/2 qts	38-1/4 lbs
WATER,BOILING	20-7/8 lbs	2 gal 2 qts	
SALT	1/4 oz	1/8 tsp	
SHORTENING,VEGETABLE,MELTED	1-1/3 lbs	3 cup	
SALT	1-7/8 oz	3 tbsp	
PEPPER,BLACK,GROUND	1/8 oz	1/3 tsp	

Method

1. Cover potatoes with boiling salted water; bring to a boil; reduce heat; simmer 15 minutes or until tender. DO NOT OVERCOOK. Drain well.
2. Spread a layer of potatoes over well greased griddle at 400 F. Cook 10 minutes or until golden brown on one side.
3. Turn potatoes; cook 10 minutes or until golden brown.
4. Sprinkle with salt and pepper. CCP: Hold at 140 F. or higher for service.

VEGETABLES No.Q 046 01

COTTAGE FRIED POTATOES

Yield 100   **Portion** 2/3 Cup

Calories	Carbohydrates	Protein	Fat	Cholesterol	Sodium	Calcium
175 cal	28 g	2 g	6 g	0 mg	242 mg	13 mg

Ingredient	Weight	Measure	Issue
POTATOES,FRESH,PEELED,SLICED	31 lbs	5 gal 2-1/2 qts	
WATER,BOILING	20-7/8 lbs	2 gal 2 qts	
SALT	1/4 oz	1/8 tsp	
SHORTENING,VEGETABLE,MELTED	1-1/3 lbs	3 cup	
SALT	1-7/8 oz	3 tbsp	
PEPPER,BLACK,GROUND	1/8 oz	1/3 tsp	

Method

1. Cut potatoes in half lengthwise. Slice 1/4-inch thick. Cover potatoes with boiling salted water; bring to a boil; reduce heat; simmer 15 minutes or until tender. DO NOT OVERCOOK. Drain well.
2. Spread a layer of potatoes over well greased 400 F. griddle. Cook 10 minutes or until golden brown on one side.
3. Turn potatoes; cook 10 minutes or until golden brown.
4. Sprinkle with salt and pepper. CCP: Hold at 140 F. or higher for service.

VEGETABLES No.Q 046 02

HASHED BROWN POTATOES (FROZEN, SHREDDED, 3 OZ)

Yield 100   Portion 1/2 Cup

Calories	Carbohydrates	Protein	Fat	Cholesterol	Sodium	Calcium
121 cal	14 g	2 g	7 g	0 mg	227 mg	8 mg

Ingredient	Weight	Measure	Issue
POTATOES,WHITE,FROZEN,SHREDDED,HASHBROWN	18 lbs	2 gal 1-3/4 qts	
SHORTENING,VEGETABLE,MELTED	1-1/3 lbs	3 cup	
SALT	1-7/8 oz	3 tbsp	
PEPPER,BLACK,GROUND	1/8 oz	1/3 tsp	

Method

2 Place layer of potatoes on well greased 400 F. griddle; cook 15 minutes; turn; brown on other side.

3 Sprinkle with salt and pepper. CCP: Hold at 140 F. or higher for service.

VEGETABLES No.Q 046 03

LYONNAISE POTATOES

Yield 100　　　　　　　　　　　　　　　　**Portion** 2/3 Cup

Calories	Carbohydrates	Protein	Fat	Cholesterol	Sodium	Calcium
204 cal	35 g	3 g	6 g	0 mg	218 mg	15 mg

Ingredient　　　　　　　　　　　　　　Weight　　　Measure　　　Issue

Ingredient	Weight	Measure	Issue
POTATOES,FRESH,PEELED,SLICED	38 lbs	6 gal 3-5/8 qts	
ONIONS,FRESH,SLICED	1-1/2 lbs	1 qts 2 cup	1-2/3 lbs
COOKING SPRAY,NONSTICK	2 oz	1/4 cup 1/3 tbsp	
OIL,SALAD	1-1/4 lbs	2-1/2 cup	
SALT	1-7/8 oz	3 tbsp	
PEPPER,BLACK,GROUND	1/8 oz	1/3 tsp	

Method

1. Mix sliced potatoes with sliced onions. Lightly spray each steam table pan with non-stick cooking spray. Place mixture in sprayed pans.
2. Add salad oil, salt and pepper. Mix lightly.
3. Using a convection oven, bake in 350 F. for 1 hour 15 minutes on high fan, closed vent or until tender. CCP: Hold for service at 140 F. or higher.

VEGETABLES No.Q 046 04

HASHED BROWN POTATOES (FROZEN, SHREDDED, 2.5 OZ)

Yield 100 Portion 1 Patty

Calories	Carbohydrates	Protein	Fat	Cholesterol	Sodium	Calcium
60 cal	13 g	2 g	0 g	0 mg	16 mg	7 mg

Ingredient	Weight	Measure	Issue
POTATOES,WHITE,FROZEN,SHREDDED,HASHBROWN	16 lbs	2 gal 5/8 qts	

Method

1. Use frozen hashed brown potatoes. DO NOT THAW. Place patties on ungreased sheet pans. DO NOT allow patties to touch each other.
2. Using a convection oven, bake at 400 F. 15 to 17 minutes or until lightly browned on high fan, open vent. CCP: Hold at 140 F. or higher for service.

VEGETABLES No.Q 047 00

HOME FRIED POTATOES

Yield 100   Portion 2/3 Cup

Calories	Carbohydrates	Protein	Fat	Cholesterol	Sodium	Calcium
214 cal	32 g	3 g	9 g	0 mg	217 mg	13 mg

Ingredient	Weight	Measure	Issue
OIL,SALAD	1-7/8 lbs	1 qts	
POTATOES,FRESH,PEELED,SLICED	35 lbs	6 gal 1-1/2 qts	
SALT	1-7/8 oz	3 tbsp	
PEPPER,BLACK,GROUND	1/8 oz	1/3 tsp	

Method

1 Spread a layer of potatoes on well greased griddle.
2 Cook on 400 F. griddle for about 25 minutes, turning occasionally to ensure even browning.
3 Sprinkle with salt and pepper. CCP: Hold at 140 F. or higher for service.

VEGETABLES No.Q 048 00

MASHED POTATOES

Yield 100　　　　　　　　　　　　　　　　　　**Portion**　1/2 Cup

Calories	Carbohydrates	Protein	Fat	Cholesterol	Sodium	Calcium
105 cal	20 g	2 g	2 g	0 mg	172 mg	20 mg

Ingredient	Weight	Measure	Issue
POTATOES,FRESH,PEELED,CUBED	22 lbs	4 gal	27-1/8 lbs
WATER	12-1/2 lbs	1 gal 2 qts	
SALT	1-1/4 oz	2 tbsp	
MARGARINE,SOFTENED	8 oz	1 cup	
PEPPER,WHITE,GROUND	1/8 oz	1/4 tsp	
MILK,NONFAT,DRY	2-2/3 oz	1-1/8 cup	
WATER,WARM	3 lbs	1 qts 1-3/4 cup	

Method

1. Cover potatoes with salted water; bring to a boil; reduce heat; simmer 25 minutes or until tender. Drain well.
2. Beat potatoes in mixer bowl at low speed until broken into smaller pieces, about 1 minute.
3. Add butter or margarine and pepper. Beat at high speed 3 to 5 minutes or until smooth.
4. Reconstitute milk; heat to a simmer; blend into potatoes at low speed. Beat at high speed 2 minutes or until light and fluffy. CCP: Internal temperature must reach 145 F. or higher for 15 seconds. Hold at 140 F. or higher for service.

VEGETABLES No.Q 048 01

GRILLED POTATO PATTIES

Yield 100 **Portion** 2 Patties

Calories	Carbohydrates	Protein	Fat	Cholesterol	Sodium	Calcium
124 cal	23 g	3 g	2 g	18 mg	203 mg	26 mg

Ingredient	**Weight**	**Measure**	**Issue**
POTATOES,FRESH,PEELED,CUBED	22 lbs	4 gal	27-1/8 lbs
WATER	12-1/2 lbs	1 gal 2 qts	
SALT	1-1/4 oz	2 tbsp	
MARGARINE,SOFTENED	8 oz	1 cup	
PEPPER,WHITE,GROUND	1/8 oz	1/4 tsp	
MILK,NONFAT,DRY	2-3/8 oz	1 cup	
WATER,WARM	2-1/8 lbs	1 qts	
EGGS,WHOLE,FROZEN	14-1/4 oz	1-5/8 cup	
BREADCRUMBS,DRY,GROUND,FINE	1-1/8 lbs	1 qts 1/2 cup	

Method

1 Cover potatoes with salted water; bring to a boil; reduce heat; simmer 25 minutes or until tender. Drain well.
2 Beat potatoes in mixer bowl at high speed until broken into smaller pieces, about 1 minute.
3 Add butter or margarine and pepper. Beat at high speed 1 minute.
4 Reconstitute milk, heat to a simmer; blend into potatoes, blend in beaten eggs at low speed. Beat at high speed 1 minute.
5 Shape into 2 ounce patties. Dredge patties in bread crumbs. Shake off excess. Grill on lightly greased 350 F. griddle 3 minutes per side or until golden brown. CCP: Hold at 140 F. or higher for service.

VEGETABLES No.Q 049 00

O'BRIEN POTATOES

Yield 100   Portion 2/3 Cup

Calories	Carbohydrates	Protein	Fat	Cholesterol	Sodium	Calcium
175 cal	29 g	3 g	6 g	0 mg	194 mg	13 mg

Ingredient

Ingredient	Weight	Measure	Issue
PEPPERS,GREEN,FRESH,CHOPPED	3 lbs	2 qts 1 cup	3-5/8 lbs
PIMIENTO,CANNED,DRAINED,CHOPPED	12-2/3 oz	1-7/8 cup	
SHORTENING,VEGETABLE,MELTED	3-5/8 oz	1/2 cup	
POTATOES,FRESH,PEELED,CUBED	31 lbs	5 gal 2-1/2 qts	38-1/4 lbs
SALT	1-2/3 oz	2-2/3 tbsp	
PEPPER,BLACK,GROUND	1/8 oz	1/8 tsp	

Method

1 Saute peppers in shortening or salad oil 5 minutes or until tender. Add pimientos; saute until heated through.
2 Fry potatoes in 365 F. deep fat in 25-portion batches 7 minutes or until lightly browned and tender.
3 Drain well in basket or on absorbent paper.
4 Combine 2-1/3 cups of sauteed vegetables with each pan of potatoes.
5 Combine salt and pepper. Sprinkle 2 teaspoons salt-pepper mixture over each batch of potatoes. Stir lightly but thoroughly.
6 Using a convection oven, bake at 350 F. for 8 to 10 minutes until thoroughly heated on high fan, open vent. CCP: Internal temperature must reach 145 F. or higher for 15 seconds. Hold at 140 F. or higher for service.

VEGETABLES No.Q 050 00

OVEN BROWNED POTATOES

Yield 100   **Portion** 1/2 Cup

Calories	Carbohydrates	Protein	Fat	Cholesterol	Sodium	Calcium
126 cal	22 g	2 g	4 g	0 mg	234 mg	10 mg

Ingredient

Ingredient	Weight	Measure	Issue
POTATOES,FRESH,CHOPPED	23-7/8 lbs	4 gal 1-1/3 qts	29-1/2 lbs
MARGARINE,MELTED	1 lbs	2 cup	
SALT	1-2/3 oz	2-2/3 tbsp	
PEPPER,BLACK,GROUND	1/8 oz	1/3 tsp	
PAPRIKA,GROUND	1/4 oz	1 tbsp	

Method

1. Place 8 pounds or 5-3/4 quarts potatoes in each steam table pan.
2. Drizzle 2/3 cup butter or margarine over potatoes in each pan; stir gently to coat potatoes well.
3. Mix salt, pepper and paprika together. Sprinkle 1-1/2 tablespoon mixture over potatoes in each pan.
4. Using a convection oven, bake in 350 F. for 25 to 30 minutes on high fan, open vent or until browned and done. Turn potatoes once during cooking. CCP: Internal temperature must reach 145 F. or higher for 15 seconds. Hold at 140 F. or higher for service.

VEGETABLES No.Q 050 01

FRANCONIA POTATOES

Yield 100   Portion   1/2 Cup

Calories	Carbohydrates	Protein	Fat	Cholesterol	Sodium	Calcium
126 cal	22 g	2 g	4 g	10 mg	231 mg	12 mg

Ingredient	Weight	Measure	Issue
POTATOES,FRESH,CHOPPED	24 lbs	4 gal 1-1/2 qts	29-5/8 lbs
WATER	16-3/4 lbs	2 gal	
BUTTER	1 lbs	2 cup	
SALT	1-2/3 oz	2-2/3 tbsp	
PEPPER,BLACK,GROUND	1/8 oz	1/3 tsp	
PAPRIKA,GROUND	1/4 oz	1 tbsp	

Method

1. Partially cook potatoes in steam-jacketed kettle or stock pot 10 minutes. Drain. Place about 7 pounds 15 ounces partially cooked potatoes in each pan.
2. Drizzle 2/3 cup butter or margarine over potatoes in each steam table pan; stir gently to coat potatoes well.
3. Mix salt, pepper, and paprika together. Sprinkle 1-1/2 tablespoons mixture over potatoes in each pan.
4. Using a convection oven, bake at 400 F. for 15 minutes on high fan, closed vent until browned and done, turning once. CCP: Internal temperature must reach 145 F. or higher for 15 seconds. Hold at 140 F. or higher for service.

VEGETABLES No.Q 050 02

OVEN-GLO POTATOES

Yield 100   **Portion** 1/2 Cup

Calories	Carbohydrates	Protein	Fat	Cholesterol	Sodium	Calcium
130 cal	23 g	2 g	4 g	10 mg	268 mg	14 mg

Ingredient	**Weight**	**Measure**	**Issue**
POTATOES,FRESH,PEELED,CUBED	23-7/8 lbs	4 gal 1-1/3 qts	29-1/2 lbs
WATER	16-3/4 lbs	2 gal	
BUTTER,MELTED	1 lbs	2 cup	
SALT	1-2/3 oz	2-2/3 tbsp	
PEPPER,BLACK,GROUND	1/8 oz	1/3 tsp	
TOMATO PASTE,CANNED	1 lbs	1-3/4 cup	
WATER	4-1/8 lbs	2 qts	
GARLIC POWDER	1/4 oz	3/8 tsp	

Method

1. Partially cook potatoes in steam-jacketed kettle or stock pot 10 minutes or partially cook potatoes in 15 pounds PSI steam cooker 5 to 7 minutes or 5 pounds PSI steam cooker, 12 to 15 minutes. Drain. Use steam table pans. Place about 7 pounds 15 ounce partially cooked potatoes in each pan.
2. Thoroughly combine butter or margarine, salt, pepper, tomato paste, hot water and garlic powder; blend thoroughly.
3. Pour 2 pounds 2 ounce mixture over potatoes in each steam table pan.
4. Using a convection oven, bake at 400 F. for 15 minutes on high fan, closed vent. CCP: Internal temperature must reach 145 F. or higher for 15 seconds. Hold at 140 F. or higher for service.

VEGETABLES No.Q 050 03

OVEN-GLO POTATOES (CANNED)

Yield 100   Portion   3/4 Cup

Calories	Carbohydrates	Protein	Fat	Cholesterol	Sodium	Calcium
87 cal	12 g	2 g	4 g	0 mg	516 mg	49 mg

Ingredient	Weight	Measure	Issue
POTATOES, CANNED, WHOLE	34 lbs	2 gal 1-5/8 qts	
TOMATO PASTE, CANNED	1 lbs	1-3/4 cup	
MARGARINE	1 lbs	2 cup	
SALT	1-2/3 oz	2-2/3 tbsp	
PEPPER, BLACK, GROUND	1/8 oz	1/3 tsp	
WATER	4-1/8 lbs	2 qts	
GARLIC POWDER	1/4 oz	3/8 tsp	

Method

1. Drain potatoes. Place 1-1/3 gal potatoes in each pan.
2. Combine tomato paste, margarine or butter, salt, garlic powder and pepper. Add hot water; blend thoroughly.
3. Pour 1 qt mixture over potatoes in each pan.
4. Using a convection oven, bake at 400 F. 15 minutes or until browned on high fan, closed vent. CCP: Temperature must reach 145 F. or higher for 15 seconds. Hold for service at 140 F. or higher.

VEGETABLES No.Q 050 04

OVEN BROWNED POTATOES (CANNED)

Yield 100   **Portion** 3/4 Cup

Calories	Carbohydrates	Protein	Fat	Cholesterol	Sodium	Calcium
95 cal	12 g	2 g	5 g	0 mg	447 mg	47 mg

Ingredient	Weight	Measure	Issue
POTATOES, CANNED, WHOLE	34 lbs	2 gal 1-5/8 qts	
MARGARINE	1-1/3 lbs	2-5/8 cup	
SALT	1-1/4 oz	2 tbsp	
PAPRIKA, GROUND	1/3 oz	1 tbsp	
PEPPER, BLACK, GROUND	1/4 oz	3/8 tsp	

Method

1. Drain potatoes. Place 1-1/3 gal potatoes in each pan.
2. Drizzle 2/3 cup margarine or butter over potatoes in each pan; stir gently to coat potatoes.
3. Mix salt, paprika and pepper together. Sprinkle about 1 tbsp over potatoes in each pan.
4. Using a convection oven, bake at 350 F. 25-30 minutes or until browned on high fan, open vent. CCP: Temperature must reach 145 F. or higher for 15 seconds. Hold for service at 140 F. or higher.

VEGETABLES No.Q 051 00

POTATOES AU GRATIN

Yield 100 **Portion** 2/3 Cup

Calories	Carbohydrates	Protein	Fat	Cholesterol	Sodium	Calcium
228 cal	30 g	6 g	10 g	28 mg	444 mg	103 mg

Ingredient	Weight	Measure	Issue
POTATOES,FRESH,PEELED,SLICED	25-1/2 lbs	4 gal 2-1/2 qts	
WATER,BOILING	18-3/4 lbs	2 gal 1 qts	
SALT	1-1/4 oz	2 tbsp	
BUTTER,MELTED	1-1/2 lbs	3 cup	
FLOUR,WHEAT,GENERAL PURPOSE	13-1/4 oz	3 cup	
MILK,NONFAT,DRY	9-5/8 oz	1 qts	
WATER,WARM	11 lbs	1 gal 1-1/4 qts	
SALT	1-1/4 oz	2 tbsp	
PEPPER,WHITE,GROUND	1/8 oz	1/3 tsp	
CHEESE,CHEDDAR,SHREDDED	1-1/2 lbs	1 qts 2 cup	
MUSTARD,DRY	1/2 oz	1 tbsp	
BREADCRUMBS,DRY,GROUND,FINE	1 lbs	1 qts	
BUTTER,MELTED	8 oz	1 cup	

Method

1. Cover potatoes with salted water; bring to a boil; cook 10 minutes or until tender.
2. Drain well. Place about 8 pounds or 1-1/2 gallon potatoes in each steam table pan. Set aside for use in Step 6.
3. Melt butter. Blend butter and flour together using wire whip; stir until smooth.
4. Reconstitute milk; bring to just below boiling. DO NOT BOIL. Add milk to flour mixture stirring constantly. Add salt and pepper. Simmer 10 to 15 minutes or until thickened. Stir as necessary.
5. Add cheese and mustard to sauce. Stir until cheese is melted.
6. Pour 2-1/3 quarts sauce evenly over potatoes in each pan.
7. Mix crumbs and butter or margarine. Sprinkle 1-1/3 cups crumbs over potatoes in each pan.
8. Using a convection oven, bake in 325 F. for 30 minutes on low fan, open vent or until browned. CCP: Hold for service at 140 F. or higher.

VEGETABLES No.Q 051 01

POTATOES AU GRATIN (DEHYDRATED, SLICES)

Yield 100   **Portion** 2/3 Cup

Calories	Carbohydrates	Protein	Fat	Cholesterol	Sodium	Calcium
152 cal	12 g	4 g	10 g	28 mg	517 mg	107 mg

Ingredient	**Weight**	**Measure**	**Issue**
POTATO,WHITE,DEHYDRATED,SLICED	5-1/2 lbs		
WATER	37-5/8 lbs	4 gal 2 qts	
SALT	1-7/8 oz	3 tbsp	
BUTTER,MELTED	1-1/2 lbs	3 cup	
FLOUR,WHEAT,GENERAL PURPOSE	13-1/4 oz	3 cup	
MILK,NONFAT,DRY	12-1/4 oz	1 qts 1-1/8 cup	
WATER,WARM	13-5/8 lbs	1 gal 2-1/2 qts	
SALT	1-1/4 oz	2 tbsp	
PEPPER,BLACK,GROUND	1/8 oz	1/3 tsp	
CHEESE,CHEDDAR,GRATED	1-1/2 lbs	1 qts 2 cup	
MUSTARD,DRY	1/2 oz	1 tbsp	
BREADCRUMBS	1 lbs	1 qts	
BUTTER,MELTED	8 oz	1 cup	

Method

1. Bring water to a boil; add salt; pour over potatoes. Cover; bring to a boil; simmer until tender.
2. Drain well; place about 6 pounds 8 ounces or 4-1/2 quarts cooked, drained potatoes in each steam table pan. Set aside for use in Step 6.
3. Melt butter. Blend butter and flour together using wire whip; stir until smooth.
4. Reconstitute milk; bring to just below boiling. DO NOT BOIL. Add milk to flour mixture stirring constantly. Add salt and pepper. Simmer 10 to 15 minutes or until thickened. Stir as necessary.
5. Add cheese and mustard to sauce. Stir until cheese is melted.
6. Pour 2-3/4 quarts sauce over potatoes in each pan.
7. Mix crumbs and butter or margarine. Sprinkle 1-1/3 cups crumbs over potatoes in each pan.
8. Using a convection oven, bake in 325 F. for 30 minutes or until browned on low fan, open vent. CCP: Hold for service at 140 F. or higher.

VEGETABLES No.Q 052 00

RISSOLE POTATOES

Yield 100   Portion 2/3 Cup

Calories	Carbohydrates	Protein	Fat	Cholesterol	Sodium	Calcium
217 cal	32 g	3 g	9 g	0 mg	8 mg	13 mg

Ingredient   Weight   Measure   Issue

POTATOES,FRESH,PEELED,SLICED   35 lbs   6 gal 1-1/2 qts

Method

1. Cook potatoes in steamer 5 to 7 minutes at 15 PSI or 12 to 15 minutes at 5 PSI. Drain.
2. Fry in deep fat until golden brown in 360 F. deep fat.
3. Drain well in basket or on absorbent paper.
4. CCP: Hold for service at 140 F. or higher.

VEGETABLES No.Q 053 00

SCALLOPED POTATOES

Yield 100 **Portion** 2/3 Cup

Calories	Carbohydrates	Protein	Fat	Cholesterol	Sodium	Calcium
152 cal	28 g	4 g	3 g	1 mg	339 mg	64 mg

Ingredient	Weight	Measure	Issue
POTATOES,FRESH,PEELED,SLICED	25-1/2 lbs	4 gal 2-1/2 qts	
WATER,BOILING	18-3/4 lbs	2 gal 1 qts	
SALT	1-1/4 oz	2 tbsp	
COOKING SPRAY,NONSTICK	2 oz	1/4 cup 1/3 tbsp	
MARGARINE,MELTED	10 oz	1-1/4 cup	
FLOUR,WHEAT,GENERAL PURPOSE	13-1/4 oz	3 cup	
MILK,NONFAT,DRY	14-3/8 oz	1 qts 2 cup	
WATER,WARM	15-2/3 lbs	1 gal 3-1/2 qts	
SALT	1-1/4 oz	2 tbsp	
PEPPER,WHITE,GROUND	1/8 oz	1/3 tsp	

Method

1. Cover potatoes with salted water; bring to a boil; cook 10 minutes or until tender.
2. Drain well. Lightly spray each steam table pan with non-stick cooking spray. Place about 8 pounds potatoes in each sprayed pan.
3. Blend butter or margarine and flour together using a wire whip. Stir until smooth.
4. Reconstitute milk; bring to just below boiling. DO NOT BOIL. Add milk to roux stirring constantly. Add salt and pepper. Simmer 10 to 15 minutes or until thickened. Stir as necessary.
5. Pour 2-3/4 quarts sauce over potatoes in each pan.
6. Using a convection oven, bake at 325 F. for 30 minutes on low fan, open vent or until browned. CCP: Hold at 140 F. or higher for service.

VEGETABLES No.Q 053 01

SCALLOPED POTATOES AND ONIONS

Yield 100   Portion   2/3 Cup

Calories	Carbohydrates	Protein	Fat	Cholesterol	Sodium	Calcium
155 cal	29 g	4 g	3 g	1 mg	336 mg	60 mg

Ingredient	Weight	Measure	Issue
POTATOES,FRESH,PEELED,SLICED	25-1/2 lbs	4 gal 2-1/2 qts	
WATER,BOILING	18-3/4 lbs	2 gal 1 qts	
SALT	1-1/4 oz	2 tbsp	
ONIONS,FRESH,SLICED	2-1/2 lbs	2 qts 2 cup	2-7/8 lbs
COOKING SPRAY,NONSTICK	2 oz	1/4 cup 1/3 tbsp	
MARGARINE,MELTED	10 oz	1-1/4 cup	
FLOUR,WHEAT,GENERAL PURPOSE	13-1/4 oz	3 cup	
MILK,NONFAT,DRY	12-5/8 oz	1 qts 1-1/4 cup	
WATER,WARM	13-5/8 lbs	1 gal 2-1/2 qts	
SALT	1-1/4 oz	2 tbsp	
PEPPER,WHITE,GROUND	1/8 oz	1/3 tsp	

Method

1. Cover potatoes with salted water; bring to a boil; cook 10 minutes or until tender.
2. Drain well. Lightly spray each steam table pan with non-stick cooking spray. Place onions in layers with potatoes. Place about 8 pounds potatoes in each sprayed pan.
3. Blend butter or margarine and flour together using a wire whip. Stir until smooth.
4. Reconstitute milk; bring to just below boiling. DO NOT BOIL. Add milk to roux stirring constantly. Add salt and pepper. Simmer 10 to 15 minutes or until thickened. Stir as necessary.
5. Pour 2-1/2 quarts sauce over potatoes in each pan.
6. Using a convection oven, bake at 325 F. for 30 minutes on low fan, open vent or until browned. CCP: Hold at 140 F. or higher for service.

VEGETABLES No.Q 054 00

HASHED BROWN POTATOES (DEHYDRATED, SLICED)

Yield 100   Portion   2/3 Cup

Calories	Carbohydrates	Protein	Fat	Cholesterol	Sodium	Calcium
77 cal	7 g	1 g	5 g	0 mg	288 mg	8 mg

Ingredient	Weight	Measure	Issue
POTATO,WHITE,DEHYDRATED,SLICED	8 lbs		
WATER,BOILING	50-1/8 lbs	6 gal	
SALT	1-7/8 oz	3 tbsp	
SHORTENING,VEGETABLE,MELTED	1-1/8 lbs	2-1/2 cup	
SALT	5/8 oz	1 tbsp	
PEPPER,BLACK,GROUND	1/8 oz	1/3 tsp	

Method

1. Add potatoes to boiling salted water. Cover. Bring quickly to a boil; reduce heat; simmer 15 to 20 minutes or until tender. Drain well.
2. Spread potatoes on greased 375 F. griddle. Sprinkle with mixture of salt and pepper. Cook 10 minutes or until golden brown. Turn potatoes; continue to cook 10 minutes or until golden brown. Proceed with remaining potatoes. CCP: Hold at 140 F. or higher for service.

VEGETABLES No.Q 054 01

LYONNAISE POTATOES (DEHYDRATED)

Yield 100   Portion   2/3 Cup

Calories	Carbohydrates	Protein	Fat	Cholesterol	Sodium	Calcium
82 cal	8 g	1 g	5 g	0 mg	288 mg	12 mg

Ingredient	Weight	Measure	Issue
ONIONS,DEHYDRATED,CHOPPED	5-1/4 oz	2-5/8 cup	
POTATO,WHITE,DEHYDRATED,SLICED	8 lbs		
WATER,BOILING	50-1/8 lbs	6 gal	
SALT	1-7/8 oz	3 tbsp	
SHORTENING,VEGETABLE,MELTED	1-1/8 lbs	2-1/2 cup	
SALT	5/8 oz	1 tbsp	
PEPPER,BLACK,GROUND	1/8 oz	1/3 tsp	

Method

1. Add potatoes to boiling salted water. Cover. Bring quickly to a boil; reduce heat; simmer 15 to 20 minutes or until tender. Drain well. Rehydrate onions. Add to cooked potatoes.
2. Spread potatoes on greased 375 F. griddle. Sprinkle with mixture of salt and pepper. Cook 10 minutes or until golden brown. Turn potatoes; continue to cook 10 minutes or until golden brown. Proceed with remaining potatoes. CCP: Hold at 140 F. or higher for service.

VEGETABLES No.Q 054 02

O'BRIEN POTATOES (DEHYDRATED, SLICED)

Yield 100 **Portion** 2/3 Cup

Calories	Carbohydrates	Protein	Fat	Cholesterol	Sodium	Calcium
78 cal	8 g	1 g	5 g	0 mg	289 mg	8 mg

Ingredient
Ingredient	Weight	Measure	Issue
POTATO,WHITE,DEHYDRATED,SLICED	8 lbs		
PEPPERS,SWEET,DICED,DEHYDRATED	1/3 oz	1-1/2 cup	
WATER,BOILING	50-1/8 lbs	6 gal	
SALT	1-7/8 oz	3 tbsp	
PIMIENTO,CANNED,DRAINED,CHOPPED	1-1/4 lbs	3 cup	
SHORTENING,VEGETABLE,MELTED	1-1/8 lbs	2-1/2 cup	
SALT	5/8 oz	1 tbsp	
PEPPER,BLACK,GROUND	1/8 oz	1/3 tsp	

Method

1 Add potatoes to boiling salted water. Cover. Bring quickly to a boil; reduce heat; simmer 15 to 20 minutes or until tender. Drain well. Rehydrate green peppers. Add peppers and pimientos to cooked potatoes.

2 Spread potatoes on greased 375 F. griddle. Sprinkle with mixture of salt and pepper. Cook 10 minutes or until golden brown. Turn potatoes; continue to cook 10 minutes or until golden brown. Proceed with remaining potatoes. CCP: Hold at 140 F. or higher for service.

VEGETABLES No.Q 054 03

HASHED BROWN POTATOES (DEHYDRATED, SHREDDED)

Yield 100　　　　　　　　　　　　　　　　　　　　**Portion**　1/2 Cup

Calories	Carbohydrates	Protein	Fat	Cholesterol	Sodium	Calcium
58 cal	5 g	0 g	4 g	0 mg	354 mg	5 mg

Ingredient   Weight   Measure   Issue

POTATO,WHITE,DEHYDRATED,SHREDDED　　　　5-5/8 lbs
WATER　　　　　　　　　　　　　　　　　　　　31-1/3 lbs　　　　3 gal 3 qts
SALT　　　　　　　　　　　　　　　　　　　　　2-1/2 oz　　　　　1/4 cup 1/3 tbsp
SHORTENING,VEGETABLE,MELTED　　　　　　　14-1/2 oz　　　　　2 cup
SALT　　　　　　　　　　　　　　　　　　　　　5/8 oz　　　　　　1 tbsp
PEPPER,BLACK,GROUND　　　　　　　　　　　　1/8 oz　　　　　　1/3 tsp

Method

1. Use dehydrated hash brown potatoes. Add hot water and salt. Stir. Let stand 20 minutes; drain.
2. Spread 1/3 layer of potatoes on greased 375 F. griddle. Sprinkle with mixture of salt and pepper. Cook 2 to 3 minutes on each side. CCP: Hold at 140 F. or higher for service.

VEGETABLES No.Q 054 04

HASHED BROWN POTATOES (DEHYDRATED, DICED)

Yield 100 **Portion** 2/3 Cup

Calories	Carbohydrates	Protein	Fat	Cholesterol	Sodium	Calcium
75 cal	7 g	1 g	5 g	0 mg	288 mg	8 mg

Ingredient	**Weight**	**Measure**	**Issue**
POTATO,WHITE,DEHYDRATED,DICED	7-1/2 lbs		
WATER,BOILING	50-1/8 lbs	6 gal	
SALT	1-7/8 oz	3 tbsp	
SHORTENING,VEGETABLE,MELTED	1-1/8 lbs	2-1/2 cup	
SALT	5/8 oz	1 tbsp	
PEPPER,BLACK,GROUND	1/8 oz	1/3 tsp	

Method

1 Add dehydrated diced potatoes to boiling salted water. Cover. Bring quickly to a boil. Reduce heat and simmer for 15 minutes.
2 Spread 1/3 layer of potatoes on greased 375 F. griddle. Sprinkle with mixture of salt and pepper. Cook 10 minutes or until golden brown. Turn potatoes; continue to cook 10 minutes or until golden brown. Proceed with remaining layers. CCP: Hold at 140 F. or higher for service.

VEGETABLES No.Q 055 00

SCALLOPED POTATOES AND ONIONS (DEHYDRATED, SLICED)

Yield 100   Portion   2/3 Cup

Calories	Carbohydrates	Protein	Fat	Cholesterol	Sodium	Calcium
95 cal	11 g	2 g	5 g	12 mg	286 mg	71 mg

Ingredient	Weight	Measure	Issue
POTATO,WHITE,DEHYDRATED,SLICED	5 lbs		
ONIONS,DEHYDRATED,CHOPPED	7-7/8 oz	1 qts	
WATER,BOILING	41-3/4 lbs	5 gal	
SALT	1-7/8 oz	3 tbsp	
COOKING SPRAY,NONSTICK	2 oz	1/4 cup 1/3 tbsp	
BUTTER,MELTED	1-1/8 lbs	2-1/4 cup	
FLOUR,WHEAT,GENERAL PURPOSE	9-7/8 oz	2-1/4 cup	
MILK,NONFAT,DRY	1 lbs	1 qts 2-5/8 cup	
WATER,WARM	17-3/4 lbs	2 gal 1/2 qts	
PEPPER,WHITE,GROUND	1/8 oz	1/3 tsp	
PAPRIKA,GROUND	1/4 oz	1 tbsp	

Method

1. Add potatoes and onions to boiling salted water. Cover. Bring quickly to a boil; reduce heat; simmer 15 to 25 minutes or until tender. Drain well.
2. Lightly spray each steam table pan with non-stick cooking spray. Place about 7 pounds or 4-3/4 quarts potato mixture into each sprayed pan.
3. Blend butter or margarine and flour together; stir until smooth using a wire whip.
4. Reconstitute milk. Heat to just below boiling. DO NOT BOIL. Add milk to roux stirring constantly. Add pepper. Simmer 10 to 15 minutes or until thickened. Stir as necessary.
5. Pour 3 quarts sauce over mixture in each pan.
6. Sprinkle 1 teaspoon paprika over mixture in each pan.
7. Using a convection oven, bake at 325 F. for 30 minutes on open vent, low fan or until lightly brown. CCP: Hold at 140 F. or higher for service.

VEGETABLES No.Q 055 01

SCALLOPED POTATOES (DEHYDRATED, SLICED)

Yield 100   **Portion** 2/3 Cup

Calories	Carbohydrates	Protein	Fat	Cholesterol	Sodium	Calcium
89 cal	10 g	2 g	5 g	12 mg	285 mg	65 mg

Ingredient   Weight   Measure   Issue

Ingredient	Weight	Measure	Issue
POTATO,WHITE,DEHYDRATED,SLICED	5-1/2 lbs		
WATER,BOILING	37-5/8 lbs	4 gal 2 qts	
SALT	1-7/8 oz	3 tbsp	
COOKING SPRAY,NONSTICK	2 oz	1/4 cup 1/3 tbsp	
BUTTER,MELTED	1-1/8 lbs	2-1/4 cup	
FLOUR,WHEAT,GENERAL PURPOSE	9-7/8 oz	2-1/4 cup	
MILK,NONFAT,DRY	1 lbs	1 qts 2-5/8 cup	
WATER,WARM	17-3/4 lbs	2 gal 1/2 qts	
PEPPER,WHITE,GROUND	1/8 oz	1/3 tsp	
PAPRIKA,GROUND	1/4 oz	1 tbsp	

Method

1. Add potatoes to boiling salted water. Cover. Bring quickly to a boil; reduce heat; simmer 15 to 25 minutes or until tender. Drain well.
2. Lightly spray each steam table pan with non-stick cooking spray. Place about 6 pounds 8 ounces or 4-1/2 quarts cooked, drained potatoes into each sprayed pan. Set aside for use in Step 5.
3. Blend butter or margarine and flour together; stir until smooth using a wire whip.
4. Reconstitute milk. Heat to just below boiling. DO NOT BOIL. Add milk to roux stirring constantly. Add pepper. Simmer 10 to 15 minutes or until thickened. Stir as necessary.
5. Pour 3 quarts sauce over mixture in each pan.
6. Sprinkle 1 teaspoon paprika over mixture in each pan.
7. Using a convection oven, bake 30 minutes or until lightly browned in 325 F. oven on open vent, low fan. CCP: Hold at 140 F. or higher for service.

VEGETABLES No.Q 055 02

SCALLOPED POTATOES (DEHYDRATED, DICED)

Yield 100 Portion 2/3 Cup

Calories	Carbohydrates	Protein	Fat	Cholesterol	Sodium	Calcium
88 cal	9 g	2 g	5 g	12 mg	285 mg	64 mg

Ingredient	Weight	Measure	Issue
POTATO,WHITE,DEHYDRATED,DICED	5 lbs		
WATER,BOILING	33-1/2 lbs	4 gal	
SALT	1-7/8 oz	3 tbsp	
COOKING SPRAY,NONSTICK	2 oz	1/4 cup 1/3 tbsp	
BUTTER,MELTED	1-1/8 lbs	2-1/4 cup	
FLOUR,WHEAT,GENERAL PURPOSE	9-7/8 oz	2-1/4 cup	
MILK,NONFAT,DRY	1 lbs	1 qts 2-5/8 cup	
WATER,WARM	17-3/4 lbs	2 gal 1/2 qts	
PEPPER,WHITE,GROUND	1/8 oz	1/3 tsp	
PAPRIKA,GROUND	1/4 oz	1 tbsp	

Method

1. Add potatoes to boiling salted water. Cover. Bring quickly to a boil; reduce heat; simmer 15 to 25 minutes or until tender. Drain well.
2. Lightly spray each steam table pan with non-stick cooking spray. Place about 7 pounds or 4-3/4 quarts cooked, drained potatoes into each sprayed pan. Set aside for use in Step 5.
3. Blend butter or margarine and flour together; stir until smooth using a wire whip.
4. Reconstitute milk. Heat to just below boiling. DO NOT BOIL. Add milk to roux stirring constantly. Add pepper. Simmer 10 to 15 minutes or until thickened. Stir as necessary.
5. Pour 3 quarts sauce over mixture in each pan.
6. Sprinkle 1 teaspoon paprika over mixture in each pan.
7. Using a convection oven, bake at 325 F. for 30 minutes on open vent, low fan or until lightly brown. CCP: Hold at 140 F. or higher for service.

VEGETABLES No.Q 056 00

GOLDEN POTATO BALLS (INSTANT)

Yield 100   Portion 3 Each

Calories	Carbohydrates	Protein	Fat	Cholesterol	Sodium	Calcium
123 cal	9 g	2 g	9 g	0 mg	226 mg	25 mg

Ingredient	Weight	Measure	Issue
ONIONS,DEHYDRATED,CHOPPED	3 oz	1-1/2 cup	
WATER,WARM	2-1/8 lbs	1 qts	
POTATO,WHITE,INSTANT,GRANULES	1 lbs	2 qts 2 cup	
MILK,NONFAT,DRY	5-3/8 oz	2-1/4 cup	
FLOUR,WHEAT,GENERAL PURPOSE	13-1/4 oz	3 cup	
SALT	1-7/8 oz	3 tbsp	
NUTMEG,GROUND	<1/16th oz	1/8 tsp	
THYME,GROUND	<1/16th oz	<1/16th tsp	
WATER,BOILING	13 lbs	1 gal 2-1/4 qts	
POTATO,WHITE,INSTANT,GRANULES	3-3/8 oz	2 cup	

Method

1. Rehydrate onions in water for 15 minutes. Drain; set aside for use in Step 5.
2. Combine potatoes, milk, flour, salt, nutmeg, and thyme; mix well.
3. Pour water into mixer bowl.
4. At low speed, rapidly add dry ingredients. Mix 1 minute or until well blended.
5. Add onions. Mix until well blended.
6. Shape mixture into balls, about 1 ounce each. Roll into potato granules.
7. Fry 3 minutes or until golden brown in 375 F. deep fat fryer.
8. Drain well in basket or on absorbent paper. CCP: Hold at 140 F. or higher for service.

VEGETABLES No.Q 057 00

MASHED POTATOES (INSTANT)

Yield 100   Portion   1/2 Cup

Calories	Carbohydrates	Protein	Fat	Cholesterol	Sodium	Calcium
150 cal	30 g	4 g	2 g	5 mg	185 mg	31 mg

Ingredient	Weight	Measure	Issue
POTATO,WHITE,INSTANT,GRANULES	4-3/4 lbs	4 gal 2-7/8 qts	
MILK,NONFAT,DRY	5-3/8 oz	2-1/4 cup	
WATER,BOILING	20-7/8 lbs	2 gal 2 qts	
BUTTER	8 oz	1 cup	
SALT	1 oz	1 tbsp	
PEPPER,WHITE,GROUND	1/8 oz	1/4 tsp	

Method

1. Blend potatoes and milk together.
2. Blend water, butter or margarine, salt and pepper in mixer bowl.
3. At low speed, using wire whip, rapidly add potato and milk mixture to liquid; mix 1/2 minute. Stop mixer; scrape down sides and bottom of bowl.
4. Whip at high speed about 2 minutes or until light and fluffy. DO NOT OVERWHIP. CCP: Hold at 140 F. or higher for service.

VEGETABLES No.Q 057 01

GRILLED POTATO CAKES

Yield 100 **Portion** 1 Cake

Calories	Carbohydrates	Protein	Fat	Cholesterol	Sodium	Calcium
79 cal	10 g	2 g	3 g	31 mg	193 mg	29 mg

Ingredient | Weight | Measure | Issue

Ingredient	Weight	Measure
POTATO,WHITE,INSTANT,GRANULES	1-3/8 lbs	3 qts 1 cup
MILK,NONFAT,DRY	6 oz	2-1/2 cup
WATER,BOILING	16-3/4 lbs	2 gal
BUTTER	12 oz	1-1/2 cup
SALT	1-1/4 oz	2 tbsp
PEPPER,WHITE,GROUND	1/8 oz	1/4 tsp
FLOUR,WHEAT,GENERAL PURPOSE	1-1/8 lbs	1 qts
EGGS,WHOLE,FROZEN	1-1/4 lbs	2-1/4 cup

Method

1. Blend potatoes and milk together.
2. Blend water, butter or margarine, salt and pepper in mixer bowl.
3. At low speed, using wire whip, rapidly add potato and milk mixture to liquid; mix 1/2 minute. Stop mixer; scrape down sides and bottom of bowl.
4. Whip at high speed about 1 minute or until light and fluffy. At low speed, blend slightly beaten whole eggs into potatoes 1 minute. Whip at medium speed 1/2 minute. DO NOT OVERWHIP. Chill mixture.
5. Shape into 4 ounce cakes.
6. Dredge cakes in sifted general purpose flour.
7. Grill on well-greased 375 F. griddle about 3-1/2 to 4 minutes per side or until golden brown. CCP: Hold at 140 F. or higher for service.

VEGETABLES   No.Q 058 00

STEWED TOMATOES

Yield 100   Portion   1/2 Cup

Calories	Carbohydrates	Protein	Fat	Cholesterol	Sodium	Calcium
23 cal	5 g	1 g	0 g	0 mg	172 mg	35 mg

Ingredient	Weight	Measure	Issue
TOMATOES,CANNED,WHOLE,PEELED,INCL LIQUIDS	25-3/8 lbs	3 gal	
ONIONS,FRESH,CHOPPED	4 oz	1/2 cup 3-1/3 tbsp	4-1/2 oz
PEPPERS,GREEN,FRESH,CHOPPED	2-1/2 oz	1/4 cup	3 oz
CELERY,FRESH,CHOPPED	4 oz	3/4 cup 3 tbsp	5-1/2 oz
PEPPER,BLACK,GROUND	1/8 oz	1/8 tsp	

Method

1 Combine tomatoes, onions, peppers, celery, and pepper. Mix well.
2 Bring to a boil to blend flavors. CCP: Hold for service at 140 F. or higher.

VEGETABLES No.Q 058 01

STEWED TOMATOES WITH CROUTONS

Yield 100   Portion   1/2 Cup

Calories	Carbohydrates	Protein	Fat	Cholesterol	Sodium	Calcium
57 cal	9 g	2 g	2 g	5 mg	227 mg	43 mg

Ingredient	Weight	Measure	Issue
TOMATOES,CANNED,WHOLE,PEELED,INCL LIQUIDS	25-3/8 lbs	3 gal	
ONIONS,FRESH,CHOPPED	4 oz	1/2 cup 3-1/3 tbsp	4-1/2 oz
PEPPERS,GREEN,FRESH,CHOPPED	2-1/2 oz	1/4 cup	3 oz
CELERY,FRESH,CHOPPED	4 oz	3/4 cup 3 tbsp	5-1/2 oz
PEPPER,BLACK,GROUND	1/8 oz	1/8 tsp	
CROUTONS		8 unit	

Method

1 Combine tomatoes, onions, peppers, celery, and pepper. Mix well.
2 Bring to a boil to blend flavors.
3 Serve with croutons. CCP: Hold for service at 145 F. or higher.

VEGETABLES No.Q 059 00

GERMAN SAUERKRAUT

Yield 100   Portion   1/2 Cup

Calories	Carbohydrates	Protein	Fat	Cholesterol	Sodium	Calcium
36 cal	7 g	1 g	1 g	1 mg	583 mg	31 mg

Ingredient	Weight	Measure	Issue
SAUERKRAUT,SHREDDED,CANNED,INCL LIQUIDS	18-3/4 lbs	2 gal 1 qts	
BACON,RAW	1 lbs		
ONIONS,FRESH,CHOPPED	3 lbs	2 qts 1/2 cup	3-1/3 lbs
APPLES,FRESH,MEDIUM,PEELED,CORED,CHOPPED	1 lbs	3-5/8 cup	1-1/4 lbs
CARAWAY SEED	3/4 oz	3 tbsp	
SUGAR,BROWN,PACKED	3-7/8 oz	3/4 cup	

Method

1 Combine sauerkraut, bacon, onions, apples, caraway seed, and brown sugar; cook 1-1/2 hours, stirring occasionally. CCP: Hold at 140 F. or higher for service.

VEGETABLES No.Q 060 00

CLUB SPINACH

Yield 100   **Portion** 1/2 Cup

Calories	Carbohydrates	Protein	Fat	Cholesterol	Sodium	Calcium
163 cal	14 g	9 g	8 g	19 mg	682 mg	265 mg

Ingredient	**Weight**	**Measure**	**Issue**
SPINACH,CANNED,INCL LIQUIDS	37-1/8 lbs	4 gal 2 qts	
CHEESE,CHEDDAR,SHREDDED	3-3/4 lbs	3 qts 3 cup	
CRACKER CRUMBS	2-1/2 lbs	2 qts 1-3/4 cup	
MARGARINE,MELTED	6 oz	3/4 cup	
BACON,SLICED,RAW	1 lbs		

Method

1. Drain spinach; chop coarsely; place about 7-1/2 pounds or 3-3/4 quarts in each steam table pan.
2. Cover spinach in each pan with 1 pounds 4 ounces or 1-1/4 quarts cheese.
3. Combine crumbs and butter or margarine; sprinkle 3 cups crumbs over cheese in pan.
4. Cook bacon according to Recipe No. L 002 00 or L 002 02. Drain fat. Finely chop bacon. Sprinkle 1/3 cup bacon over mixture in each pan.
5. Using a convection oven, bake at 325 F. for 30 minutes on low fan, open vent or until thoroughly heated. CCP: Hold at 140 F. or higher for service.

Notes

1. In Step 3, DO NOT substitute bread crumbs for cracker crumbs.

VEGETABLES No.Q 061 00

BAKED HUBBARD SQUASH

Yield 100   Portion 3-1/2 Ounces

Calories	Carbohydrates	Protein	Fat	Cholesterol	Sodium	Calcium
97 cal	14 g	3 g	4 g	10 mg	118 mg	23 mg

Ingredient   **Weight**   **Measure**   **Issue**

SQUASH,HUBBARD,FRESH   29 lbs   7 gal 3/8 qts   32-5/8 lbs
WATER,WARM   3-7/8 lbs   1 qts 3-1/2 cup
BUTTER,MELTED   1 lbs   2 cup
WATER   8-1/3 oz   1 cup
SUGAR,BROWN,PACKED   10-7/8 oz   2-1/8 cup
CINNAMON,GROUND   1/4 oz   1 tbsp
SALT   5/8 oz   1 tbsp

Method

1 Cut squash in half; remove seeds. Cut into 4-1/2 ounce pieces.
2 Place squash cut side up in steam table pans.
3 Add 1-1/2 cups water to each pan. Cover pans.
4 Using a convection oven, bake at 350 F. 1 hour on high fan, closed vent or until tender.
5 Combine butter or margarine, water, cinnamon, brown sugar and salt; mix well. Simmer about 5 minutes or until heated thoroughly in steam-jacketed kettle or stock pot.
6 Pour brown sugar sauce over squash in each pan. CCP: Hold at 140 F. or higher for service.

VEGETABLES No.Q 062 00

CREOLE SUMMER SQUASH

Yield 100   Portion   2/3 Cup

Calories	Carbohydrates	Protein	Fat	Cholesterol	Sodium	Calcium
44 cal	8 g	2 g	1 g	0 mg	252 mg	33 mg

Ingredient	Weight	Measure	Issue
ONIONS,FRESH,CHOPPED	3-1/8 lbs	2 qts 1 cup	3-1/2 lbs
OIL,SALAD	2-7/8 oz	1/4 cup 2-1/3 tbsp	
SQUASH,FRESH,SUMMER	20 lbs		21 lbs
WATER,BOILING	1 lbs	2 cup	
TOMATOES,CANNED,CRUSHED,INCL LIQUIDS	6-5/8 lbs	3 qts	
SALT	1-7/8 oz	3 tbsp	
SUGAR,GRANULATED	2-2/3 oz	1/4 cup 2-1/3 tbsp	
PEPPER,BLACK,GROUND	1/4 oz	1 tbsp	
GARLIC POWDER	1/8 oz	1/8 tsp	
PARSLEY,FRESH,BUNCH,CHOPPED	2 oz	3/4 cup 3 tbsp	2-1/8 oz

Method

1. Saute onions in salad oil until tender.
2. Combine sauteed onions, squash and water.
3. Bring to a boil. Cover; reduce heat. Simmer 10 minutes.
4. Add tomatoes, salt, sugar, pepper, garlic and parsley.
5. Bring to a boil. Reduce heat; simmer 5 minutes. CCP: Hold at 140 F. or higher for service.

VEGETABLES No. Q 063 00

TANGY SPINACH

Yield 100 **Portion** 1/2 Cup

Calories	Carbohydrates	Protein	Fat	Cholesterol	Sodium	Calcium
31 cal	5 g	3 g	1 g	0 mg	187 mg	122 mg

Ingredient	Weight	Measure	Issue
SPINACH,FROZEN	18 lbs	2 gal 2-3/4 qts	
OIL,SALAD	1-1/2 oz	3 tbsp	
ONIONS,FRESH,CHOPPED	1-3/4 lbs	1 qts 1 cup	2 lbs
VINEGAR,DISTILLED	1-1/8 lbs	2-1/4 cup	
SALT	1 oz	1 tbsp	
PEPPER,BLACK,GROUND	1/2 oz	2 tbsp	

Method

1. Cook spinach for 4 to 6 minutes. Drain.
2. Saute onions in oil until tender.
3. Stir in vinegar, salt and pepper; simmer 3 minutes.
4. Pour vinegar-onion mixture over spinach. CCP: Hold at 140 F. or higher for service.

VEGETABLES No.Q 064 00

LOUISIANA STYLE SMOTHERED SQUASH

Yield 100 Portion 1/2 Cup

Calories	Carbohydrates	Protein	Fat	Cholesterol	Sodium	Calcium
57 cal	8 g	1 g	3 g	7 mg	241 mg	25 mg

Ingredient	Weight	Measure	Issue
ONIONS,FRESH,CHOPPED	1-3/8 lbs	1 qts	1-5/8 lbs
PEPPERS,GREEN,FRESH,CHOPPED	6 oz	1-1/8 cup	7-1/3 oz
CELERY,FRESH,CHOPPED	6 oz	1-3/8 cup	8-1/4 oz
BUTTER	12 oz	1-1/2 cup	
SQUASH,FRESH,SUMMER	24 lbs		25-1/4 lbs
WATER,BOILING	1 lbs	2 cup	
SUGAR,GRANULATED	7 oz	1 cup	
SALT	1-7/8 oz	3 tbsp	
PEPPER,BLACK,GROUND	1/8 oz	1/3 tsp	

Method

1 Saute onions, peppers and celery in butter or margarine until tender. Set aside for use in Step 3.
2 Add squash to water; cook, covered, in steam-jacketed kettle or stock pot about 5 minutes or until just tender.
3 Add sugar, salt and pepper to squash. Add sauteed vegetables; mix lightly.
4 Cook, covered, about 5 minutes, or until just heated through, stirring occasionally. CCP: Hold at 140 F. or higher for service.

Notes

1 Prepare in batches of 25 as needed. Do not peel squash.

VEGETABLES No.Q 064 01

SAVORY SUMMER SQUASH

Yield 100   Portion 1/2 Cup

Calories	Carbohydrates	Protein	Fat	Cholesterol	Sodium	Calcium
35 cal	7 g	1 g	1 g	0 mg	212 mg	26 mg

Ingredient	Weight	Measure	Issue
ONIONS,FRESH,SLICED	2 lbs	2 qts	2-1/4 lbs
OIL,SALAD	2-3/8 oz	1/4 cup 1-1/3 tbsp	
SQUASH,FRESH,SUMMER	24 lbs		25-1/4 lbs
WATER,BOILING	1 lbs	2 cup	
SUGAR,GRANULATED	3-1/2 oz	1/2 cup	
SALT	1-7/8 oz	3 tbsp	
BASIL,SWEET,WHOLE,CRUSHED	1/3 oz	2 tbsp	
PEPPER,BLACK,GROUND	1/8 oz	1/3 tsp	

Method

1 Saute onions in salad oil or melted shortening until tender. Set aside for use in Step 3.
2 Add squash to water; cook, covered, in steam-jacketed kettle or stock pot about 5 minutes or until just tender.
3 Add sugar, salt and pepper to squash. Add basil if desired. Add sauteed vegetables; mix lightly.
4 Cook, covered, about 5 minutes, or until just heated through, stirring occasionally. CCP: Hold at 140 F. or higher for service.

Notes

1 Prepare in batches of 25 as needed. DO NOT peel squash.

VEGETABLES No.Q 065 00

HERBED BROCCOLI

Yield 100							Portion 1/2 Cup

Calories	Carbohydrates	Protein	Fat	Cholesterol	Sodium	Calcium
27 cal	5 g	3 g	0 g	0 mg	23 mg	52 mg

Ingredient	Weight	Measure	Issue
WATER	10-1/2 lbs	1 gal 1 qts	
BROCCOLI,FROZEN,CUT	20 lbs	3 gal 2-1/2 qts	
ONIONS,FRESH,CHOPPED	7 oz	1-1/4 cup	7-7/8 oz
MARJORAM,SWEET,GROUND	1/4 oz	1/4 cup 1/3 tbsp	
BASIL,DRIED,CRUSHED	1/2 oz	3 tbsp	

Method

1. Bring water to a boil.
2. Add broccoli, onions, marjoram and basil to boiling water.
3. Return to boil; cover.
4. Reduce heat; cook 7 to 9 minutes or until tender.
5. Drain; reserve 1 quart liquid to pour over vegetables. CCP: Hold at 140 F. or higher for service.

VEGETABLES No.Q 066 00

BAKED SWEET POTATOES

Yield 100   Portion 1 Each

Calories	Carbohydrates	Protein	Fat	Cholesterol	Sodium	Calcium
191 cal	44 g	3 g	0 g	0 mg	24 mg	40 mg

Ingredient   Weight   Measure   Issue

SWEET POTATOES, FRESH   40 lbs   8 gal 2-1/8 qts

Method

1. Scrub potatoes well; dry; remove any blemishes; place on sheet pans.
2. Prick skin with fork to allow steam to escape.
3. Using a convection oven, bake at 400 F. for 40 to 45 minutes on high fan, closed vent or until done. CCP: Hold at 140 F. or higher for service.

VEGETABLES No.Q 067 00

CANDIED SWEET POTATOES

Yield 100   **Portion** 1/2 Cup

Calories	Carbohydrates	Protein	Fat	Cholesterol	Sodium	Calcium
179 cal	34 g	1 g	4 g	10 mg	253 mg	29 mg

Ingredient   Weight   Measure   Issue

Ingredient	Weight	Measure	Issue
SWEET POTATOES,CANNED,W/SYRUP	24-1/8 lbs	3 gal	
COOKING SPRAY,NONSTICK	2 oz	1/4 cup 1/3 tbsp	
BUTTER,MELTED	1 lbs	2 cup	
SUGAR,BROWN,PACKED	2-1/2 lbs	2 qts	
SALT	1-1/2 oz	2-1/3 tbsp	
ORANGE,FRESH,SLICED	9-1/4 oz	2 each	

Method

1. Drain potatoes. Lightly spray each steam table pan with non-stick cooking spray. Place potatoes in single layer in each sprayed steam table pan.
2. Pour 1/2 cup butter or margarine over potatoes in each pan.
3. Combine brown sugar and salt. Sprinkle 2 cups mixture over potatoes in each pan.
4. Using a convection oven, bake at 325 F. 20 minutes on low fan, closed vent or until thoroughly heated. CCP: Internal temperature must reach 145 F. or higher for 15 seconds. Hold at 140 F. or higher for service.
5. Garnish with orange slices before serving.

VEGETABLES No.Q 067 01

GLAZED SWEET POTATOES

Yield 100   Portion   1/2 Cup

Calories	Carbohydrates	Protein	Fat	Cholesterol	Sodium	Calcium
176 cal	34 g	1 g	4 g	10 mg	253 mg	27 mg

Ingredient	Weight	Measure	Issue
SWEET POTATOES,CANNED,W/SYRUP	24-1/8 lbs	3 gal	
COOKING SPRAY,NONSTICK	2 oz	1/4 cup 1/3 tbsp	
CORNSTARCH	4-1/2 oz	1 cup	
RESERVED LIQUID	4-1/8 lbs	2 qts	
BUTTER,MELTED	1 lbs	2 cup	
SUGAR,BROWN,PACKED	2-1/8 lbs	1 qts 2-1/2 cup	
SALT	1-1/2 oz	2-1/3 tbsp	
ORANGE,FRESH,SLICED	9-1/4 oz	2 each	

Method

1 Lightly spray each steam table pan with non-stick cooking spray. Drain potatoes and reserve 2 quarts of liquid for use in Step 2. Place potatoes in single layer in each sprayed pan.
2 Combine cornstarch with liquid from potatoes and water. Add melted butter or margarine.
3 Combine sugar and salt. Add to cornstarch mixture. Bring to a boil; cook 5 minutes. Pour 3-1/4 cup sauce over potatoes in each pan.
4 Using a convection oven, bake at 325 F. 20 minutes on low fan, closed vent or until thoroughly heated. CCP: Internal temperature must reach 145 F. or higher for 15 seconds. Hold at 140 F. or higher for service.
5 Garnish with orange slices before serving.

VEGETABLES No.Q 067 02

GLAZED SWEET POTATOES (SYRUP)

Yield 100 **Portion** 1/2 Cup

Calories	Carbohydrates	Protein	Fat	Cholesterol	Sodium	Calcium
226 cal	47 g	1 g	4 g	10 mg	280 mg	23 mg

Ingredient | Weight | Measure | Issue

Ingredient	Weight	Measure	Issue
SWEET POTATOES,CANNED,W/SYRUP	24-1/8 lbs	3 gal	
COOKING SPRAY,NONSTICK	2 oz	1/4 cup 1/3 tbsp	
CORNSTARCH	4-1/2 oz	1 cup	
BUTTER,MELTED	1 lbs	2 cup	
SYRUP	6-3/4 lbs	2 qts 1-3/4 cup	
SALT	1-1/2 oz	2-1/3 tbsp	
ORANGE,FRESH,SLICED	9-1/4 oz	2 each	

Method

1. Drain potatoes and reserve liquid for use in Step 2. Combine reserved liquid with water to equal 2 quarts. Lightly spray each pan with non-stick cooking spray. Place potatoes in single layer in each sprayed pan.
2. Combine cornstarch with syrup from potatoes or with water. Add melted butter or margarine.
3. Add salt. Add syrup to cornstarch mixture. Bring to a boil; cook about 5 minutes. Pour 3-1/2 cups sauce over potatoes in each pan.
4. Using a convection oven, bake at 325 F. 20 minutes on low fan, closed vent or until thoroughly heated. CCP: Internal temperature must reach 145 F. or higher for 15 seconds. Hold at 140 F. or higher for service.
5. Garnish with orange slices before serving.

VEGETABLES No.Q 068 00

TEMPURA VEGETABLES

Yield 100　　　　　　　　　　　　　　　　　　　Portion 3-1/2 Ounces

Calories	Carbohydrates	Protein	Fat	Cholesterol	Sodium	Calcium
143 cal	19 g	5 g	6 g	36 mg	420 mg	72 mg

Ingredient

Ingredient	Weight	Measure	Issue
BROCCOLI,FRESH,CHOPPED	5 lbs	1 gal 2-1/2 qts	8-1/4 lbs
CAULIFLOWER FLORETS,FRESH	5 lbs	1 gal 1-2/3 qts	
SQUASH,ZUCCHINI,FRESH,SLICED	5 lbs	1 gal 1 qts	5-1/4 lbs
FLOUR,WHEAT,GENERAL PURPOSE	4-1/2 lbs	1 gal 1/8 qts	
BAKING POWDER	2-2/3 oz	1/4 cup 2 tbsp	
SALT	2-7/8 oz	1/4 cup 2/3 tbsp	
EGGS,WHOLE,FRESH	1-7/8 lbs	17 Eggs	
WATER,COLD	6-2/3 lbs	3 qts 3/4 cup	

Method

1. Wash and trim vegetables. Set aside for use in Step 6.
2. Sift together flour, baking powder, and salt in mixer bowl. Set aside for use in Step 4.
3. Separate eggs. Beat egg yolks. Set egg whites aside for use in Step 5.
4. Add half of ice water to egg yolks. Add to dry mixture beating at low speed until blended. Add remaining ice water; whip at high speed until smooth.
5. Whip egg whites until stiff but not dry. Fold into batter.
6. Dip dry vegetables into batter.
7. Fry about 3 to 5 minutes or until golden brown in 365 F. deep fat fryer.
8. Drain well in basket or on absorbent paper. CCP: Hold at 140 F. or higher for service.

VEGETABLES No.Q 069 00

MASHED SWEET POTATOES

Yield 100   **Portion** 1/2 Cup

Calories	Carbohydrates	Protein	Fat	Cholesterol	Sodium	Calcium
158 cal	32 g	2 g	3 g	5 mg	227 mg	35 mg

Ingredient	**Weight**	**Measure**	**Issue**
MILK,NONFAT,DRY	3-5/8 oz	1-1/2 cup	
WATER,WARM	4-1/2 lbs	2 qts 1/2 cup	
SWEET POTATOES,CANNED,W/SYRUP	31-1/8 lbs	3 gal 3-1/2 qts	
SALT	1-1/4 oz	2 tbsp	
BUTTER,MELTED	8 oz	1 cup	
SUGAR,GRANULATED	7 oz	1 cup	
COOKING SPRAY,NONSTICK	2 oz	1/4 cup 1/3 tbsp	

Method

1. Reconstitute milk in mixer bowl.
2. Add sweet potatoes; beat at low speed 2 minutes or until smooth.
3. Add salt, melted butter or margarine and sugar; blend at medium speed.
4. Scrape bowl down; beat at medium speed 2 minutes.
5. Lightly spray each steam table pan with non-stick cooking spray. Place 7-1/2 quarts potatoes in each sprayed pan; cover pan.
6. Using a convection oven, bake at 325 F. 30 minutes on high fan, closed vent or until heated thoroughly. CCP: Internal temperature must reach 145 F. or higher for 15 seconds. Hold at 140 F. or higher for service.

VEGETABLES No.Q 069 01

SWEET POTATOES SOUTHERN STYLE

Yield 100   Portion   1/2 Cup

Calories	Carbohydrates	Protein	Fat	Cholesterol	Sodium	Calcium
156 cal	33 g	2 g	2 g	4 mg	223 mg	39 mg

Ingredient	Weight	Measure	Issue
MILK,NONFAT,DRY	3-5/8 oz	1-1/2 cup	
WATER,WARM	4-1/2 lbs	2 qts 1/2 cup	
SWEET POTATOES,CANNED,W/SYRUP	31-1/8 lbs	3 gal 3-1/2 qts	
SALT	1-1/4 oz	2 tbsp	
BUTTER,MELTED	6 oz	3/4 cup	
SUGAR,BROWN,PACKED	8-1/2 oz	1-5/8 cup	
CINNAMON,GROUND	1/2 oz	2 tbsp	
NUTMEG,GROUND	1/4 oz	1 tbsp	
COOKING SPRAY,NONSTICK	2 oz	1/4 cup 1/3 tbsp	

Method

1. Reconstitute milk in mixer bowl.
2. Add sweet potatoes; beat at low speed 2 minutes or until smooth.
3. Add salt, melted butter or margarine and brown sugar; blend at medium speed. If desired, add cinnamon and nutmeg.
4. Scrape bowl down; beat at medium speed 2 minutes.
5. Lightly spray each pan with non-stick cooking spray. Place 7-1/2 quarts potatoes in each sprayed pan; cover.
6. Using a convection oven, bake at 325 F. 30 minutes on high fan, closed vent or until heated thoroughly. CCP: Internal temperature must reach 145 F. or higher for 15 seconds. Hold at 140 F. or higher for service.

VEGETABLES No.Q 069 02

MARSHMALLOW SWEET POTATOES

Yield 100   Portion   1/2 Cup

Calories	Carbohydrates	Protein	Fat	Cholesterol	Sodium	Calcium
172 cal	36 g	2 g	3 g	5 mg	229 mg	35 mg

Ingredient   Weight   Measure   Issue

Ingredient	Weight	Measure
MILK,NONFAT,DRY	3-5/8 oz	1-1/2 cup
WATER,WARM	4-1/2 lbs	2 qts 1/2 cup
SWEET POTATOES,CANNED,W/SYRUP	31-1/8 lbs	3 gal 3-1/2 qts
SALT	1-1/4 oz	2 tbsp
BUTTER,MELTED	8 oz	1 cup
SUGAR,GRANULATED	7 oz	1 cup
COOKING SPRAY,NONSTICK	2 oz	1/4 cup 1/3 tbsp
MARSHMALLOWS,MINIATURE	1 lbs	2 qts 1 cup

Method

1. Reconstitute milk in mixer bowl.
2. Add sweet potatoes; beat at low speed 2 minutes or until smooth.
3. Add salt, melted butter or margarine and sugar; blend at medium speed.
4. Scrape bowl down; beat at medium speed 2 minutes.
5. Lightly spray each pan with non-stick cooking spray. Place 7-1/2 quarts potatoes in each sprayed pan; cover.
6. Using a convection oven, bake at 400 F. 10 minutes on high fan, closed vent or until heated thoroughly. After potatoes are heated through, sprinkle marshmallows over potatoes. Bake until marshmallows are lightly browned. CCP: Internal temperature must reach 145 F. or higher for 15 seconds. Hold at 140 F. or higher for service.

VEGETABLES No.Q 070 00

GARLIC ROASTED POTATO WEDGES

Yield 100   Portion   4 Wedges

Calories	Carbohydrates	Protein	Fat	Cholesterol	Sodium	Calcium
105 cal	23 g	2 g	1 g	0 mg	192 mg	12 mg

Ingredient	Weight	Measure	Issue
GARLIC POWDER	2-3/8 oz	1/2 cup	
SALT	1-2/3 oz	2-2/3 tbsp	
ONION POWDER	5/8 oz	2-2/3 tbsp	
PAPRIKA,GROUND	5/8 oz	2-1/3 tbsp	
PARSLEY,DEHYDRATED,FLAKED	1/3 oz	1/4 cup 3 tbsp	
PEPPER,BLACK,GROUND	1/8 oz	1/3 tsp	
POTATOES,WHITE,FRESH,WEDGED	24-3/4 lbs	4 gal 2 qts	
COOKING SPRAY,NONSTICK	2 oz	1/4 cup 1/3 tbsp	

Method

1. Combine salt, garlic powder, onion powder, paprika, parsley and pepper.
2. Wash potatoes and dry; toss well with seasonings.
3. Lightly spray sheet pans with cooking spray.
4. Place 8-1/2 pounds or 1-3/4 gallons seasoned potatoes on each pan. Lightly spray potatoes with cooking spray.
5. Using a convection oven, bake 20 minutes at 350 F. on high fan, closed vent. Lightly spray potatoes. Bake 15 minutes longer or until tender and light brown. CCP: Hold at 140 F. or higher for serving.

VEGETABLES No.Q 071 00

ROSEMARY ROASTED POTATO WEDGES

Yield 100   **Portion** 4 Wedges

Calories	Carbohydrates	Protein	Fat	Cholesterol	Sodium	Calcium
112 cal	25 g	2 g	1 g	0 mg	193 mg	23 mg

Ingredient	**Weight**	**Measure**	**Issue**
SALT	1-2/3 oz	2-2/3 tbsp	
ROSEMARY,GROUND	7/8 oz	1/2 cup	
ONION POWDER	9-3/4 oz	2-5/8 cup	
PEPPER,BLACK,GROUND	1/8 oz	1/3 tsp	
POTATOES,WHITE,FRESH,WEDGED	24-3/4 lbs	4 gal 2 qts	
COOKING SPRAY,NONSTICK	2 oz	1/4 cup 1/3 tbsp	

Method

1. Combine salt, rosemary, onion powder and pepper.
2. Wash potatoes and dry, toss with seasonings.
3. Lightly spray sheet pans with cooking spray. Place 8-1/2 pounds or 1-3/4 gallons seasoned potatoes on each pan. Lightly spray potatoes with cooking spray.
4. Using a convection oven, bake 20 minutes at 350 F. on high fan, closed vent. Lightly spray potatoes. Bake 15 minutes longer or until tender and light brown. CCP: Hold at 140 F. or higher for serving.

VEGETABLES No.Q 072 00

SESAME GLAZED GREEN BEANS

Yield 100   Portion 3/4 Cup

Calories	Carbohydrates	Protein	Fat	Cholesterol	Sodium	Calcium
90 cal	14 g	4 g	3 g	0 mg	412 mg	62 mg

Ingredient	Weight	Measure	Issue
SESAME SEEDS	1-1/8 lbs	3-1/2 cup	
JUICE,APPLE,CANNED	1-7/8 lbs	3-1/2 cup	
SOY SAUCE	1-5/8 lbs	2-1/2 cup	
SUGAR,BROWN,PACKED	5-1/8 oz	1 cup	
GARLIC POWDER	7-1/8 oz	1-1/2 cup	
CORNSTARCH	1-1/8 oz	1/4 cup 1/3 tbsp	
PEPPER,BLACK,GROUND	1/2 oz	2 tbsp	
GINGER,GROUND	1/2 oz	2-1/3 tbsp	
WATER	18-3/4 lbs	2 gal 1 qts	
BEANS,GREEN,FROZEN,WHOLE	25-1/8 lbs	5 gal 3 qts	

Method

1 Place sesame seeds on sheet pan in single layer. Using a convection oven, bake in 350 F. for 10 minutes on low fan, open vent. Transfer immediately to another sheet pan to cool.
2 Combine apple juice, soy sauce, brown sugar, garlic powder, cornstarch, ginger and pepper in steam-jacketed kettle or stock pot. Stir well to dissolve cornstarch. Bring to a boil; reduce heat; simmer 5 minutes or until thick and clear.
3 Bring water to a boil in steam-jacketed kettle or stock pot. Add beans. Stir well. Return to a boil. Cook 3 to 4 minutes or until almost tender, stirring occasionally. Drain well.
4 Add sauce to beans; stir-cook 2 to 3 minutes to evenly coat and thoroughly heat the beans. Do not over cook.
5 Remove to serving pans. Sprinkle 3-1/2 ounces or 3/4 cup sesame seeds over each 25 portions of beans. Toss to distribute sesame seeds. CCP: Hold at 140 F. or higher for service.

VEGETABLES No.Q 073 00

JAPANESE VEGETABLE STIR FRY

Yield 100   Portion   3/4 Cup

Calories	Carbohydrates	Protein	Fat	Cholesterol	Sodium	Calcium
81 cal	13 g	4 g	2 g	0 mg	475 mg	63 mg

Ingredient	Weight	Measure	Issue
CARROTS,FRESH,SLICED	7-3/4 lbs	1 gal 2-7/8 qts	9-1/2 lbs
ONIONS,FRESH,SLICED	2 lbs	1 qts 3-7/8 cup	2-1/4 lbs
SOY SAUCE	1-3/4 lbs	2-3/4 cup	
JUICE,PINEAPPLE,CANNED,UNSWEETENED	11-3/4 oz	1-3/8 cup	
VINEGAR,DISTILLED	1-3/4 oz	3-1/3 tbsp	
JUICE,LEMON	1-1/8 oz	2 tbsp	
SUGAR,BROWN,PACKED	3-7/8 oz	3/4 cup	
ONIONS,FRESH,GRATED	1-7/8 oz	1/4 cup 1-2/3 tbsp	2-1/8 oz
GINGER,GROUND	1/2 oz	2-1/3 tbsp	
GARLIC POWDER	1/8 oz	1/8 tsp	
WATER	14-5/8 oz	1-3/4 cup	
CORNSTARCH	7/8 oz	3 tbsp	
OIL, CANOLA	7-2/3 oz	1 cup	
BROCCOLI,FROZEN,CUT	16-1/2 lbs	3 gal	
BEANS,GREEN,FROZEN,WHOLE	4-3/8 lbs	1 gal	
ONIONS,FRESH,CHOPPED	1-3/4 lbs	1 qts 1 cup	2 lbs

Method

1. Wash and trim fresh vegetables.
2. Combine soy sauce, pineapple juice, vinegar, and lemon juice.
3. Add brown sugar, minced onions, ginger, and garlic to soy sauce mixture. Mix until well blended. Bring to a boil, reduce heat, simmer 1 minute.
4. Blend cornstarch with water to make a smooth paste. Slowly add paste to soy sauce mixture stirring constantly, simmer 5 minutes or until lightly thickened. Remove from heat.
5. Stir-fry vegetables in 50 portion batches in salad oil as follows: Carrots, 3 minutes; add green beans and onions, 2 minutes; add broccoli, 2 minutes. Do not overcook.
6. Pour approximately 3-1/2 cups of sauce over each batch of vegetables and garnish with 1 quart of green onions.
7. CCP: Hold at 140 F. or higher for service.

VEGETABLES No.Q 074 00

SQUASH AND CARROT MEDLEY (FRESH)

Yield 100　　　　　　　　　　　　　　　　　Portion   3/4 Cup

Calories	Carbohydrates	Protein	Fat	Cholesterol	Sodium	Calcium
42 cal	9 g	2 g	0 g	0 mg	132 mg	38 mg

Ingredient	Weight	Measure	Issue
CARROTS,FRESH,SLICED	8 lbs	1 gal 3-1/8 qts	9-3/4 lbs
SQUASH,FRESH,SUMMER	13-1/2 lbs		14-1/4 lbs
SQUASH,ZUCCHINI,FRESH,JULIENNE	13-1/2 lbs	3 gal 1-5/8 qts	14-1/4 lbs
COOKING SPRAY,NONSTICK	3/4 oz	1 tbsp	
GARLIC POWDER	1-5/8 oz	1/4 cup 1-2/3 tbsp	
SALT	1 oz	1 tbsp	
BASIL,DRIED,CRUSHED	7/8 oz	1/4 cup 1-2/3 tbsp	
PEPPER,BLACK,GROUND	1/4 oz	1 tbsp	
ONIONS,FRESH,CHOPPED	13-1/2 oz	2-3/8 cup	15 oz

Method

1 Wash and trim fresh vegetables.
2 Lightly spray steam-jacketed kettle with non-stick spray; add carrots; stir-fry carrots 5 minutes.
3 Add yellow squash, zucchini, garlic, salt, basil, and pepper; stir well; cover; cook 5 minutes; uncover; stir-cook 3 to 4 minutes or until squash is tender-crisp. Do not overcook.
4 Transfer to serving pans; garnish with green onions. CCP: Heat to 145 F. or higher for 15 seconds. Hold at 140 F. or higher for service.

VEGETABLES No.Q 074 01

SQUASH AND CARROT MEDLEY (FROZEN)

Yield 100   Portion 3/4 Cup

Calories	Carbohydrates	Protein	Fat	Cholesterol	Sodium	Calcium
40 cal	8 g	2 g	0 g	0 mg	140 mg	40 mg

Ingredient	**Weight**	**Measure**	**Issue**
COOKING SPRAY,NONSTICK	3/4 oz	1 tbsp	
CARROTS,FROZEN,SLICED	8 lbs	1 gal 3-1/8 qts	
SQUASH,ZUCCHINI,FROZEN	27 lbs		
GARLIC POWDER	1-5/8 oz	1/4 cup 1-2/3 tbsp	
SALT	1 oz	1 tbsp	
BASIL,DRIED,CRUSHED	7/8 oz	1/4 cup 1-2/3 tbsp	
PEPPER,BLACK,GROUND	1/4 oz	1 tbsp	
ONIONS,FRESH,CHOPPED	13-1/2 oz	2-3/8 cup	15 oz

Method

1 Lightly spray steam-jacketed kettle with non-stick spray; add carrots; cook 10 minutes.
2 Add zucchini, garlic, salt, basil, and pepper; stir well; cover; cook 5 minutes. Do not overcook. CCP: Heat to 145 F. or higher for 15 seconds. Hold at 140 F. or higher for service.
3 Transfer to serving pans; garnish with green onions.

VEGETABLES No.Q 075 00

DEVILED OVEN FRIES

Yield 100 Portion 4 Pieces

Calories	Carbohydrates	Protein	Fat	Cholesterol	Sodium	Calcium
113 cal	24 g	2 g	1 g	0 mg	227 mg	19 mg

Ingredient	Weight	Measure	Issue
CHILI POWDER,DARK,GROUND	4-1/4 oz	1 cup	
GARLIC POWDER	3 oz	1/2 cup 2 tbsp	
SALT	1-7/8 oz	3 tbsp	
MUSTARD,DRY	4 oz	1/2 cup 2 tbsp	
POTATOES,WHITE,FRESH,WEDGED	24-3/4 lbs	4 gal 2 qts	
COOKING SPRAY,NONSTICK	2 oz	1/4 cup 1/3 tbsp	

Method

1. Combine chili powder, garlic powder, mustard, and salt.
2. Divide potatoes into 3 equal batches. Toss each well dried batch with 2/3 cup seasoning mixture.
3. Lightly spray sheet pans with cooking spray. Place 8-1/2 pounds or 1-3/4 gallons of seasoned potatoes, skin side down on each pan. Lightly spray potatoes with cooking spray.
4. Using a convection oven, bake 20 minutes at 350 F.; lightly spray potatoes; bake about 15 minutes longer or until tender and light brown on high fan, open vent. CCP: Hold at 140 F. or higher for serving.

VEGETABLES No.Q 076 00

SOUTH OF THE BORDER BROCCOLI

Yield 100   **Portion** 2/3 Cup

Calories	Carbohydrates	Protein	Fat	Cholesterol	Sodium	Calcium
40 cal	7 g	4 g	0 g	0 mg	421 mg	57 mg

Ingredient   Weight   Measure   Issue

Ingredient	Weight	Measure	Issue
BROCCOLI,FROZEN,CUT	20 lbs	3 gal 2-1/2 qts	
SALT	7/8 oz	1 tbsp	
WATER,BOILING	8 lbs	3 qts 3-3/8 cup	
SAUCE,SALSA	11-1/4 lbs	1 gal 1-1/4 qts	
CILANTRO,DRY	1/2 oz	1/2 cup	

Method

1. Add broccoli to boiling salted water; return to a boil. Cook uncovered, 5 minutes or until tender-crisp. Drain; place 5 pounds broccoli in each steam table pan.
2. Mix Salsa and cilantro. Heat to 145 F. or higher for 15 seconds. Pour 5 cups Salsa over broccoli in each pan. Mix lightly. CCP: Hold for service at 140 F. or higher.
3. Use batch preparation techniques. Toss salsa and broccoli just before service to prevent discoloration of the broccoli.

VEGETABLES No.Q 076 01

SOUTH OF THE BORDER MEDLEY

Yield 100   Portion 2/3 Cup

Calories	Carbohydrates	Protein	Fat	Cholesterol	Sodium	Calcium
33 cal	6 g	2 g	0 g	0 mg	417 mg	37 mg

Ingredient | Weight | Measure | Issue

Ingredient	Weight	Measure	Issue
BROCCOLI,FROZEN,CUT	8-1/4 lbs	1 gal 2 qts	
CAULIFLOWER,FROZEN	10 lbs		
WATER,BOILING	8-1/3 lbs	1 gal	
SALT	7/8 oz	1 tbsp	
SAUCE,SALSA	11-1/4 lbs	1 gal 1-1/4 qts	
CILANTRO,DRY	1/2 oz	1/2 cup	

Method

1. Add broccoli and cauliflower to boiling salted water. Return to a boil; cook uncovered 5 minutes or until tender-crisp. Drain; place 5 pounds of evenly distributed vegetable mixture in each steam table pan.
2. Mix salsa and cilantro. Heat to 145 F. or higher for 15 seconds. Pour 5 cups salsa over broccoli and cauliflower in each pan, mix lightly. CCP: Hold for service at 140 F. or higher.
3. Use batch preparation techniques. Toss salsa and vegetable medley just before service to prevent discoloration of the broccoli.

VEGETABLES No.Q 077 00

BAKED POTATO PANCAKES (FROZEN SHREDDED POTATO)

Yield 100   **Portion** 1 Cake

Calories	Carbohydrates	Protein	Fat	Cholesterol	Sodium	Calcium
129 cal	22 g	6 g	2 g	30 mg	230 mg	54 mg

Ingredient	Weight	Measure	Issue
POTATOES,WHITE,FROZEN,SHREDDED,HASHBROWN	25-7/8 lbs	3 gal 2 qts	
CHEESE,MONTEREY JACK,REDUCED FAT,SHREDDED	2 lbs	2 qts	
ONIONS,GREEN,FRESH,CHOPPED	8-7/8 oz	2-1/2 cup	9-3/4 oz
ONIONS,FRESH,CHOPPED	5-5/8 oz	1 cup	6-1/4 oz
EGGS,WHOLE,FROZEN	1-3/8 lbs	2-5/8 cup	
GARLIC POWDER	1-5/8 oz	1/4 cup 1-2/3 tbsp	
SALT	1-1/4 oz	2 tbsp	
PEPPER,WHITE,GROUND	1/8 oz	1/4 tsp	
COOKING SPRAY,NONSTICK	2 oz	1/4 cup 1/3 tbsp	

Method

2. Add cheese, green onions, and fresh onions to potato; mix well.
3. Add garlic powder, salt, and pepper to egg; stir to blend.
4. Add egg mixture to potatoes; stir lightly to combine all ingredients.
5. Lightly spray each steam table pan with non-stick cooking spray.
6. Shape potato mixture into 4-ounce balls. Place balls in rows of 3x5 on each pan. Flatten into 4x1/2-inch thick cakes; lightly spray tops with non-stick cooking spray.
7. Using a convection oven, bake 30 to 35 minutes at 375 F. oven on high fan, open vent or until well browned. CCP: Internal temperature must reach 145 F. or higher for 15 seconds. Hold at 140 F. or higher for service.

VEGETABLES No.Q 077 01

BAKED POTATO PANCAKES

Yield 100   **Portion** 1 Cake

Calories	Carbohydrates	Protein	Fat	Cholesterol	Sodium	Calcium
107 cal	18 g	5 g	2 g	30 mg	208 mg	50 mg

Ingredient	Weight	Measure	Issue
POTATO,WHITE,DEHYDRATED,SHREDDED	19 lbs		
CHEESE,MONTEREY JACK,REDUCED FAT,SHREDDED	2 lbs	2 qts	
ONIONS,GREEN,FRESH,CHOPPED	8-7/8 oz	2-1/2 cup	9-3/4 oz
ONIONS,FRESH,CHOPPED	5-5/8 oz	1 cup	6-1/4 oz
EGGS,WHOLE,FROZEN	1-3/8 lbs	2-5/8 cup	
GARLIC POWDER	1-5/8 oz	1/4 cup 1-2/3 tbsp	
SALT	1-1/4 oz	2 tbsp	
PEPPER,WHITE,GROUND	1/8 oz	1/4 tsp	
COOKING SPRAY,NONSTICK	2 oz	1/4 cup 1/3 tbsp	

Method

1. Rehydrate potatoes according to package directions.
2. Add cheese, green onions, and fresh onions to potato; mix well.
3. Add garlic powder, salt, and pepper to egg; stir to blend.
4. Add egg mixture to potatoes; stir lightly to combine all ingredients.
5. Lightly spray each pan with non-stick cooking spray.
6. Shape potato mixture into 4-ounce balls. Place balls in rows of 3x5 on each pan. Flatten into 4x1/2-inch thick cakes; lightly spray tops with non-stick cooking spray.
7. Using a convection oven, bake 30 to 35 minutes in 375 F. oven on high fan, open vent or until well browned. CCP: Internal temperature must reach 145 F. or higher for 15 seconds. Hold at 140 F. or higher for service.

VEGETABLES No.Q 078 00

POTATOES AND HERBS

Yield 100 **Portion** 2/3 Cup

Calories	Carbohydrates	Protein	Fat	Cholesterol	Sodium	Calcium
132 cal	31 g	3 g	0 g	0 mg	57 mg	21 mg

Ingredient	**Weight**	**Measure**	**Issue**
GARLIC POWDER	3/4 oz	2-2/3 tbsp	
PARSLEY,DEHYDRATED,FLAKED	1/2 oz	3/4 cup	
DILL WEED,DRIED	1/3 oz	3 tbsp	
PEPPER,BLACK,GROUND	1/8 oz	1/3 tsp	
THYME LEAVES,DRIED	1/8 oz	1 tbsp	
POTATOES,FRESH,PEELED,SLICED	31 lbs	5 gal 2-1/2 qts	
ONIONS,FRESH,CHOPPED	4-1/4 lbs	3 qts	4-2/3 lbs
VEGETABLE BROTH		1 gal 1/2 qts	

Method

1. Thoroughly combine garlic powder, parsley, dillweed, pepper and thyme.
2. Place 7 pounds or 5-1/2 quarts of potatoes, 3 cups onions, and 1/3 cup of herb mixture in each steam table pan.
3. Gently and thoroughly toss potatoes with onions and herbs to evenly distribute onions and herbs.
4. Prepare vegetable broth according to package directions. Pour 4-1/2 cups broth around edges of potato mixture in each pan.
5. Using a convection oven, bake 60 to 65 minutes at 350 F. or until potatoes are tender and most of the stock is absorbed. CCP: Heat to 145 F. or higher for 15 seconds. Hold for service at 140 F. or higher.

VEGETABLES No.Q 079 00

HACIENDA POTATOES

Yield 100   Portion   3/4 Cup

Calories	Carbohydrates	Protein	Fat	Cholesterol	Sodium	Calcium
145 cal	33 g	4 g	1 g	0 mg	498 mg	46 mg

Ingredient	Weight	Measure	Issue
POTATOES,FRESH,PEELED,CUBED	24-1/8 lbs	4 gal 1-1/2 qts	29-3/4 lbs
WATER	16-3/4 lbs	2 gal	
TOMATOES,CANNED,DICED,DRAINED	17-5/8 lbs	2 gal	
PEPPERS,GREEN,FRESH,CHOPPED	2 lbs	1 qts 2 cup	2-3/8 lbs
ONIONS,FRESH,CHOPPED	2-1/8 lbs	1 qts 2 cup	2-1/3 lbs
SUGAR,GRANULATED	4-2/3 oz	1/2 cup 2-2/3 tbsp	
CHILI POWDER,DARK,GROUND	4-1/4 oz	1 cup	
SALT	3-3/8 oz	1/4 cup 1-2/3 tbsp	
GARLIC POWDER	1-1/4 oz	1/4 cup 1/3 tbsp	
CUMIN,GROUND	3/8 oz	2 tbsp	
PEPPER,BLACK,GROUND	1/4 oz	1 tbsp	
WATER	1 lbs	2 cup	
FLOUR,WHEAT,GENERAL PURPOSE	8-7/8 oz	2 cup	

Method

1 Add potatoes to water. Bring to a boil. Reduce heat. Simmer 20 minutes or until potatoes are just tender.
2 Combine tomatoes, green peppers, onions, sugar, chili powder, salt, garlic powder, cumin, and black pepper in steam-jacketed kettle. Bring to a boil; reduce heat; cover; simmer 5 minutes.
3 Blend water and flour to make a smooth paste. Add to sauce. Stir to combine. Simmer 5 minutes or until thickened, stirring occasionally.
4 Add potatoes to sauce. Stir to evenly distribute ingredients. Cover; bring to a boil, stirring occasionally until mixture comes to a complete boil. Uncover; reduce heat. Simmer 10 minutes, stirring occasionally until potatoes are thoroughly heated. CCP: Heat to 145 F. or higher for 15 seconds. Hold for service at 140 F. or higher.

VEGETABLES No.Q 080 00

HACIENDA CORN AND BLACK BEANS

Yield 100   **Portion** 3/4 Cup

Calories	Carbohydrates	Protein	Fat	Cholesterol	Sodium	Calcium
160 cal	34 g	7 g	1 g	0 mg	484 mg	46 mg

Ingredient	**Weight**	**Measure**	**Issue**
TOMATOES,CANNED,DICED,DRAINED	16 lbs	1 gal 3-1/4 qts	
PEPPERS,GREEN,FRESH,CHOPPED	2 lbs	1 qts 2 cup	2-3/8 lbs
ONIONS,FRESH,CHOPPED	2-1/8 lbs	1 qts 2 cup	2-1/3 lbs
SUGAR,GRANULATED	4-2/3 oz	1/2 cup 2-2/3 tbsp	
CHILI POWDER,DARK,GROUND	4-1/4 oz	1 cup	
SALT	3-3/8 oz	1/4 cup 1-2/3 tbsp	
GARLIC POWDER	1-1/4 oz	1/4 cup 1/3 tbsp	
CUMIN,GROUND	3/8 oz	2 tbsp	
PEPPER,BLACK,GROUND	1/4 oz	1 tbsp	
WATER	1 lbs	2 cup	
FLOUR,WHEAT,GENERAL PURPOSE	8-1/4 oz	1-7/8 cup	
CORN,FROZEN,WHOLE KERNEL	16 lbs	2 gal 3-1/8 qts	
BEANS,BLACK,CANNED,DRAINED	8 lbs	3 qts 2-1/8 cup	

Method

1. Combine tomatoes, green peppers, onions, sugar, chili powder, salt, garlic powder, cumin, and black pepper in a steam-jacketed kettle. Bring to a boil; reduce heat; cover; simmer 5 minutes.
2. Blend water and flour to make a smooth paste. Add to sauce. Stir to combine. Simmer 5 minutes or until thickened, stirring constantly.
3. Add corn and black beans to sauce. Stir to evenly distribute ingredients. Cover; bring to a boil, stirring occasionally until mixture comes to complete boil. Uncover; reduce heat, simmer 15 minutes, stirring occasionally until corn and black beans are thoroughly heated. CCP: Heat to 145 F. or higher for 15 seconds. Hold for service at 140 F. or higher.

VEGETABLES No.Q 081 00

HACIENDA GREEN BEANS

Yield 100   **Portion** 3/4 Cup

Calories	Carbohydrates	Protein	Fat	Cholesterol	Sodium	Calcium
87 cal	20 g	4 g	1 g	0 mg	494 mg	81 mg

Ingredient	Weight	Measure	Issue
TOMATOES,CANNED,DICED,DRAINED	17-5/8 lbs	2 gal	
PEPPERS,GREEN,FRESH,CHOPPED	2 lbs	1 qts 2 cup	2-3/8 lbs
ONIONS,FRESH,CHOPPED	2-1/8 lbs	1 qts 2 cup	2-1/3 lbs
SUGAR,GRANULATED	4-2/3 oz	1/2 cup 2-2/3 tbsp	
CHILI POWDER,DARK,GROUND	4-1/4 oz	1 cup	
SALT	3-3/8 oz	1/4 cup 1-2/3 tbsp	
GARLIC POWDER	1-1/4 oz	1/4 cup 1/3 tbsp	
CUMIN,GROUND	3/8 oz	2 tbsp	
PEPPER,BLACK,GROUND	1/4 oz	1 tbsp	
WATER	1 lbs	2 cup	
FLOUR,WHEAT,GENERAL PURPOSE	8-1/4 oz	1-7/8 cup	
BEANS,GREEN,FROZEN,WHOLE	24 lbs	5 gal 2 qts	

Method

1. Combine tomatoes, green peppers, onions, sugar, chili powder, salt, garlic powder, cumin, and black pepper in a steam-jacketed kettle. Bring to a boil; reduce heat; cover; simmer for 5 minutes; stirring occasionally.
2. Blend water and flour to make a smooth paste. Add to sauce. Stir to combine. Simmer 5 minutes or until thickened, stirring constantly.
3. Add green beans to sauce. Stir to evenly distribute ingredients. Cover, bring to a boil, stirring occasionally until mixture comes to complete boil. Uncover; reduce heat; simmer 20 minutes, stirring occasionally until green beans are thoroughly heated and just tender. CCP: Heat to 145 F. or higher for 15 seconds. Hold for service at 140 F. or higher.

VEGETABLES No.Q 082 00

HONEY DIJON VEGETABLES

Yield 100   **Portion** 3/4 Cup

Calories	Carbohydrates	Protein	Fat	Cholesterol	Sodium	Calcium
62 cal	14 g	3 g	0 g	0 mg	121 mg	38 mg

Ingredient

Ingredient	Weight	Measure	Issue
CARROTS,FROZEN,SLICED	12 lbs	2 gal 2-5/8 qts	
CAULIFLOWER,FROZEN	12 lbs		
BRUSSELS SPROUTS,FROZEN	6 lbs	1 gal 3/8 qts	
COOKING SPRAY,NONSTICK	1/8 oz	1/8 tsp	
ONIONS,FRESH,CHOPPED	8-1/2 oz	1-1/2 cup	9-3/8 oz
VEGETABLE BROTH		3 qts	
MUSTARD,DIJON	12-3/4 oz	1-1/2 cup	
HONEY	10-1/2 oz	3/4 cup 2 tbsp	
PEPPER,RED,GROUND	<1/16th oz	1/8 tsp	
CORNSTARCH	4 oz	3/4 cup 2 tbsp	

Method

1. Cook carrots for 10 to 13 minutes, cauliflower for 4 to 8 minutes and brussels sprouts for 7 to 9 minutes. Use progressive cooking techniques for optimal vegetable texture.
2. Stir-cook onions in a lightly sprayed steam jacketed kettle about 5 minutes or until tender, stirring constantly.
3. Prepare vegetable broth according to package directions. Reserve 2 cups vegetable broth for use in Step 5.
4. Add remaining vegetable broth, mustard, honey and pepper to onions in steam jacketed kettle. Stir to blend. Bring to a simmer.
5. Blend reserved broth and cornstarch until smooth. Add to hot liquid mixture stirring constantly. Bring to a boil. Cook gently 2 to 3 minutes, stirring occasionally.
6. Pour glaze evenly over vegetables. Toss lightly until well coated. CCP: Heat to 145 F. or higher for 15 seconds. Hold for service at 140 F. or higher.

VEGETABLES No.Q 083 00

CORN AND GREEN BEAN CASSEROLE

Yield 100   Portion 3/4 Cup

Calories	Carbohydrates	Protein	Fat	Cholesterol	Sodium	Calcium
361 cal	29 g	8 g	25 g	19 mg	431 mg	120 mg

Ingredient	Weight	Measure	Issue
SALAD DRESSING,MAYONNAISE TYPE	9-3/8 lbs	1 gal 3/4 qts	
CHEESE,CHEDDAR,LOWFAT,SHREDDED	4 lbs	1 gal	
CELERY,FRESH,CHOPPED	3 lbs	2 qts 3-3/8 cup	4-1/8 lbs
ONIONS,FRESH,CHOPPED	1-1/2 lbs	1 qts 1/4 cup	1-2/3 lbs
GARLIC POWDER	1-1/4 oz	1/4 cup 1/3 tbsp	
PEPPER,WHITE,GROUND	3/8 oz	1 tbsp	
BEANS,GREEN,FROZEN,CUT	15 lbs	3 gal 1-3/4 qts	
CORN,FROZEN,WHOLE KERNEL	15 lbs	2 gal 2-3/8 qts	
BREADCRUMBS	1 lbs	1 qts 1/4 cup	
MARGARINE,MELTED	8 oz	1 cup	

Method

1. Combine salad dressing, cheese, celery, onions, garlic powder and white pepper in a mixer bowl. Mix at medium speed 1 minute.
2. Combine green beans and corn. Add salad dressing mixture. Mix lightly but thoroughly until all ingredients are blended.
3. Pour approximately 5-3/4 quart of mixture into steam table pans. Spread evenly.
4. Mix crumbs and margarine. Sprinkle 1 cup of crumb mixture evenly over mixture in each pan.
5. Using a convection oven, bake 45 minutes at 325 F. on high fan, closed vent. CCP: Internal temperature must reach 145 F. or higher for 15 seconds. Hold for service at 140 F. or higher.

VEGETABLES No.Q 084 00

GARLIC-LEMON ROASTED POTATO WEDGES

Yield 100   **Portion** 4 Wedges

Calories	Carbohydrates	Protein	Fat	Cholesterol	Sodium	Calcium
146 cal	33 g	3 g	1 g	0 mg	183 mg	18 mg

Ingredient	**Weight**	**Measure**	**Issue**
GARLIC POWDER	3 oz	1/2 cup 2 tbsp	
SALT	1-5/8 oz	2-1/3 tbsp	
SEASONING,LEMON N' HERB	7/8 oz	3 tbsp	
ONION POWDER	2/3 oz	3 tbsp	
PEPPER,BLACK,GROUND	1/4 oz	1 tbsp	
OREGANO,CRUSHED	1/2 oz	3 tbsp	
POTATOES,WHITE,FRESH,WEDGES	35 lbs		
COOKING SPRAY,NONSTICK	2 oz	1/4 cup 1/3 tbsp	

Method

1. Combine garlic powder, salt, lemon n' herb seasoning, onion powder, pepper and oregano.
2. Wash potatoes and dry; toss with seasonings.
3. Spray sheet pans with non-stick cooking spray. Place 8-1/2 pounds seasoned potatoes on each pan. Lightly spray potatoes with cooking spray.
4. Using a convection oven, bake 35 minutes or until tender and light brown on high fan, open vent. Spray with cooking spray as needed. CCP: Hold for service at 140 F. or higher.

VEGETABLES No.Q 100 00

ASPARAGUS (FROZEN)

Yield 100　　　　　　　　　　　　　　　　　　**Portion** 3/4 Cup

Calories	Carbohydrates	Protein	Fat	Cholesterol	Sodium	Calcium
23 cal	4 g	2 g	0 g	0 mg	74 mg	20 mg

Ingredient	**Weight**	**Measure**	**Issue**
ASPARAGUS,FROZEN,SPEARS,SLICED	18 lbs	2 gal 3-1/3 qts	
WATER	8-1/3 lbs	1 gal	
SALT	5/8 oz	1 tbsp	

Method

1 Cook asparagus for 5 to 8 minutes. Bring water to a boil in a steam-jacketed kettle or stock pot.
2 Add salt. Return to a boil. Cover.
3 Place asparagus in serving pan. CCP: Heat to 145 F. or higher for 15 seconds. Hold at 140 F. or higher for service.

VEGETABLES No.Q 100 01

ASPARAGUS (CANNED)

Yield 100 **Portion** 3/4 Cup

Calories	Carbohydrates	Protein	Fat	Cholesterol	Sodium	Calcium
22 cal	4 g	3 g	0 g	0 mg	409 mg	22 mg

Ingredient   Weight   Measure   Issue

ASPARAGUS,CANNED,SPEARS,INCL LIQUIDS   31-3/4 lbs   3 gal 2-3/4 qts

Method

1. Pour off half the liquid.
2. Place asparagus in steam-jacketed kettle or stock pot.
3. Heat to a simmer. Simmer about 10 minutes, stirring gently. DO NOT BOIL. CCP: Heat to 145 F. or higher for 15 seconds. Hold at 140 F. or higher for service.

VEGETABLES No.Q 100 02

ASPARAGUS (FRESH)

Yield 100   **Portion** 4 Spears

Calories	Carbohydrates	Protein	Fat	Cholesterol	Sodium	Calcium
21 cal	4 g	2 g	0 g	0 mg	73 mg	20 mg

Ingredient	Weight	Measure	Issue
WATER	12-1/2 lbs	1 gal 2 qts	
SALT	5/8 oz	1 tbsp	
ASPARAGUS,FRESH,WASHED & TRIMMED	20 lbs	4 gal 7/8 qts	37-3/4 lbs

Method

1 Bring water to a boil in steam-jacketed kettle or stock pot.
2 Add salt.
3 Add asparagus; bring water back to a boil. Cover; cook 10 to 20 minutes.
4 Place asparagus in serving pans. CCP: Heat to 145 F. or higher for 15 seconds. Hold at 140 F. or higher for service.

VEGETABLES No.Q 101 00

BEANS, GREEN (FROZEN)

Yield 100 **Portion** 3/4 Cup

Calories	Carbohydrates	Protein	Fat	Cholesterol	Sodium	Calcium
36 cal	8 g	2 g	0 g	0 mg	76 mg	47 mg

Ingredient	Weight	Measure	Issue
WATER	18-3/4 lbs	2 gal 1 qts	
SALT	5/8 oz	1 tbsp	
BEANS,GREEN,FROZEN,CUT	24 lbs	5 gal 2 qts	

Method

1. Bring water to a boil in a steam-jacketed kettle or stock pot.
2. Add salt.
3. Add beans; stir well.
4. Return to a boil; cover. Cook beans 5 to 8 minutes.
5. Place beans in serving pan. CCP: Heat to 145 F. or higher for 15 seconds. Hold at 140 F. or higher for service.

VEGETABLES No.Q 101 01

BEANS, GREEN (CANNED)

Yield 100　　　　　　　　　　　　　　　　　　　　　Portion 3/4 Cup

Calories	Carbohydrates	Protein	Fat	Cholesterol	Sodium	Calcium
26 cal	6 g	1 g	0 g	0 mg	443 mg	41 mg

Ingredient　　　　　　　　　　　　　　　**Weight**　　　　**Measure**　　　　**Issue**

BEANS,GREEN,CANNED　　　　　　　　37-3/4 lbs　　　4 gal 1-7/8 qts

Method

1. Pour off half the liquid.
2. Place green beans in steam-jacketed kettle or stock pot.
3. Heat to a simmer. Simmer about 10 minutes, stirring gently. DO NOT BOIL. CCP: Heat to 145 F. or higher for 15 seconds.
4. Place in serving pans. Garnish if desired. CCP: Hold at 140 F. or higher for service.

VEGETABLES No.Q 101 02

BEANS, WAX (CANNED)

Yield 100　　　　　　　　　　　　　　　　　　　　　　**Portion** 3/4 Cup

Calories	Carbohydrates	Protein	Fat	Cholesterol	Sodium	Calcium
26 cal	6 g	1 g	0 g	0 mg	443 mg	41 mg

Ingredient　　　　　　　　　　　　　　　Weight　　　　　Measure　　　　　Issue

BEANS,WAX,CANNED,INCL LIQUIDS　　　　37-3/4 lbs　　　4 gal 1-7/8 qts

Method

1. Pour off half the liquid.
2. Place green beans in steam-jacketed kettle or stock pot.
3. Heat to a simmer. Simmer about 10 minutes, stirring gently. DO NOT BOIL. CCP: Heat to 145 F. or higher for 15 seconds.
4. Place in serving pans. Garnish if desired. CCP: Hold at 140 F. or higher for service.

VEGETABLES No.Q 101 03

BEANS, FRENCH STYLE CUT (FROZEN)

Yield 100   **Portion** 3/4 Cup

Calories	Carbohydrates	Protein	Fat	Cholesterol	Sodium	Calcium
36 cal	8 g	2 g	0 g	0 mg	76 mg	47 mg

Ingredient	**Weight**	**Measure**	**Issue**
WATER	18-3/4 lbs	2 gal 1 qts	
SALT	5/8 oz	1 tbsp	
BEANS,GREEN,FROZEN,FRENCH STYLE	24 lbs	5 gal 2 qts	

Method

1. Bring water to a boil in a steam-jacketed kettle or stock pot.
2. Add salt.
3. Add beans; stir well.
4. Return to a boil; cover. Cook beans 5 to 8 minutes.
5. Place beans in serving pan. CCP: Heat to 145 F. or higher for 15 seconds. Hold at 140 F. or higher for service.

VEGETABLES No.Q 101 04

BEANS, WAX (FROZEN)

Yield 100   Portion   3/4 Cup

Calories	Carbohydrates	Protein	Fat	Cholesterol	Sodium	Calcium
30 cal	7 g	2 g	0 g	0 mg	82 mg	55 mg

| **Ingredient** | **Weight** | **Measure** | **Issue** |

WATER — 18-3/4 lbs — 2 gal 1 qts
SALT — 5/8 oz — 1 tbsp
BEANS,WAX,FROZEN — 24 lbs

Method

1 Bring water to a boil in a steam-jacketed kettle or stock pot.
2 Add salt.
3 Add beans; stir well.
4 Return to a boil; cover. Cook beans 5 to 8 minutes.
5 Place beans in serving pan. CCP: Heat to 145 F. or higher for 15 seconds. Hold at 140 F. or higher for service.

VEGETABLES No.Q 102 00

BEANS, LIMA (FROZEN)

Yield 100 Portion 3/4 Cup

Calories	Carbohydrates	Protein	Fat	Cholesterol	Sodium	Calcium
162 cal	31 g	9 g	0 g	0 mg	136 mg	45 mg

Ingredient	**Weight**	**Measure**	**Issue**
WATER	18-3/4 lbs	2 gal 1 qts
SALT	5/8 oz	1 tbsp
BEANS,LIMA,FROZEN	27 lbs	4 gal 2-2/3 qts

Method

1. Bring water to a boil in a steam-jacketed kettle or stock pot.
2. Add salt.
3. Add lima beans; stir well. Return to a boil; cover.
4. Reduce heat; cook for 6 to 12 minutes.
5. Place lima beans in serving pan. CCP: Heat to 145 F. or higher for 15 seconds. Hold for service at 140 F. or higher.

VEGETABLES No.Q 102 01

BEANS, LIMA (CANNED)

Yield 100   Portion 3/4 Cup

Calories	Carbohydrates	Protein	Fat	Cholesterol	Sodium	Calcium
126 cal	24 g	7 g	0 g	0 mg	449 mg	50 mg

Ingredient	Weight	Measure	Issue
BEANS,LIMA,CANNED,INCL LIQUIDS	39-1/4 lbs	4 gal 2 qts	

Method

1. Pour off half the liquid.
2. Place lima beans in a steam-jacketed kettle or stock pot.
3. Heat to a simmer. Simmer about 10 minutes, stirring gently. DO NOT BOIL. CCP: Heat to 145 F. or higher for 15 seconds.
4. Place in serving pans. Garnish if desired. CCP: Hold for service at 140 F. or higher.

VEGETABLES No.Q 103 01

BEANS, WHITE IN TOMATO SAUCE (CANNED)

Yield 100   Portion   3/4 Cup

Calories	Carbohydrates	Protein	Fat	Cholesterol	Sodium	Calcium
198 cal	37 g	10 g	3 g	13 mg	775 mg	99 mg

Ingredient | Weight | Measure | Issue

BEANS,BAKED,W/PORK,CANNED   41-1/4 lbs   4 gal 2-1/2 qts

Method

1. Place in steam-jacketed kettle or stock pot.
2. Heat to a simmer. Simmer about 10 minutes, stirring gently. DO NOT BOIL. CCP: Internal temperature must reach 145 F. or higher for 15 seconds.
3. Place in serving pans. Garnish if desired. CCP: Hold for service at 140 F. or higher.

VEGETABLES No.Q 104 01

BEETS (CANNED)

Yield 100　　　　　　　　　　　　　　　　**Portion** 3/4 Cup

Calories	Carbohydrates	Protein	Fat	Cholesterol	Sodium	Calcium
50 cal	12 g	1 g	0 g	0 mg	446 mg	23 mg

Ingredient	**Weight**	**Measure**	**Issue**
BEETS,CANNED,SLICED,INCL LIQUIDS	39 lbs	4 gal 2 qts	

Method

1. Pour off half the liquid.
2. Place beets in steam-jacketed kettle or stock pot.
3. Heat to a simmer. Simmer about 10 minutes, stirring gently. DO NOT BOIL. CCP: Heat to 145 F. or higher for 15 seconds.
4. Place in serving pans. Garnish if desired. CCP: Hold for service at 140 F. or higher.

VEGETABLES No.Q 105 00

BROCCOLI (FROZEN)

Yield 100　　　　　　　　　　　　　　　　　　　　Portion 3/4 Cup

Calories	Carbohydrates	Protein	Fat	Cholesterol	Sodium	Calcium
38 cal	7 g	4 g	0 g	0 mg	105 mg	71 mg

Ingredient　　　　　　　　　　　　　　　　**Weight**　　**Measure**　　**Issue**

WATER　　　　　　　　　　　　　　　　　　18-3/4 lbs　　2 gal 1 qts
SALT　　　　　　　　　　　　　　　　　　　5/8 oz　　　1 tbsp
BROCCOLI,FROZEN,CHOPPED　　　　　　　30 lbs　　　5 gal

Method

1. Bring water to a boil in a steam-jacketed kettle or stock pot.
2. Add salt.
3. Add broccoli; stir well. Return to a boil; cover.
4. Reduce heat; cook broccoli for 6 to 8 minutes. CCP: Heat to 145 F. or higher for 15 seconds.
5. Place broccoli in serving pan. CCP: Hold for service at 140 F. or higher.

VEGETABLES No.Q 105 02

BROCCOLI (FRESH)

Yield 100   **Portion** 3 Stalks

Calories	Carbohydrates	Protein	Fat	Cholesterol	Sodium	Calcium
38 cal	7 g	4 g	0 g	0 mg	112 mg	69 mg

Ingredient

Ingredient	Weight	Measure	Issue
WATER	37-5/8 lbs	4 gal 2 qts	
SALT	5/8 oz	1 tbsp	
BROCCOLI,FRESH,CHOPPED	30 lbs	9 gal 2-5/8 qts	49-1/8 lbs

Method

1 Bring water to a boil in steam-jacketed kettle or stock pot.
2 Add salt.
3 Add broccoli; bring water back to a boil. Cover; cook for 10 to 15 minutes.
4 Place broccoil in serving pans. CCP: Heat to 145 F. or higher for 15 seconds for service. Hold for service at 140 F. or higher.

VEGETABLES No.Q 106 00

BRUSSELS SPROUTS (FROZEN)

Yield 100   **Portion**   3/4 Cup

Calories	Carbohydrates	Protein	Fat	Cholesterol	Sodium	Calcium
57 cal	11 g	5 g	0 g	0 mg	104 mg	34 mg

Ingredient | Weight | Measure | Issue

WATER — 18-3/4 lbs — 2 gal 1 qts
SALT — 5/8 oz — 1 tbsp
BRUSSELS SPROUTS,FROZEN — 30 lbs — 5 gal 2 qts

Method

1. Bring water to a boil in a steam-jacketed kettle or stock pot.
2. Add salt.
3. Add brussels sprouts; stir well. Return to a boil; cover.
4. Reduce heat; cook brussels sprouts for 7 to 9 minutes.
5. Place brussels sprouts in serving pans. CCP: Heat to 145 F. or higher for 15 seconds. Hold for service at 140 F. or higher.

VEGETABLES No.Q 107 02

CABBAGE (FRESH)

Yield 100   **Portion** 4-1/2 Ounces

Calories	Carbohydrates	Protein	Fat	Cholesterol	Sodium	Calcium
33 cal	7 g	2 g	0 g	0 mg	99 mg	67 mg

Ingredient	Weight	Measure	Issue
WATER	37-5/8 lbs	4 gal 2 qts	
SALT	5/8 oz	1 tbsp	
CABBAGE,GREEN,FRESH,CHOPPED	30 lbs	12 gal 5/8 qts	37-1/2 lbs

Method

1. Bring water to a boil in steam-jacketed kettle or stock pot.
2. Add salt.
3. Add cabbage; bring water back to a boil. Cover; cook cabbage for 10 minutes.
4. Place cabbage in serving pans. CCP: Heat to 145 F. or higher for 15 seconds. Hold for service at 140 F. or higher.

VEGETABLES   No.Q 108 00

CARROT SLICES (FROZEN)

Yield 100   Portion 3/4 Cup

Calories	Carbohydrates	Protein	Fat	Cholesterol	Sodium	Calcium
53 cal	12 g	2 g	0 g	0 mg	152 mg	45 mg

Ingredient   **Weight**   **Measure**   **Issue**
WATER   12-1/2 lbs   1 gal 2 qts
SALT   5/8 oz   1 tbsp
CARROTS,FROZEN,SLICED   30 lbs   6 gal 2-5/8 qts

Method

1 Bring water to a boil in a steam-jacketed kettle or stock pot.
2 Add salt.
3 Add carrots; stir well. Return to a boil; cover.
4 Reduce heat; cook carrots 10 to 13 minutes. CCP: Heat to 145 F. or higher for 15 seconds.
5 Place carrots in serving pan. CCP: Hold for service at 140 F. or higher.

VEGETABLES No.Q 108 01

CARROT SLICES (CANNED)

Yield 100　　　　　　　　　　　　　　　　　　　　　Portion　3/4 Cup

Calories	Carbohydrates	Protein	Fat	Cholesterol	Sodium	Calcium
41 cal	10 g	1 g	0 g	0 mg	427 mg	55 mg

Ingredient	Weight	Measure	Issue
CARROTS,CANNED,SLICED,INCL LIQUIDS	39-1/4 lbs	4 gal 2-1/8 qts	

Method

1. Pour off half the liquid.
2. Place carrots in steam-jacketed kettle or stock pot.
3. Heat to a simmer. Simmer about 10 minutes, stirring gently. DO NOT BOIL. CCP: Heat to 145 F. or higher for 15 seconds.
4. Place in serving pans. CCP: Hold for service at 140 F. or higher.

VEGETABLES No.Q 108 02

CARROTS (1/4 INCH SLICES) (FRESH)

Yield 100 **Portion** 3/4 Cup

Calories	Carbohydrates	Protein	Fat	Cholesterol	Sodium	Calcium
59 cal	14 g	1 g	0 g	0 mg	121 mg	39 mg

Ingredient	**Weight**	**Measure**	**Issue**
WATER	25-1/8 lbs	3 gal	
SALT	5/8 oz	1 tbsp	
CARROTS,FRESH,SLICED	30 lbs	6 gal 2-5/8 qts	36-5/8 lbs

Method

1 Bring water to a boil in steam-jacketed kettle or stock pot.
2 Add salt.
3 Add carrots; bring water back to a boil. Cover; cook carrots for 15 to 25 minutes.
4 Place carrots in serving pans. CCP: Heat to 145 F. or higher for 15 seconds. Hold for service at 140 F. or higher.

VEGETABLES No.Q 108 03

CARROT STRIPS (FRESH)

Yield 100 **Portion** 3/4 Cup

Calories	Carbohydrates	Protein	Fat	Cholesterol	Sodium	Calcium
53 cal	12 g	1 g	0 g	0 mg	117 mg	36 mg

Ingredient	**Weight**	**Measure**	**Issue**
WATER	29-1/4 lbs	3 gal 2 qts	
SALT	5/8 oz	1 tbsp	
CARROTS,FRESH,2" STRIPS	27 lbs	6 gal 3-7/8 qts	32-7/8 lbs

Method

1. Bring water to a boil in steam-jacketed kettle or stock pot.
2. Add salt.
3. Add carrots; bring water back to a boil. Cover; cook carrots for 15 minutes.
4. Place carrots in serving pans. CCP: Heat to 145 F. or higher for 15 seconds. Hold for service at 140 F. or higher.

VEGETABLES No.Q 109 00

CAULIFLOWER (FROZEN)

Yield 100   Portion 3/4 Cup

Calories	Carbohydrates	Protein	Fat	Cholesterol	Sodium	Calcium
26 cal	5 g	2 g	0 g	0 mg	98 mg	26 mg

Ingredient	**Weight**	**Measure**	**Issue**
WATER	27-7/8 lbs	3 gal 1-1/3 qts	
SALT	5/8 oz	1 tbsp	
CAULIFLOWER, FROZEN	30 lbs		

Method

1. Bring water to a boil in a steam-jacketed kettle or stock pot.
2. Add salt.
3. Add cauliflower; stir well.
4. Return to a boil; cover.
5. Reduce heat; cook cauliflower for 4 to 8 minutes. CCP: Heat to 145 F. or higher for 15 seconds.
6. Place cauliflower in serving pan. CCP: Hold for service at 140 F. or higher.

VEGETABLES No.Q 109 02

CAULIFLOWER (FRESH)

Yield 100   Portion   5 Flowerets

Calories	Carbohydrates	Protein	Fat	Cholesterol	Sodium	Calcium
34 cal	7 g	3 g	0 g	0 mg	116 mg	33 mg

Ingredient	Weight	Measure	Issue
WATER	37-5/8 lbs	4 gal 2 qts	
SALT	5/8 oz	1 tbsp	
CAULIFLOWER,FRESH	30 lbs	8 gal 2 qts	36-1/8 lbs

Method

1 Bring water to a boil in steam-jacketed kettle or stock pot.
2 Add salt.
3 Add cauliflower; bring water back to a boil. Cover; cook for 12 minutes. CCP: Heat to 145 F. or higher for 15 seconds.
4 Place cauliflower in serving pans. CCP: Hold for service at 140 F. or higher.

VEGETABLES No.Q 110 00

CORN, WHOLE KERNEL (FROZEN)

Yield 100 Portion 3/4 Cup

Calories	Carbohydrates	Protein	Fat	Cholesterol	Sodium	Calcium
108 cal	26 g	4 g	1 g	0 mg	75 mg	6 mg

Ingredient	Weight	Measure	Issue
WATER	12-1/2 lbs	1 gal 2 qts	
SALT	5/8 oz	1 tbsp	
CORN,FROZEN,WHOLE KERNEL	27 lbs	4 gal 2-2/3 qts	

Method

1. Bring water to a boil in a steam-jacketed kettle or stock pot.
2. Add salt.
3. Add corn; stir well. Return to a boil; cover.
4. Reduce heat; cook corn 4 to 6 minutes. CCP: Heat to 145 F. or higher for 15 seconds.
5. Place corn in serving pans. CCP: Hold for service at 140 F. or higher.

VEGETABLES No.Q 110 01

CORN, WHOLE KERNEL (CANNED)

Yield 100   Portion   3/4 Cup

Calories	Carbohydrates	Protein	Fat	Cholesterol	Sodium	Calcium
115 cal	28 g	4 g	1 g	0 mg	384 mg	7 mg

Ingredient   Weight   Measure   Issue
CORN,CANNED,WHOLE KERNEL,INCL LIQUIDS   39-3/4 lbs   4 gal 1-5/8 qts

Method
1. Pour off half the liquid.
2. Place corn in steam-jacketed kettle or stock pot.
3. Heat to a simmer. Simmer about 10 minutes, stirring gently. DO NOT BOIL. CCP: Heat to 145 F. or higher for 15 seconds.
4. Place corn in serving pans. Garnish if desired. CCP: Hold for service at 140 F. or higher.

VEGETABLES No.Q 111 00

CORN ON THE COB (FROZEN)

Yield 100   Portion   1 Ear

Calories	Carbohydrates	Protein	Fat	Cholesterol	Sodium	Calcium
123 cal	29 g	4 g	1 g	0 mg	83 mg	10 mg

Ingredient	**Weight**	**Measure**	**Issue**
WATER	50-1/8 lbs	6 gal	
SALT	5/8 oz	1 tbsp	
CORN ON THE COB, FROZEN	27-1/2 lbs	100 each	

Method

1. Bring water to a boil in a steam-jacketed kettle or stock pot.
2. Add salt.
3. Add corn; stir well. Return to a boil; cover.
4. Reduce heat; cook corn 5 to 10 minutes. CCP: Heat to 145 F. or higher for 15 seconds.
5. Place corn in serving pans. CCP: Hold for service at 140 F. or higher.

VEGETABLES No.Q 111 02

CORN ON THE COB (FRESH)

Yield 100   Portion 1 Each

Calories	Carbohydrates	Protein	Fat	Cholesterol	Sodium	Calcium
148 cal	34 g	5 g	2 g	0 mg	100 mg	7 mg

Ingredient   **Weight**   **Measure**   **Issue**

WATER   50-1/8 lbs   6 gal
SALT   5/8 oz   1 tbsp
CORN ON THE COB, FRESH   55 lbs   5 gal 7/8 qts   60-3/8 each

Method

1. Bring water to a boil in steam-jacketed kettle or stock pot.
2. Add salt.
3. Add corn; bring water back to a boil. Cover; corn 5 to 10 minutes. CCP: Heat to 145 F. or higher for 15 seconds.
4. Place corn in serving pans. CCP: Hold for service at 140 F. or higher.

VEGETABLES No.Q 112 01

CREAM STYLE CORN (CANNED)

Yield 100	**Portion** 3/4 Cup

Calories	Carbohydrates	Protein	Fat	Cholesterol	Sodium	Calcium
130 cal	33 g	3 g	1 g	0 mg	514 mg	5 mg

Ingredient
CORN,CANNED,CREAM STYLE

Weight 39-3/4 lbs

Measure 4 gal 1-5/8 qts

Issue

Method

1. Place corn in steam-jacketed kettle or stock pot.
2. Heat corn to a simmer. Simmer about 10 minutes, stirring gently. DO NOT BOIL. CCP: Heat to 145 F. or higher for 15 seconds.
3. Place in serving pans. Garnish if desired. CCP: Hold for service at 140 F. or higher.

VEGETABLES No.Q 113 00

GREENS, COLLARD (FROZEN)

Yield 100   **Portion** 3/4 Cup

Calories	Carbohydrates	Protein	Fat	Cholesterol	Sodium	Calcium
88 cal	17 g	7 g	1 g	0 mg	774 mg	517 mg

Ingredient

Ingredient	Weight	Measure	Issue
WATER	25-1/8 lbs	3 gal	
SALT	5/8 oz	1 tbsp	
GREENS,COLLARD,FROZEN	54 lbs	9 gal <1/16th qts	

Method

1. Bring water to a boil in a steam-jacketed kettle or stock pot.
2. Add salt.
3. Add greens; stir well. Return to a boil; cover.
4. Reduce heat; cook greens 15 to 30 minutes. CCP: Heat to 145 F. or higher for 15 seconds.
5. Place greens in serving pans. CCP: Hold for service at 140 F. or higher. Garnish if desired.

VEGETABLES No.Q 113 02

GREENS, COLLARD (FRESH)

Yield 100 **Portion** 3/4 Cup

Calories	Carbohydrates	Protein	Fat	Cholesterol	Sodium	Calcium
35 cal	7 g	3 g	0 g	0 mg	85 mg	164 mg

Ingredient	Weight	Measure	Issue
WATER	25-1/8 lbs	3 gal	
SALT	5/8 oz	1 tbsp	
GREENS,COLLARD,FRESH	30 lbs	4 gal 1-7/8 qts	40-1/2 lbs

Method

1. Bring water to a boil in steam-jacketed kettle or stock pot.
2. Add salt.
3. Add greens; bring water back to a boil. Cover; cook greens 20 to 30 minutes. CCP: Heat to 145 F. or higher for 15 seconds.
4. Place greens in serving pans. CCP: Hold for service at 140 F. or higher. Garnish if desired.

VEGETABLES No.Q 113 03

GREENS, KALE (FRESH)

Yield 100 Portion 3/4 Cup

Calories	Carbohydrates	Protein	Fat	Cholesterol	Sodium	Calcium
41 cal	8 g	3 g	1 g	0 mg	106 mg	111 mg

Ingredient	Weight	Measure	Issue
WATER	6-1/4 lbs	3 qts	
SALT	5/8 oz	1 tbsp	
GREENS,KALE,FRESH	18 lbs	7 gal 2-1/2 qts	25-1/3 lbs

Method

1. Bring water to a boil in steam-jacketed kettle or stock pot.
2. Add salt.
3. Add greens; bring water back to a boil. Cover; cook greens 10 to 12 minutes.
4. Place greens in serving pans.
5. Garnish as desired. CCP: Hold for service at 140 F. or higher.

VEGETABLES No.Q 114 00

OKRA (FROZEN)

Yield 100   Portion 3/4 Cup

Calories	Carbohydrates	Protein	Fat	Cholesterol	Sodium	Calcium
37 cal	8 g	2 g	0 g	0 mg	75 mg	100 mg

Ingredient   **Weight**   **Measure**   **Issue**

WATER   12-1/2 lbs   1 gal 2 qts
SALT   5/8 oz   1 tbsp
OKRA,FROZEN,CUT   27 lbs   4 gal 5/8 qts

Method

1 Bring water to a boil in a steam-jacketed kettle or stock pot.
2 Add salt.
3 Add okra; stir well. Return to a boil; cover.
4 Reduce heat; cook okra 4 to 7 minutes. CCP: Heat to 145 F. or higher for 15 seconds.
5 Place okra in serving pan. CCP: Hold for service at 140 F. or higher.

VEGETABLES No.Q 114 01

OKRA (CANNED)

Yield 100 **Portion** 3/4 Cup

Calories	Carbohydrates	Protein	Fat	Cholesterol	Sodium	Calcium
35 cal	8 g	1 g	0 g	0 mg	556 mg	111 mg

Ingredient	**Weight**	**Measure**	**Issue**
OKRA,CANNED,INCL LIQUIDS	38 lbs	4 gal 1-3/8 qts	

Method

1. Pour off half the liquid.
2. Place okra in steam-jacketed kettle or stock pot.
3. Heat to a simmer. Simmer about 10 minutes, stirring gently. DO NOT BOIL. CCP: Heat to 145 F. or higher for 15 seconds.
4. Place okra in serving pans. Garnish if desired. CCP: Hold for service at 140 F. or higher.

VEGETABLES No.Q 115 01

ONIONS (CANNED)

Yield 100   **Portion** 3/4 Cup

Calories	Carbohydrates	Protein	Fat	Cholesterol	Sodium	Calcium
33 cal	7 g	2 g	0 g	0 mg	635 mg	77 mg

Ingredient	**Weight**	**Measure**	**Issue**
ONIONS,CANNED,WHOLE,TINY,INCL LIQUIDS	37-3/4 lbs	4 gal 3-1/8 qts	

Method

1. Pour off half the liquid.
2. Place onions in steam-jacketed kettle or stock pot.
3. Heat to a simmer. Simmer about 10 minutes, stirring gently. DO NOT BOIL. CCP: Heat to 145 F. or higher for 15 seconds.
4. Place onions in serving pans. Garnish if desired. CCP: Hold for service at 140 F. or higher.

VEGETABLES No.Q 115 02

ONIONS (FRESH)

Yield 100   **Portion** 3/4 Cup

Calories	Carbohydrates	Protein	Fat	Cholesterol	Sodium	Calcium
52 cal	12 g	2 g	0 g	0 mg	79 mg	31 mg

Ingredient	**Weight**	**Measure**	**Issue**
WATER	37-5/8 lbs	4 gal 2 qts	
SALT	5/8 oz	1 tbsp	
ONIONS,FRESH,SLICED	30 lbs	7 gal 1-5/8 qts	33-1/3 lbs

Method

1. Bring water to a boil in steam-jacketed kettle or stock pot.
2. Add salt.
3. Add onions; bring water back to a boil. Cover; cook onions 15 to 25 minutes. CCP: Heat to 145 F. or higher for 15 seconds.
4. Place vegetables in serving pans. Garnish as desired. CCP: Hold for service at 140 F. or higher.

VEGETABLES No.Q 116 00

PEAS (FROZEN)

Yield 100 **Portion** 3/4 Cup

Calories	Carbohydrates	Protein	Fat	Cholesterol	Sodium	Calcium
103 cal	19 g	7 g	0 g	0 mg	76 mg	35 mg

Ingredient | Weight | Measure | Issue

Ingredient	Weight	Measure	Issue
WATER	15-2/3 lbs	1 gal 3-1/2 qts	
SALT	5/8 oz	1 tbsp	
PEAS, GREEN, FROZEN	27 lbs	5 gal 1-1/4 qts	

Method

1. Bring water to a boil in a steam-jacketed kettle or stock pot.
2. Add salt.
3. Add peas; stir well. Return to a boil; cover.
4. Reduce heat; cook peas 7 minutes. CCP: Heat to 145 F. or higher for 15 seconds.
5. Place vegetables in serving pan. CCP: Hold for service at 140 F. or higher.

VEGETABLES No.Q 116 01

PEAS (CANNED)

Yield 100   **Portion** 3/4 Cup

Calories	Carbohydrates	Protein	Fat	Cholesterol	Sodium	Calcium
95 cal	18 g	6 g	0 g	0 mg	448 mg	32 mg

Ingredient	Weight	Measure	Issue
PEAS,GREEN,CANNED,INCL LIQUIDS	39-1/2 lbs	4 gal 2-1/8 qts	

Method

1. Pour off half the liquid.
2. Place peas in steam-jacketed kettle or stock pot.
3. Heat to a simmer. Simmer about 10 minutes, stirring gently. DO NOT BOIL. CCP: Heat to 145 F. or higher for 15 seconds.
4. Place peas in serving pans. Garnish if desired. CCP: Hold for service at 140 F. or higher.

VEGETABLES No.Q 117 01

BLACK-EYED PEAS (CANNED)

Yield 100 **Portion** 3/4 Cup

Calories	Carbohydrates	Protein	Fat	Cholesterol	Sodium	Calcium
139 cal	25 g	8 g	1 g	0 mg	539 mg	36 mg

Ingredient	**Weight**	**Measure**	**Issue**
PEAS,BLACKEYE,CANNED,INCL LIQUIDS	39-3/4 lbs	4 gal 2-7/8 qts	

Method

1. Place black eyed peas in steam-jacketed kettle or stock pot.
2. Heat to a simmer. Simmer about 10 minutes, stirring gently. DO NOT BOIL. CCP: Heat to 145 F. or higher for 15 seconds.
3. Place black eyed peas in serving pans. Garnish if desired. CCP: Hold for service at 140 F. or higher.

VEGETABLES No.Q 118 01

POTATOES, SWEET (CANNED)

Yield 100 **Portion** 3/4 Cup

Calories	Carbohydrates	Protein	Fat	Cholesterol	Sodium	Calcium
161 cal	38 g	2 g	0 g	0 mg	80 mg	27 mg

Ingredient   Weight   Measure   Issue

SWEET POTATOES,CANNED,W/SYRUP   40 lbs   4 gal 3-7/8 qts

Method

1. Pour off half the liquid.
2. Place sweet potatoes in steam-jacketed kettle or stock pot.
3. Heat to a simmer. Simmer about 10 minutes, stirring gently. DO NOT BOIL. CCP: Heat to 145 F. or higher for 15 seconds.
4. Place sweet potatoes in serving pans. Garnish if desired. CCP: Hold for service at 140 F. or higher.

VEGETABLES No.Q 118 02

POTATOES, SWEET (FRESH)

Yield 100 **Portion** 3 Pieces

Calories	Carbohydrates	Protein	Fat	Cholesterol	Sodium	Calcium
164 cal	38 g	3 g	0 g	0 mg	95 mg	38 mg

Ingredient | Weight | Measure | Issue

Ingredient	Weight	Measure	Issue
SWEET POTATOES, FRESH	34-1/2 lbs	7 gal 1-3/8 qts	
WATER	37-5/8 lbs	4 gal 2 qts	
SALT	5/8 oz	1 tbsp	

Method

1 Cut sweet potatoes into 1 inch pieces.
2 Add salt to water. Bring water to a boil in steam-jacketed kettle or stock pot.
3 Add sweet potatoes; bring water back to a boil. Cover; cook 25 to 35 minutes. CCP: Heat to 145 F. or higher for 15 seconds.
4 Place sweet potatoes in serving pans. Garnish as desired. CCP: Hold for service at 140 F. or higher.

VEGETABLES No.Q 119 01

POTATOES, WHITE (CANNED)

Yield 100 **Portion** 3/4 Cup

Calories	Carbohydrates	Protein	Fat	Cholesterol	Sodium	Calcium
76 cal	17 g	2 g	0 g	0 mg	376 mg	68 mg

Ingredient	**Weight**	**Measure**	**Issue**
POTATOES,CANNED,DICED,WHITE,INCL LIQUIDS	38-1/4 lbs	3 gal 2-1/2 qts	

Method

1. Pour off half the liquid.
2. Place potatoes in steam-jacketed kettle or stock pot.
3. Heat to a simmer. Simmer about 10 minutes, stirring gently. DO NOT BOIL. CCP: Heat to 145 F. or higher for 15 seconds.
4. Place potatoes in serving pans. Garnish if desired. CCP: Hold for service at 140 F. or higher.

VEGETABLES No.Q 119 02

POTATOES, WHITE (FRESH)

Yield 100   Portion 3/4 Cup

Calories	Carbohydrates	Protein	Fat	Cholesterol	Sodium	Calcium
137 cal	32 g	3 g	0 g	0 mg	83 mg	16 mg

Ingredient	Weight	Measure	Issue
WATER	37-5/8 lbs	4 gal 2 qts	
SALT	5/8 oz	1 tbsp	
POTATOES,WHITE,FRESH	35 lbs	6 gal 1-1/2 qts	

Method

1. Bring water to a boil in steam-jacketed kettle or stock pot.
2. Add salt.
3. Add potatoes; bring water back to a boil. Cover; cook potatoes 20 to 25 minutes. CCP: Heat to 145 F. or higher for 15 seconds.
4. Place potatoes in serving pans. Garnish as desired. CCP: Hold for service at 140 F. or higher.

VEGETABLES No.Q 120 01

SAUERKRAUT (CANNED)

Yield 100   **Portion** 3/4 Cup

Calories	Carbohydrates	Protein	Fat	Cholesterol	Sodium	Calcium
32 cal	7 g	2 g	0 g	0 mg	1113 mg	51 mg

Ingredient   Weight   Measure   Issue

SAUERKRAUT,SHREDDED,CANNED,INCL LIQUIDS   37-1/8 lbs   4 gal 1-7/8 qts

Method

1. Place sauerkraut in steam-jacketed kettle or stock pot.
2. Heat to a simmer. Simmer about 10 minutes, stirring gently. DO NOT BOIL. CCP: Heat to 145 F. or higher for 15 seconds.
3. Place sauerkraut in serving pans. Garnish if desired. CCP: Hold for service at 140 F. or higher.

VEGETABLES No.Q 121 00

SPINACH (FROZEN)

Yield 100   **Portion** 3/4 Cup

Calories	Carbohydrates	Protein	Fat	Cholesterol	Sodium	Calcium
34 cal	6 g	4 g	0 g	0 mg	176 mg	179 mg

Ingredient	**Weight**	**Measure**	**Issue**
WATER	6-1/4 lbs	3 qts	
SALT	5/8 oz	1 tbsp	
SPINACH, FROZEN	27 lbs	4 gal 1/8 qts	

Method

1. Bring water to a boil in a steam-jacketed kettle or stock pot.
2. Add salt.
3. Add spinach; stir well. Return to a boil; cover.
4. Reduce heat; cook spinach 4 to 6 minutes. CCP: Heat to 145 F. or higher for 15 seconds.
5. Place spinach in serving pans. CCP: Hold for service at 140 F. or higher.

VEGETABLES No.Q 121 01

SPINACH (CANNED)

Yield 100   Portion 3/4 Cup

Calories	Carbohydrates	Protein	Fat	Cholesterol	Sodium	Calcium
32 cal	5 g	4 g	1 g	0 mg	532 mg	138 mg

Ingredient	Weight	Measure	Issue
SPINACH,CANNED,INCL LIQUIDS	36-3/4 lbs	4 gal 1-7/8 qts	

Method

1. Pour off half the liquid.
2. Place spinach in steam-jacketed kettle or stock pot.
3. Heat to a simmer. Simmer about 10 minutes, stirring gently. DO NOT BOIL. CCP: Heat to 145 F. or higher for 15 seconds.
4. Place spinach in serving pans. Garnish if desired. CCP: Hold for service at 140 F. or higher.

VEGETABLES No.Q 121 02

SPINACH (FRESH)

Yield 100   Portion 3/4 Cup

Calories	Carbohydrates	Protein	Fat	Cholesterol	Sodium	Calcium
27 cal	4 g	4 g	0 g	0 mg	167 mg	122 mg

Ingredient	Weight	Measure	Issue
WATER	6-1/4 lbs	3 qts	
SALT	5/8 oz	1 tbsp	
SPINACH,FRESH,BUNCH	27 lbs	25 gal 2 qts	29-1/3 lbs

Method

1. Bring water to a boil in steam-jacketed kettle or stock pot.
2. Add salt.
3. Chop spinach. Add chopped spinach; bring water back to a boil. Cover; cook spinach 3 to 10 minutes. CCP: Heat to 145 F. or higher for 15 seconds.
4. Place spinach in serving pans. Garnish as desired. CCP: Hold for service at 140 F. or higher.

VEGETABLES No.Q 122 00

SQUASH, SUMMER (FROZEN)

Yield 100 **Portion** 3/4 Cup

Calories	Carbohydrates	Protein	Fat	Cholesterol	Sodium	Calcium
46 cal	10 g	3 g	0 g	0 mg	76 mg	49 mg

Ingredient	Weight	Measure	Issue
WATER	3-1/8 lbs	1 qts 2 cup	
SALT	5/8 oz	1 tbsp	
SQUASH,ZUCCHINI,FROZEN,CHOPPED	60 lbs		

Method

1 Bring water to a boil in a steam-jacketed kettle or stock pot.
2 Add salt.
3 Add squash; stir well. Return to a boil; cover.
4 Reduce heat; cook squash 7 to 9 minutes. CCP: Heat to 145 F. or higher for 15 seconds.
5 Place squash in serving pans. CCP: Hold for service at 140 F. or higher.

VEGETABLES No. Q 122 02

SQUASH, SUMMER (FRESH)

Yield 100 **Portion** 3/4 Cup

Calories	Carbohydrates	Protein	Fat	Cholesterol	Sodium	Calcium
33 cal	7 g	2 g	0 g	0 mg	73 mg	33 mg

Ingredient	Weight	Measure	Issue
WATER	1-5/8 lbs	3 cup	
SALT	5/8 oz	1 tbsp	
SQUASH,FRESH,SUMMER,SLICED	36 lbs	9 gal 1/8 qts	37-7/8 lbs

Method

1. Bring water to a boil in steam-jacketed kettle or stock pot.
2. Add salt.
3. Add squash; bring water back to a boil. Cover; cook squash 10 to 20 minutes. CCP: Heat to 145 F. or higher for 15 seconds.
4. Place squash in serving pans. Garnish as desired. CCP: Hold for service at 140 F. or higher.

VEGETABLES No.Q 123 02

SQUASH, FALL AND WINTER (FRESH)

Yield 100 **Portion** 3/4 Cup

Calories	Carbohydrates	Protein	Fat	Cholesterol	Sodium	Calcium
54 cal	12 g	3 g	1 g	0 mg	81 mg	20 mg

Ingredient

Ingredient	Weight	Measure	Issue
WATER	9-3/8 lbs	1 gal 1/2 qts	
SALT	5/8 oz	1 tbsp	
SQUASH,HUBBARD,FRESH	30 lbs	7 gal 1-1/3 qts	33-3/4 lbs

Method

1 Bring water to a boil in steam-jacketed kettle or stock pot.
2 Add salt.
3 Add squash; bring water back to a boil. Cover; cook squash 15 to 30 minutes. CCP: Heat to 145 F. or higher for 15 seconds.
4 Place squash in serving pans. Garnish as desired. CCP: Hold for service at 140 F. or higher.

VEGETABLES No.Q 124 00

SUCCOTASH (FROZEN)

Yield 100   Portion 3/4 Cup

Calories	Carbohydrates	Protein	Fat	Cholesterol	Sodium	Calcium
114 cal	24 g	5 g	1 g	0 mg	127 mg	21 mg

Ingredient	Weight	Measure	Issue
WATER	18-3/4 lbs	2 gal 1 qts	
SALT	5/8 oz	1 tbsp	
SUCCOTASH,FROZEN	27 lbs	4 gal 3-5/8 qts	

Method

1. Bring water to a boil in a steam-jacketed kettle or stock pot.
2. Add salt.
3. Add succotash; stir well. Return to a boil; cover.
4. Reduce heat; cook succotash 6 to 12 minutes. CCP: Heat to 145 F. or higher for 15 seconds.
5. Place succotash in serving pans. CCP: Hold for service at 140 F. or higher.

VEGETABLES No.Q 125 01

TOMATOES (CANNED)

Yield 100   Portion 3/4 Cup

Calories	Carbohydrates	Protein	Fat	Cholesterol	Sodium	Calcium
33 cal	8 g	2 g	0 g	0 mg	257 mg	52 mg

Ingredient

TOMATOES,CANNED,INCL LIQUIDS

Weight 38-1/4 lbs

Measure 4 gal 5/8 qts

Issue

Method

1. Place tomatoes in steam-jacketed kettle or stock pot.
2. Heat to a simmer. Simmer about 10 minutes, stirring gently. DO NOT BOIL. CCP: Heat to 145 F. or higher for 15 seconds.
3. Place tomatoes in serving pans. Garnish if desired. CCP: Hold for service at 140 F. or higher.

VEGETABLES No.Q 126 00

MIXED VEGETABLES (FROZEN)

Yield 100 Portion 3/4 Cup

Calories	Carbohydrates	Protein	Fat	Cholesterol	Sodium	Calcium
78 cal	16 g	4 g	1 g	0 mg	130 mg	32 mg

Ingredient	**Weight**	**Measure**	**Issue**
WATER	18-3/4 lbs	2 gal 1 qts	
SALT	5/8 oz	1 tbsp	
VEGETABLES,MIXED,FROZEN	27 lbs	4 gal 7/8 qts	

Method

1 Bring water to a boil in a steam-jacketed kettle or stock pot.
2 Add salt.
3 Add mixed vegetables; stir well. Return to a boil; cover.
4 Reduce heat; cook mixed vegetables 7 minutes. CCP: Heat to 145 F. or higher for 15 seconds.
5 Place mixed vegetables in serving pan. CCP: Hold for service at 140 F. or higher.

VEGETABLES No.Q 127 00

PEAS AND CARROTS (FROZEN)

Yield 100   **Portion** 3/4 Cup

Calories	Carbohydrates	Protein	Fat	Cholesterol	Sodium	Calcium
65 cal	14 g	4 g	1 g	0 mg	169 mg	35 mg

Ingredient   Weight   Measure   Issue

Ingredient	Weight	Measure	Issue
WATER	15-2/3 lbs	1 gal 3-1/2 qts	
SALT	5/8 oz	1 tbsp	
PEAS & CARROTS, FROZEN	27 lbs	5 gal 1-7/8 qts	

Method

1. Bring water to a boil in a steam-jacketed kettle or stock pot.
2. Add salt.
3. Add peas; stir well. Return to a boil; cover.
4. Reduce heat; cook peas 7 minutes. CCP: Heat to 145 F. or higher for 15 seconds.
5. Place vegetables in serving pan. CCP: Hold for service at 140 F. or higher.

VEGETABLES No.Q 128 00

RUTABAGAS (FRESH)

Yield 100   Portion 3/4 Cup

Calories	Carbohydrates	Protein	Fat	Cholesterol	Sodium	Calcium
49 cal	11 g	2 g	0 g	0 mg	98 mg	65 mg

Ingredient	Weight	Measure	Issue
WATER	9-3/8 lbs	1 gal 1/2 qts	
SALT	5/8 oz	1 tbsp	
RUTABAGAS,FRESH	30 lbs	6 gal 1/4 qts	35-1/4 lbs

Method

1 Bring water to a boil in steam-jacketed kettle or stock pot.
2 Add salt.
3 Add rutabagas; bring water back to a boil. Cover; cook rutabagas 15 to 30 minutes. CCP: Heat to 145 F. or higher for 15 seconds.
4 Place rutabagas in serving pans. Garnish as desired. CCP: Hold for service at 140 F. or higher.

VEGETABLES No.Q 129 00

TURNIPS (FRESH)

Yield 100   Portion   3/4 Cup

Calories	Carbohydrates	Protein	Fat	Cholesterol	Sodium	Calcium
37 cal	8 g	1 g	0 g	0 mg	162 mg	42 mg

Ingredient	Weight	Measure	Issue
WATER	9-3/8 lbs	1 gal 1/2 qts	
SALT	5/8 oz	1 tbsp	
TURNIPS, FRESH	30 lbs	6 gal 2-1/8 qts	37 lbs

Method

1. Bring water to a boil in steam-jacketed kettle or stock pot.
2. Add salt.
3. Add turnips; bring water back to a boil. Cover; cook turnips 15 to 30 minutes. CCP: Heat to 145 F. or higher for 15 seconds.
4. Place turnips in serving pans. Garnish as desired. CCP: Hold for service at 140 F. or higher.

VEGETABLES No.Q 500 00

GARLIC CHEESE POTATOES

Yield 100   Portion 1/2 Cup

Calories	Carbohydrates	Protein	Fat	Cholesterol	Sodium	Calcium
121 cal	23 g	4 g	2 g	3 mg	234 mg	81 mg

Ingredient	Weight	Measure	Issue
POTATOES,WHITE,FRESH,WEDGED	23-7/8 lbs	4 gal 1-3/8 qts	
WATER	12-1/2 lbs	1 gal 2 qts	
MARGARINE	2 oz	1/4 cup 1/3 tbsp	
MILK,NONFAT,DRY	4-1/2 oz	1-7/8 cup	
WATER	5 lbs	2 qts 1-1/2 cup	
SALT	1-1/4 oz	2 tbsp	
GARLIC POWDER	1-1/4 oz	1/4 cup 1/3 tbsp	
CHEESE,PARMESAN,GRATED	14-1/8 oz	1 qts	

Method

1. Wash vegetables thoroughly. In large stock pot, cover peeled, quartered potatoes with cold water; bring to rapid boil; reduce heat; simmer and cook until potatoes are tender throughout.
2. Reconstitute milk. Heat margarine, milk, salt, and garlic until just hot. Place potatoes in a large mixer, add 2/3 of the milk mixture, whip until potatoes are just combined. Add remainder of the milk if necessary for a fluffy, not dry, consistency.
3. Fold in parmesan cheese. CCP: Hold at 140 F. or higher for service.

VEGETABLES No.Q 500 01

GARLIC CHEESE POTATOES (INSTANT)

Yield 100　　　　　　　　　　　　　　　　**Portion** 1/2 Cup

Calories	Carbohydrates	Protein	Fat	Cholesterol	Sodium	Calcium
117 cal	19 g	4 g	3 g	8 mg	244 mg	82 mg

Ingredient　　　　　　　　　　　　　　Weight　　　Measure　　　Issue

Ingredient	Weight	Measure
POTATO,WHITE,INSTANT,GRANULES	4-3/4 lbs	2 gal 3-1/4 qts
MILK,NONFAT,DRY	5-3/8 oz	2-1/4 cup
WATER,BOILING	20-7/8 lbs	2 gal 2 qts
BUTTER	8 oz	1 cup
SALT	1 oz	1 tbsp
PEPPER,WHITE,GROUND	1/8 oz	1/4 tsp
GARLIC POWDER	1-1/4 oz	1/4 cup 1/3 tbsp
CHEESE,PARMESAN,GRATED	14-1/8 oz	1 qts

Method

1. Blend instant potatoes and nonfat dry milk together.
2. Blend water, butter or margarine, salt, pepper, and garlic powder in mixer bowl.
3. At low speed, using wire whip, rapidly add potato and milk mixture to liquid; mix 1/2 minute. Stop mixer; scrape down sides and bottom of bowl.
4. Whip at high speed about 2 minutes or until light and fluffy. DO NOT OVERWHIP. Fold in parmesan cheese. CCP: Hold at 140 F. or higher for service.

VEGETABLES No.Q 502 00

ITALIAN ROASTED POTATOES

Yield 100   Portion   1/2 Cup

Calories	Carbohydrates	Protein	Fat	Cholesterol	Sodium	Calcium
129 cal	29 g	3 g	1 g	0 mg	237 mg	26 mg

Ingredient	Weight	Measure	Issue
POTATOES,FRESH,PEELED,CUBED	30 lbs	5 gal 1-7/8 qts	37 lbs
OIL, CANOLA	1-7/8 oz	1/4 cup 1/3 tbsp	
OREGANO,CRUSHED	5/8 oz	1/4 cup 1/3 tbsp	
BASIL,DRIED,CRUSHED	1/3 oz	2 tbsp	
ROSEMARY,GROUND	1/8 oz	1 tbsp	
THYME,GROUND	1/8 oz	1 tbsp	
GARLIC POWDER	1/3 oz	1 tbsp	
SALT	1-7/8 oz	3 tbsp	
PEPPER,BLACK,GROUND	1/2 oz	2 tbsp	
PARSLEY,FRESH,BUNCH	7-1/3 oz	3 cup	7-2/3 oz
TOMATOES,CANNED,DICED,DRAINED	3-1/3 lbs	1 qts 2 cup	

Method

1. Combine diced potatoes, oil, oregano, basil, rosemary, thyme, garlic, salt and pepper. Toss until thoroughly blended.
2. Place 8 pounds potatoes in each sheet pan. Roast at 400 F. in conventional oven for 25 minutes or until potatoes are browned and cooked through.
3. Add 1 pound drained tomatoes to each pan of potatoes, add parsley and toss. Return to oven and heat until heated through. CCP: Heat to 145 F. or higher for 15 seconds. Hold for service at 140 F. or higher.

VEGETABLES No.Q 503 00

OKRA MELANGE

Yield 100   Portion 1/2 Cup

Calories	Carbohydrates	Protein	Fat	Cholesterol	Sodium	Calcium
47 cal	10 g	2 g	1 g	0 mg	196 mg	73 mg

Ingredient	Weight	Measure	Issue
COOKING SPRAY,NONSTICK	2 oz	1/4 cup 1/3 tbsp	
ONIONS,FRESH,CHOPPED	2 lbs	1 qts 1-5/8 cup	2-1/4 lbs
GARLIC POWDER	2-3/8 oz	1/2 cup	
CELERY,FRESH,CHOPPED	1 lbs	3-3/4 cup	1-3/8 lbs
PEPPERS,GREEN,FRESH,CHOPPED	2 lbs	1 qts 2-1/8 cup	2-1/2 lbs
OKRA,FROZEN,CUT	14-2/3 lbs	2 gal 1 qts	
TOMATOES,CANNED,DICED,DRAINED	7-1/8 lbs	3 qts 1 cup	
SALT	1-1/4 oz	2 tbsp	
PEPPER,BLACK,GROUND	1/4 oz	1 tbsp	
JUICE,LEMON	8-5/8 oz	1 cup	
PARSLEY,DEHYDRATED,FLAKED	1/8 oz	1/4 cup 1/3 tbsp	
BREADCRUMBS,DRY,GROUND,FINE	3-3/4 oz	1 cup	

Method

1. Spray steam jacketed kettle with non-stick cooking spray. Saute onions and garlic in a steam-jacketed kettle until translucent. Add celery and green peppers. Cook an additional 5 minutes.
2. Add thawed okra and drained tomatoes. Cook okra.
3. Season with salt, pepper, lemon juice and parsley. Place in baking pans. Sprinkle with bread crumbs. Using a convection oven, bake at 400 F. for 10 minutes or until bread crumbs are brown. CCP: Heat to 145 F. or higher for 15 seconds. Hold at 140 F. or higher for service.

VEGETABLES No. Q 504 00

ROASTED PEPPER POTATOES

Yield 100 Portion 2/3 Cup

Calories	Carbohydrates	Protein	Fat	Cholesterol	Sodium	Calcium
103 cal	19 g	2 g	2 g	0 mg	170 mg	19 mg

Ingredient	Weight	Measure	Issue
POTATOES,FRESH,PEELED,CUBED	20 lbs	3 gal 2-1/2 qts	24-2/3 lbs
MILK,NONFAT,DRY	1-3/4 oz	3/4 cup	
WATER	2 lbs	3-3/4 cup	
PIMIENTO,CANNED,INCL LIQUIDS	1-1/4 lbs	3 cup	
MARGARINE	8 oz	1 cup	
SALT	1-1/4 oz	2 tbsp	
PARSLEY,FRESH,BUNCH	7-1/3 oz	3 cup	7-2/3 oz
PEPPER,BLACK,GROUND	7/8 oz	1/4 cup 1/3 tbsp	
COOKING SPRAY,NONSTICK	2 oz	1/4 cup 1/3 tbsp	

Method

1. Steam peeled, cubed potatoes for 15 minutes or until tender.
2. Reconstitute milk. Scald milk.
3. Whip the hot potatoes; add pimentos, milk, margarine, salt, pepper, and parsley. Mix on medium low speed for 3 to 4 minutes or until thoroughly whipped.
4. Divide potatoes evenly among steam table pans. Spray top of potatoes with cooking spray. Using a convection oven, bake at 400 F. 25 minutes or until potatoes are slightly brown. CCP: Hold at 140 F. or higher for service.

VEGETABLES No.Q 504 01

ROASTED PEPPER POTATOES (INSTANT)

Yield 100   Portion 2/3 Cup

Calories	Carbohydrates	Protein	Fat	Cholesterol	Sodium	Calcium
104 cal	19 g	2 g	2 g	0 mg	174 mg	30 mg

Ingredient	**Weight**	**Measure**	**Issue**
POTATO,WHITE,INSTANT,GRANULES	4-3/4 lbs	2 gal 3-1/4 qts	
MILK,NONFAT,DRY	5-3/8 oz	2-1/4 cup	
WATER,BOILING	20-7/8 lbs	2 gal 2 qts	
MARGARINE	8 oz	1 cup	
SALT	1 oz	1 tbsp	
PEPPER,WHITE,GROUND	1/8 oz	1/4 tsp	
PIMIENTO,CANNED,INCL LIQUIDS	1-1/4 lbs	3 cup	
PARSLEY,FRESH,BUNCH	7 oz	3 cup	
COOKING SPRAY,NONSTICK	2 oz	1/4 cup 1/3 tbsp	

Method

1. Blend potatoes and milk together.
2. Blend water, butter or margarine, salt and pepper in mixer bowl.
3. At low speed, using wire whip, rapidly add potato and milk mixture to liquid; mix 1/2 minute. Stop mixer; scrape down sides and bottom of bowl.
4. Whip at high speed about 2 minutes or until light and fluffy. DO NOT OVERWHIP. Fold in pimentos and parsley.
5. Divide potatoes evenly among steam table pans. Spray top of potatoes with cooking spray. Using a convection oven, bake at 400 F. 25 minutes or until potatoes are slightly brown. CCP: Hold at 140 F. or higher for service.

www.ingramcontent.com/pod-product-compliance
Lightning Source LLC
Chambersburg PA
CBHW080014090526
44578CB00013B/690